Proceedings of
THE FOUNDING CONVENTION OF THE MARS SOCIETY

Part II

Proceedings of
THE FOUNDING CONVENTION OF THE MARS SOCIETY

Part II

Edited by
Robert M. Zubrin
Maggie Zubrin

Proceedings of the Founding Convention of The Mars Society held August 13-16, 1998, Boulder, Colorado.

Published for The Mars Society by
Univelt, Incorporated, P.O. Box 28130,
San Diego, California 92198

Copyright 1999

by

Univelt, Incorporated, Publishers
P.O. Box 28130
San Diego, California 92198

First Printing 1999

ISBN 0-912183-12-8 (Part I - Soft Cover)
ISBN 0-912183-13-6 (Part II - Soft Cover)
ISBN 0-912183-14-4 (Part III - Soft Cover)

*Published for The Mars Society
by Univelt, Incorporated, P.O. Box 28130, San Diego, California 92198*

Printed and Bound in the U.S.A.

CONTENTS

	Page
Foreword, Robert and Maggie Zubrin	xv
The Founding Convention of the Mars Society, Richard Wagner	xvii
The Founding Declaration of the Mars Society	xxi

Part I

CHAPTER 1: PLENARY TALKS   1

Opening Address to The Founding Convention of The Mars Society (MAR 98-001)
    Robert Zubrin 3

Mars: The Case for Life (MAR 98-003)
    Christopher P. McKay 17

CHAPTER 2:
HISTORICAL AND PHILOSOPHICAL SIGNIFICANCE OF MARS   25

Civilizations at the Crossroads: Spain and China in the 15th Century (MAR 98-004)
    Wayne H. Bowen 27

On the Legitimization, Importance and Duty of Colonizing Mars (MAR 98-005)
    Josef Oehmen 33

The Race to Settle Mars - Earth Having a Baby - What's it Worth to the Parent Civilization? (MAR 98-006)
    Peter Perrine 37

Mars Exploration: The Survival of Our Civilization (MAR 98-007)
    E. G. Petrakakis 47

Reassessing the Human Condition: Philosophical Aspects of Mars Exploration (MAR 98-008)
    Richard L. Poss 51

CHAPTER 3: HISTORICAL LESSONS   61

The Outward Course of Empire: The Hard, Cold Lessons from Euro-American Involvement in the Terrestrial Polar Regions (MAR 98-009)
    Marilyn Dudley-Rowley 63

	Page
A Shining City on a Higher Hill: Lessons From the Last Colonization of a 'New World' (MAR 98-010)	
James D. Heiser	73

CHAPTER 4: MOBILIZING THE PUBLIC — 83

Is There a Short-Term Economic and Social Justification for Human Exploration and Settlement of Mars? (MAR 98-011)
  Robert E. Becker 85

Mars: Fostering Public Support (MAR 98-012)
  Sam Burbank 101

Planet Mars Home Page©: Some Results After Three Years Operation (MAR 98-013)
  Thomas A. Gunn 109

How (Why and Under Which Conditions) Could International Cooperation Reinforce the Case for Mars (MAR 98-014)
  Richard Heidmann 117

Why NASA Might Never Launch a Manned Mission to Mars - The Devil's Advocate (MAR 98-015)
  Fred Kelly 127

Mars Needs Guitars (MAR 98-016)
  M. R. Jardin 133

Yes, But Will the People Support Us? - Engaging Our Customers in the Mars Exploration Adventure (MAR 98-017)
  Humboldt C. Mandell, Jr. 143

Human Mission from Planet Earth: Technology Assessment and Social Forecasting of Moon/Mars Synergies (MAR 98-018)
  Eligar Sadeh and Evan Vlachos 151

Making it Happen (MAR 98-019)
  Jonathan Stabb 169

A Socially Supportable Mars Colonization Program: Earth Mars Ambassadors (MAR 98-020)
  Philip A. Turek 175

The Gen-X Rallying Cry? To Mars? Gen-X Needs a Cause, and That Cause Should Be Space (MAR 98-021)
  George T. Whitesides 187

CHAPTER 5: VOICES OF YOUTH — 191

Our Future on Mars from My Perspective as a Twelve Year Old (MAR 98-022)
  Kathleen B. Bohné 193

	Page
The Hakluyt Prize Letter (MAR 98-023) Adrian Hon	195

CHAPTER 6: EDUCATION AND THE ARTS — 197

Secondary Mars Education: Faster, Better, Cheaper (MAR 98-024)
    Thomas W. Becker 199

Growing the Future (MAR 98-025)
    Charmin P. Gerardy 223

The Frontier of Mars as an Agriscience Classroom: Terraforming Mars (MAR 98-026)
    Larry Payne 231

Teaching from Mars (MAR 98-027)
    Gabriel F. Rshaid 237

Preparing for the Journey: An Introduction to Mars Education (MAR 98-028)
    Donald M. Scott 243

From Bradbury to Blamont: The Science of Mars in the Arts (MAR 98-029)
    Michael Carroll 253

CHAPTER 7: PRIVATE FUNDING FOR MARS MISSIONS — 257

A Sponsoring Concept for Manned Missions to Mars (MAR 98-030)
    Michael Bosch 259

Conducting Mars Exploration on a Private Basis by Reviving the "East India Company" Financing Model (MAR 98-031)
    John Q. Coston 269

Private Sector Mars Wake-Up Call for Non-Government Participation (MAR 98-032)
    Thomas A. Gunn 275

Thinking About Martian Economics (MAR 98-033)
    Edward L. Hudgins 285

The Business of Commercializing Space (MAR 98-034)
    David M. Livingston 289

Funding The First Human Expedition to Mars (MAR 98-035)
    George Osorio 299

Promoting Privately Funded Settlement of Mars (MAR 98-036)
    Alan B. Wasser 311

CHAPTER 8: ROBOTIC EXPLORATION — 315

Navigation and Mobility Systems Architecture for Planetary Rovers (MAR 98-037)
    Pablo Flores 317

	Page
On the Development of Airborne Science Platforms for Martian Exploration (MAR 98-038)	
David W. Hall and Robert W. Parks	323
Reducing Risk and Complexity of Rover and Robotic Operations on Mars (MAR 98-039)	
Russell R. Mellon and Thomas R. Meyer	337
Soil Sampling on Mars (MAR 98-040)	
John L. Paterson	343
Marsplane - Flying on Mars with Existing Aircraft (MAR 98-041)	
Fabrizio Pirondini	353
Autonomous Rovers for Human Exploration of Mars (MAR 98-042)	
John Bresina, Gregory A. Dorais, Keith Golden, David E. Smith and Richard Washington	369

Part II

	Page

CHAPTER 9: SOFTWARE AND AUTONOMY — 379

Using COTS Software for Mars Missions (MAR 98-043)
 Ned Chapin 381

Adjustable Autonomy for Human-Centered Autonomous Systems on Mars (MAR 98-044)
 Gregory A. Dorais, R. Peter Bonasso, David Kortenkamp,
 Barney Pell and Debra Schreckenghost 397

Model-Based Autonomy for Robust Mars Operations (MAR 98-045)
 James A. Kurien, P. Pandurang Nayak and Brian C. Williams 421

CHAPTER 10: THE QUESTION OF LIFE — 435

Life on Mars: Evidence Within Martian Meteorites (MAR 98-046)
 Everett K. Gibson, Jr., David S. McKay and Kathie Thomas-Keprta . . . 437

ESA Exobiology Activities (MAR 98-047)
 Gerhard Kminek 449

Preserving Possible Martian Life (MAR 98-048)
 Mark Lupisella 457

The Ethical Ramifications of Discovering Life on Mars (MAR 98-049)
 Katherine Osborne 481

Interplanetary Biological Transfer of Bacteria Entrapped in Small Meteorites: Analysis of Bacterial Resistance to Impact in Ballistic Experiments (MAR 98-050)
 C.-A. H. Roten, A. Galluser, G. D. Borruat, S. D. Udry, G. Niederhäuser,
 A. Croxatto, O. Blanc, S. De Carlo, C. K. Mubenga-Kabambi
 and D. Karamata 485

CHAPTER 11: TECHNOLOGIES FOR HUMAN EXPLORATION — 501

MARSSAT: Assured Communication with Mars (MAR 98-051)
 Thomas Gangale 503

Mobility of Large Manned Rovers on Mars (MAR 98-052)
 George William Herbert 515

Boosters for Manned Missions to Mars, Past and Present (MAR 98-053)
 Scott Lowther 529

	Page
An RLV / Shuttle Compatible Habitation System (MAR 98-054) Kurt Anthony Micheels	555
Aresam: Student Concept of Future Mars Space Station (MAR 98-055) Jonathon Smith and Jim Bishop	567
Current Progress in Water Reclamation Technology (MAR 98-056) Bradley S. Tice	573
Design of a Nuclear-Powered Rover for Lunar or Martian Exploration (MAR 98-057) Holly R. Trellue, Rachelle Trautner, Michael G. Houts, David I. Poston, Kenji Giovig, J. A. Baca and R. J. Lipinski	577

CHAPTER 12: POWER ON MARS — 585

Surviving on Mars Without Nuclear Energy (MAR 98-058)
George James, Gregory Chamitoff and Donald Barker 587

Near-Term, Low-Cost Space Fission Systems (MAR 98-059)
Michael G. Houts, David I. Poston, Marc V. Berte and
William J. Emrich, Jr. 611

CHAPTER 13: ACCESSING MARTIAN RESOURCES — 621

Artesian Basins on Mars: Implications for Settlement, Life-Search and Terraforming (MAR 98-060)
Martyn J. Fogg 623

Drilling Operations to Support Human Mars Missions (MAR 98-061)
Brian M. Frankie, Frank E. Tarzian, Scott Lowther and Trevor Wende . . . 637

Extraction of Atmospheric Water on Mars in Support of the Mars Reference Mission (MAR 98-062)
M. R. Grover, M. O. Hilstad, L. M. Elias, K. G. Carpenter,
M. A. Schneider, C. S. Hoffman, S. Adan-Plaza and A. P. Bruckner . . . 659

A Comparison of *In Situ* Resource Utilization Options for the First Human Mars Missions (MAR 98-063)
Kristian Pauly 681

Producing a Brick from a Martian Soil Simulate (MAR 98-064)
David Seymour 695

The Case for a Mars Base ISRU Refinery (MAR 98-097)
Kelly R. McMillen and Thomas R. Meyer 699

CHAPTER 14: HUMAN FACTORS — 709

Coping With Effects of Enforced Intimacy on Long Duration Space Flight (MAR 98-065)
Lara Battles 711

	Page
On Our Best Behavior: Optimization of Group Functioning on the Early Mars Missions (MAR 98-066)	
Vadim I. Gushin and Marilyn Dudley-Rowley	717
Mars Mission Operations (MAR 98-067)	
Kenneth E. Peek	723
At What Risk is it Acceptable to Commit to a Manned Mars Mission? (MAR 98-068)	
Dennis G. Pelaccio and Joseph R. Fragola	729
The Case for Nurses as Key Contributors to Mars Exploration Teams (MAR 98-069)	
Mary Ellen Symanski	743
Man and Extended Space Flight: Mental and Physiological Factors (MAR 98-070)	
Bradley S. Tice	751
Who Should Go to Mars (MAR 98-071)	
Paul VanSteensburg	755

Part III

	Page
CHAPTER 15: MEDICAL ISSUES	761

No Means They Can Go (MAR 98-072)
   Thomas J. Burke and Michael C. Trachtenberg 763

The Effects of Variable Gravity on the Life Cycle of Tenebrio Molitor (MAR 98-073)
   Amy M. Davis 767

Nutritional Supplements as Radioprotectors: A Review and Proposal (MAR 98-074)
   Anthony C. Muscatello 773

Running to Mars: Exercise Countermeasures for Mars Astronauts (MAR 98-075)
   Erik Seedhouse 787

CHAPTER 16: MISSION STRATEGIES   791

New Directions: Reevaluating the Lunar Refueling Option (MAR 98-076)
   J. D. Beegle and H. L. Beegle 793

A Stepped Approach to the Moon and Mars (MAR 98-077)
   James A. Bickford 799

A Novel Space Transportation Concept Designed to Reduce Per-Mission Costs for Repeated Travel to/from a Celestial Body (MAR 98-078)
   Stephen Heppe 809

OneWay and Back: An Introduction to Comparative Missionology (MAR 98-079)
   George William Herbert 817

Phobos on the Cheap (MAR 98-080)
   George William Herbert 823

Free Return Trajectories for Mars Missions (MAR 98-081)
   Chris Hirata 831

Polar Landing Site for a First Mars Expedition (MAR 98-082)
   Geoffrey A. Landis 835

Page

CHAPTER 17: LAW AND SOCIETY ON MARS   **843**

Martian Law (MAR 98-083)
   Edward L. Hudgins 845

Legislation and Space Law Concepts Proposed for the Eventual
Industrialization of Mars by Man (MAR 98-084)
   James J. Hurtak 853

Martian Equality (MAR 98-085)
   Richard A. Jones 865

Mars Governance (MAR 98-086)
   Declan J. O'Donnell 873

The Politics of a Mars Colony (MAR 98-087)
   Kevin Archbold, Randall Hessler and Blaine Thompson 881

The Rights of Mars (MAR 98-088)
   Robert Zubrin 889

CHAPTER 18: TERRAFORMING MARS   **893**

Physiological Ecology of Terrestrial Microbes on a Terraformed Mars
(MAR 98-089)
   James M. Graham and Linda E. Graham 895

Successional Stages in Terraforming Mars (MAR 98-090)
   James M. Graham and Linda E. Graham 901

Terraformation of Mars (MAR 98-091)
   Charles R. Hancox 905

An Ecological Approach to Terraforming, Mapping the Dream
(MAR 98-092)
   Richard W. Miller 937

Ethics of Terraforming: A Practical System (MAR 98-093)
   George A. Smith 985

Terraforming Mars - Waterfield Reservoir Management (MAR 98-094)
   Patrick Whittome 1003

CHAPTER 19: CALENDARS AND TIMEKEEPING   **1019**

The Darian Calendar (MAR 98-095)
   Thomas Gangale 1021

The Millennium Mars Calendar (MAR 98-096)
   James M. Graham and Kandis Elliot 1031

	Page
APPENDIX	**1035**
Conference Sponsors	1036
Conference Schedule	1037
Conference Abstracts	1044
INDEX	**1123**
Numerical Index	1125
Author Index	1130

FOREWORD

Between the 13th and 16th of August, 1998, something special happened in Boulder Colorado. From all over the world, people came to be part of the beginning of a great enterprise – the founding of a society dedicated to opening Mars to humanity. The event was unique, everyone attending could feel it – there was something electric in the air. It was more than a turning point; it was akin to being present at the creation.

Seven hundred people came, and over 180 talks were given. Every aspect of the profoundly human endeavor of the development of a new world was discussed and debated -from engineering to economics to ethics – and from nearly every point of view. There was no consensus, nor could there be, for new life requires no consensus, but the riotous play of diversity seeking light through myriads of unexpected paths.

No proceedings could do this conference justice. The exciting debates in the halls, the breakout sessions, and the late-night bars, cannot be found in these pages. Most profoundly absent are the proceedings of the general meeting of Mars Society members, which occurred after the Saturday night banquet, during which the Founding Declaration was ratified and the Society formed. To get the full experience of the conference, you had to be there, or failing that, to speak with those who were.

Registration was furious on the first day. Maggie Zubrin (far right) keeps things moving.

Nevertheless, what paper, ink, and the best efforts of nearly 100 contributors can do to bring this conference to you and preserve it for the future, has been done. If nearly half the presentations are unrecorded, more than half are published here, as are some of the most important plenary talks. The article by Richard Wagner gives the blow-by-blow history of the convention, and accompanied by a fair number of photos, gives some feel for the actual flavor of the event.

As a personal note, we would like to say that the organizing of this conference was one of the most engaging and moving events of our lives. Going into the conference, we were a handful of people with no organization and no staff. On the evening of August 12, when the first large numbers of attendees showed up to pre-register, total chaos appeared unavoidable. But it took just a moment for many of those who were there to register for the conference to move to the other side of the registration tables as volunteers, and the pattern of volunteers stepping forward to fill in the innumerable gaps remained constant for the next four hectic days. Those who helped are too numerous to name here, but we thank you all. We also wish to thank the generous donations of Fisher Space Pen, the Bushnell Corporation, Robert Bigelow, the Longview Foundation, and the National Geographic Society, without which the event would have been impossible.

Since the convention, the Mars Society has grown exponentially. More volunteers with skills of every description have stepped forward, and the Society is making its presence known in over 30 countries. The seed planted in Boulder has sprouted, and a new force for life is loose in the world.

It gives us great pleasure to be able to present to the interested reader these proceedings of the Founding Convention of the Mars Society.

Robert and Maggie Zubrin
June 30, 1999

THE FOUNDING CONVENTION OF THE MARS SOCIETY

Richard Wagner

Against a backdrop of the Front Range's ruddy face, close to 700 Mars enthusiasts filled — and sometimes overflowed — the University of Colorado's Memorial Center in Boulder for the Founding Convention of the Mars Society. Over the course of four days, they heard from researchers and scientists involved in Mars exploration and shared their own ideas concerning humanity's future on the Red Planet. But, more than that, they helped create a popular movement that will promote the cause of human Mars exploration.

The buzz rocketing through the Glen Miller ballroom on the morning of August 13 was undeniable. While all at the conference were well aware of their personal, private passion for Mars, few were perhaps ready for the display of shared passion they found on walking through the ballroom doors. More than 600 people filled the seats of the auditorium. With camera crews scurrying about in back and on stage, audience members gazed around in something close to wonder at the multitude at hand. In due time, Robert Zubrin took to the stage and opened the convention with the simple words, "Hello, I'm Robert Zubrin, president of the Mars Society, and am I glad to see you," which the audience greeted with an eruption of raucous applause and cheers.

The Convention begins.

Anyone reviewing the convention schedule was well aware that this was a nearly overwhelming affair — four days of meetings with mornings devoted to Plenary sessions, afternoons to five different subject tracks, and evenings to further plenaries and society events. With more than 200 talks by roughly as many speakers on subjects ranging from terraforming to the art of Martian landscapes, it was sometimes hard to keep up, but no one seemed to mind. It added to the energy, the insights, and the sheer exuberant fun of the convention. The mix of researchers and engineers mingling with the likes of artists, writers, professionals, and students produced surprises throughout the convention. While it possibly came as no surprise that Zubrin's convention opening talk, "Humans to Mars within a Decade," brought the audience to its feet for an extended standing ovation, those lucky enough to hear Kathleen Bohné's talk "Our Future on Mars from My Perspective as a 12 Year Old" were, as one audience member remarked, "blown away" by the honesty and vision of her presentation.

Morning plenary sessions allowed all to come together to hear from world-class Mars researchers, astronauts, and others. Thursday morning's opening plenary offered up a good example of what was to come in the following days. In addition to Zubrin, the morning featured Everett Gibson of the ALH 84001 "Mars Meteorite Team" presenting a vigorous defense of recent attacks on the team's biologic interpretation of the stone's more intriguing features; Pascal Lee of NASA Ames Research Center with an overview of recent research in the Haughton Impact Crater on Devon Island in the Canadian Arctic; Rob Manning of the Jet Propulsion Lab with a review of NASA's current robotic Mars exploration program; and Jacques Blamont of the French space agency CNES with an overview of a U.S./French program to explore Mars by balloon. Afternoons were given over to the various tracks, which offered a smorgasbord of subjects for both those interested in the technical aspects of Mars exploration as well as the social aspects, while evenings were reserved for Society events or further plenary sessions.

In addition to offering a forum for Mars researchers and aficionados, the Founding Convention provided a sense of place and purpose for establishing an organization and movement dedicated to the human exploration of Mars. On Saturday afternoon, members of the Mars Society Steering Committee, along with interested convention attendees, wrestled with the questions of how to build a movement from the energies unleashed by the convention, and what projects to undertake to forward the cause of human Mars exploration.

The afternoon meeting was fast-paced and focused, with all agreeing that the Mars Society had to be a results oriented organization focusing on both politics and programs. By consensus, members of the steering committee mapped a project task list that focuses on the near, mid, and long-term. The first approved project is the design and construction of a full-scale Mars base simulation in the Canadian Arctic. Pascal Lee's aforementioned talk on the Haughton Impact Crater and the Haughton-Mars 1998 Project proved the impetus for this concept. Field methods and hab designs will be tested at this base, while providing a long-term base of operations for geological explorations in the crater. The Mars Society aims to have the base up and running by the summer of 2000. Second on the project list, a payload to Mars during the 2003 opportunity. While there is a wide range of options for this project, most envisioned it as a hitchhiker payload on a scheduled Mars mission or Ariane launch. The Society will release a request for proposals with an eye to selecting a mission that is affordable, quick, and sure to catch public attention. Finally, Steering Committee members agreed that a long term goal should be its own mission to Mars.

On an organizational level, there was quick agreement on the necessity of raising $100,000 within the next few months to enable the hiring of an executive director for the Society. In addition, the near future will see a series of local chapters and task forces established to provide both grassroots and managerial direction for the Society. Finally, all agreed that funding for the Arctic project will be a priority item over the course of the next year, as will organizing for a Mars Society convention in 1999.

Everything came together during Saturday evening's banquet. Highlights of the evening included some folk and filk singing entertainment (bravely led, at one point, by Bob Zubrin) and words and inspiration from Adrian Hon (winner of the Society's Hakluyt Prize) and Kathleen Bohné (an addition to the schedule, owing to the rousing reception her track talk received). Society President Zubrin later read the Founding Declaration of the Mars Society and, when he asked for acceptance of the document by acclamation, the energy and enthusiasm rippling through the convention burst forth in roar of approval. With that, the evening was transformed into an open discussion on the future of the Mars Society with a stream of individuals approaching two microphones to offer suggestions and ideas, ranging from project ideas to management suggestions to long-term goals. While offering ideas is relatively painless, volunteering to execute them is something else entirely. Yet, time and again, folks at the microphone would offer a suggestion and then volunteer to see it through. It became obvious that Society members were ready to take action as the final fifteen minutes of the evening involved on-the-spot chapter organization as members from various regions across the country came together to meet one another and set early plans for activities back home after the Convention.

By Sunday's final session, it became apparent that the Founding Convention had been more than just a meeting of Mars enthusiasts. True, the convention offered memorable appearances by such luminaries as astronaut John Young and Mars Pathfinder project scientists Matt Golombek; it provided reams of information for anyone interested in the planet Mars and its exploration; it gave Mars enthusiasts a forum in which to meet one another and swap ideas and tales. But just about any decent conference on any subject will offer up a similar collection of services. The Mars Society Founding Convention did more, though, by allowing individuals to become part of an organization; by allowing them to channel their energies and their own talents into a movement that will promote human Mars exploration. It would be hard to find more appropriate words to end the Convention than the ones spoken by Robert Zubrin to close the final session of the day.

"We're gonna win. On to Mars!"

Former NASA Associate Administrator for Exploration, Mike Griffin, signs the Founding Declaration of The Mars Society.

FOUNDING DECLARATION OF THE MARS SOCIETY

The time has come for humanity to journey to Mars.

We're ready. Though Mars is distant, we are far better prepared today to send humans to Mars than we were to travel to the Moon at the commencement of the space age. Given the will, we could have our first teams on Mars within a decade.

The reasons for going to Mars are powerful.

We must go for the knowledge of Mars. Our robotic probes have revealed that Mars was once a warm and wet planet, suitable for hosting life's origin. But did it? A search for fossils on the Martian surface or microbes in groundwater below could provide the answer. If found, they would show that the origin of life is not unique to the Earth, and, by implication, reveal a universe that is filled with life and probably intelligence as well. From the point of view learning our true place in the universe, this would be the most important scientific enlightenment since Copernicus.

We must go for the knowledge of Earth. As we begin the twenty-first century, we have evidence that we are changing the Earth's atmosphere and environment in significant ways. It has become a critical matter for us better to understand all aspects of our environment. In this project, comparative planetology is a very powerful tool, a fact already shown by the role Venusian atmospheric studies played in our discovery of the potential threat of global warming by greenhouse gases. Mars, the planet most like Earth, will have even more to teach us about our home world. The knowledge we gain could be key to our survival.

We must go for the challenge. Civilizations, like people, thrive on challenge and decay without it. The time is past for human societies to use war as a driving stress for technological progress. As the world moves towards unity, we must join together, not in mutual passivity, but in common enterprise, facing outward to embrace a greater and nobler challenge than that which we previously posed to each other. Pioneering Mars will provide such a challenge. Furthermore, a cooperative international exploration of Mars would serve as an example of how the same joint-action could work on Earth in other ventures.

We must go for the youth. The spirit of youth demands adventure. A humans-to-Mars program would challenge young people everywhere to develop their minds to participate in the pioneering of a new world. If a Mars program were to inspire just a single extra percent of today's youth to scientific educations, the net result would be tens of millions more scientists, engineers, inventors, medical researchers and doctors. These people will make innovations that create new industries, find new medical cures, increase income, and benefit the world in innumerable ways to provide a return that will utterly dwarf the expenditures of the Mars program.

We must go for the opportunity. The settling of the Martian New World is an opportunity for a noble experiment in which humanity has another chance to shed old baggage and begin the world anew; carrying forward as much of the best of our heritage as possible and leaving the worst behind. Such chances do not come often, and are not to be disdained lightly.

We must go for our humanity. Human beings are more than merely another kind of animal, - we are life's messenger. Alone of the creatures of the Earth, we have the ability to continue the work of creation by bringing life to Mars, and Mars to life. In doing so, we shall make a profound statement as to the precious worth of the human race and every member of it.

We must go for the future. Mars is not just a scientific curiosity; it is a world with a surface area equal to all the continents of Earth combined, possessing all the elements that are needed to support not only life, but technological society. It is a New World, filled with history waiting to be made by a new and youthful branch of human civilization that is waiting to be born. We must go to Mars to make that potential a reality. We must go, not for us, but for a people who are yet to be. We must do it for the Martians.

Believing therefore that the exploration and settlement of Mars is one of the greatest human endeavors possible in our time, we have gathered to found this Mars Society, understanding that even the best ideas for human action are never inevitable, but must be planned, advocated, and achieved by hard work. We call upon all other individuals and organizations of like-minded people to join with us in furthering this great enterprise. No nobler cause has ever been. We shall not rest until it succeeds.

www.marssociety.org

Chapter 9
SOFTWARE AND AUTONOMY

Taking a break in the refreshment room.

USING COTS SOFTWARE FOR MARS MISSIONS

Ned Chapin*

The traditional way of providing software to support space missions has been to build custom systems from scratch and then modify them to accommodate changes (the process of "software maintenance"). With the improving availability of Commercial Off-The-Shelf (COTS) software components, using COTS components extensively in building and maintaining systems is a new alternative. While we have nearly always used some COTS software components, now we can use them much more in systems associated with the exploration and settlement of Mars.

Using COTS software components has both pros and cons. The major pros can include:

1. Lowers the cost to build and maintain the system.
2. Shortens the time to build the system.
3. Encourages the use of open standards in building and maintaining the system.
4. Eases using "spare parts" during maintenance, and reuse during development.
5. Makes component interactions more explicit.
6. Encourages encapsulation, table-drive, and instrumentation.

The major cons can include:

1. Requires either customizing the software to fit the way people want to work, or making the way people work fit the COTS software.
2. Constrains the kinds and forms of changes in systems.
3. Slows making changes in the system.
4. Adds to the diversity of styles, conventions, and practices in the software.
5. Increases the difficulty and cost of making functional enhancements.
6. Lengthens the time to find and fix performance faults and shortfalls.

Given Mars mission requirements and eventualities, achieving a balance between the pros and cons in using COTS software components involves making tradeoffs. Implementing them includes the use of tailoring code, wrappers and middleware, reuse, certification, and effective management.

TRADITIONAL TECHNIQUES

The aerospace industry has frequently lead the way in seeking, refining, and adopting improved techniques for the development, use, and maintenance of computer software. Since World War II, the aerospace industry has pioneered time and again with a wide range of techniques from project management, computer-controlled machine tools, heads-up display support, simulations, configuration management, and IV&V (independent verification and validation) to virtual reality. Aerospace organizations have been among those at the forefront in work to raise the level of maturity in computer software work.[8] Software support has become critical in the performance of aerospace missions in general, and Mars missions are not an exception. Yet software has received relatively little explicit attention in the Mars literature. For example, the

---

*   Information Systems Consultant, InfoSci Inc., Box 7117, Menlo Park California 94026-7115.
     E-mail: NedChapin@aol.com.

Proceedings of even the 1988 *NASA Mars Conference* were effectively devoid of any explicit coverage of software and its critical role.[16]

Computer programmers and analysts had already recognized by 1952 that developing computer applications involved a mix of repetition and innovation.[17] Whenever the analysts and programmers developed new applications, and especially in new domains, the analysts faced the challenge of innovation in harnessing the power of the computer to accomplish work in those new applications. Orbital mechanics and commercial banking applications were not the same, for example. Yet at the same time, when developing new applications the analysts and programmers found themselves handling again and again many situations that were nearly identical. Keeping track of past conditions and limiting error in a series of multiplications were common to orbital mechanics and commercial banking applications, for example.

Pioneers in computer software introduced three major techniques to reduce the cost of handling the repetition.

- One technique was the introduction of programming languages of increasing power and sophistication. Some commands in such languages enabled the programmers to direct with a single command what took many commands in assembly-level or machine-level languages, as for example, "Sort".[17]

- A second technique was to recognize and prepare routines and subroutines for functions of various sizes. These are pre-written and tested applications that can be reused to handle frequently encountered functions, such as "Cosine," "Find Square Root," or "Sort." Some can be easily tailored or customized to fit varying requirements.

- A third technique used primarily by analysts in designing applications was to modularize the applications in terms of the known inventory of available language commands, routines, and subroutines. For example, if the implementing programming language to be used lacked a "Sort" command, then a needed sorting function was isolated into a module for the purpose of implementing it with a subroutine. The modules then became effectively components of the application software.

Some companion practices also arose gradually. This started with basing a new application on the software developed for a prior (and usually different) application, by copying the prior application and then while leaving some components unchanged or lightly modified, discarding some components of it, replacing other components, and introducing some new components. This led to the recognition of the potential use of customizable software templates, either for components of applications or for entire applications.[15] Furthermore, it lead to a recognition of the potential advantages of a formal reuse of software components.[11]

Commercial firms saw the opportunity to market applications implemented in software ("packaged applications"), usually focused on specific industries and their more common applications, such as production planning for organizations doing manufacturing. More recently, commercial firms have seen an opportunity to market reusable components for use in building and maintaining computer applications. These reusable components are usually the equivalent of subroutines, although sometimes of templates, and more rarely, specific packaged applications. In their modern forms, these components from commercial sources are becoming known as Commercial Off-The-Shelf (COTS) software components. Their use and any more than incidental use of software components generally is becoming known as Component-Based Development (CBD) and Component-Based Maintenance (CBM).[12]

World-wide, the last half-century has seen billions of lines of computer software designed, written, tested, put into use, and frequently modified to provide increased functionality and robustness in the use of the software. The majority of that software was custom-developed in high-level and assembly-level languages, with packaged applications coming in as a distant second. The primary reuse of software has been of subroutines, but their contribution has been but a few percentage points. Template usage has contributed less than one percent, although the concepts have been used since the mid-1950s. These general patterns apply also to aerospace software in general and Mars missions in particular, but with less use of packaged applications and more use of subroutines than the world-wide pattern.

The newcomer is CBD and CBM, especially with COTS software components, which have been touted as applicable to aerospace applications in general and to Mars missions in particular, with their modern emphasis on cheaper, faster, and better. This paper is not limited to the offerings of any one or several vendors,[7] as it examines the pros and cons of using CBD and CBM in systems associated with the exploration or settlement of Mars. By "developing" is meant the specification, analysis, design, implementation, testing, and initial deployment of the software portion of any system proposed to be involved in or affected by the exploration or settlement of Mars. By "maintaining" is meant the aggregate of all post-initial-deployment activities associated with any existing system proposed to be or involved in or affected by the exploration or settlement of Mars. The post-deployment activities may include evaluation of existing and proposed systems and their interactions, specification, analysis, design, implementation, testing, redeployment, and performance monitoring.[3,9] In practice, Mars projects like most aerospace projects spend about four times more for the maintenance of the software of a system than they do for the development of that software.

THE PRO SIDE OF USING COMPONENTS

Five Basic Factors

Proponents of using CBD and CBM list many supporting arguments. Three general and two specific COTS factors underlie many of these arguments.

- One general factor is that the software development and maintenance work done by the potential user or customer (hereafter called "user") has to be done in a manner congenial with functional modularization. This is a practice consistent with the principles of software engineering,[14] and has been pointed to as a desirable practice in software intended for the support of Mars missions.[2] Other than being used trivially or incidentally, the use of software components is rarely a viable alternative without modularization in the software.

- The second general factor is the potential of CBD and CBM. Component vendors and suppliers, COTS and non-COTS alike, have the opportunity and the capability to make their components broadly usable on many computer platforms. Further, they can implement their components in any language of their choice, yet make their components work in most operating system environments.[12]

- The third general factor is that the creators of software components can make them be easily and broadly customizable so that prospective users can easily tailor them to fit their specific needs. Again, not all creators of components do this without burdening the prospective user with complexity and slow performance.

- The first specific factor is that the commercial creators of COTS software components can concentrate enough effort in design, implementation, and testing of their components to make them more robust and reliable than they would be if developed in-house by a prospective user—although not all do so.

- The second specific factor is that the creators of COTS software components can provide a comprehensive package including detailed specifications, full documentation, test data, test procedures, test results, a built-in help facility, and a telephone-in help service. Many COTS component providers do not provide such a comprehensive package because it raises their costs, costs that they usually must recover from the price paid by users.

These five factors have the same significance for Mars missions as for other users of software components in CBD and CBM. They are among the variables to be weighed in deciding what components to use, but their import is not of special importance to Mars missions. Hence, they are only lightly considered further in this paper.

Lower Cost to Build and Maintain

"Cheaper" and "better" are key concepts in planning and executing Mars missions. Using software components offers the potential for realizing both of them, compared to traditional and likely software development and maintenance practices.[19] Creating software components in-house using the organization's usual development practices (instead of acquiring the components from a vendor or supplier) is subject to all the usual difficulties and costs of developing software. High quality software is difficult to achieve reliably in the absence of the opportunity for tight specialization. A vendor or supplier of a component may have the opportunity for a tight specialization, and by spreading its costs over many users, may have the opportunity to offer a price lower than the customers' in-house costs of producing an equivalent-quality software component themselves.

The arguments presented for a lower cost to build and maintain software by using software components include the following:

- The quality of the component software is higher when the component vendor or supplier has built quality into the component, than could likely be achieved by the user attempting to develop or maintain the component in-house for itself;

- The quality of the software system into which the component is integrated by the user is usually higher because the software system now includes a component that is of high quality;

- The cost of software component is less than the cost of an in-house developed or maintained component, assuming the user has chosen the lower-cost alternative; and

- The cost of the software system into which the component is integrated by the user is lower because the software system includes a new component that is of lower cost. This assumes that the cost to integrate the component is not greater than would have been the cost of integrating an in-house developed or maintained component in its place.

Shorter Time to Build the System

"Faster" is a key concept in planning and executing Mars missions. The arguments presented for a shorter time to develop software by using software components include the following:

- An available software component can be delivered ready to use to the user quickly, typically in a few minutes to a few days, in contrast to the usual development time in-house of from a few weeks to a few months;

- The time to integrate a well-chosen software component from a vendor or supplier into a software system should be no longer, excluding any tailoring or customization, than the time to integrate an in-house developed component in its place;

- The time to establish the requirements of a component, including obtaining reviews and approvals, should be the same for a software component from a vendor or supplier and a proposed in-house developed component, since checking the character and availability of suitable components should normally be part of the requirements process anyway; and

- The contracting process to obtain a software component from a vendor or supplier should be reduced to a brief routine process. While this may incur a cost and some time to set up, its routine use should, like shipping, be included as part of the normal cost and time estimates for components from vendors or suppliers, just as personnel overheads (such as vacation) are included as part of the in-house development time and costs.

Use of Open and International Standards

The use of standards in software work helps make the software components fit together, such as the use of standards for pipe threads enables pipe and pipe fittings (such as "Ts" and "Els") to be used regardless of the manufacturer. Since software components must be fitted or integrated into a software system, having them conform to any applicable standards helps them fit together quickly and at low cost.

Open standards set by groups of firms (such as "CORBA"), standards set by professional associations (such as by the IEEE), and standards set by nations and internationally (such as ISO standards) are considered to be especially helpful by smaller vendors and suppliers of components because the standards give the potential users the widest choice of suppliers. Proprietary standards, such as set by a prominent or dominant firm in the industry (such as IBM or Microsoft) help software specifically designed for those environments to fit together easily, but limit the user choices to what works in those environments.

For COTS components for use in Mars mission software, standards of particular importance are those about programming languages (such as for FORTRAN), documentation methodologies such as the Integrated Data Engineering Facility (IDEF) and the Univeral Modeling Language (UML), data interchange codes such as the American Standard Code for Information Interchange (ASCII) and communication protocols (such as ICP/TP), and software management practices such as Software Configuration Management (SCM) and Software Quality Assurance (SQA).

Reuse and Spare Parts

While software components can be first introduced during the maintenance process, they are most easily used in maintenance when they were also used in development, although different components may be used during the different stages. For the development process, software reuse characteristics dominate; for the maintenance process, a "spare parts" approach is the objective for advocates of component usage. Because any one Mars mission is a member of a family or series of Mars missions, reuse and "spare parts" offer attractive ways of saving cost and time, and of keeping software quality high.

Software reuse is the process of using software components and other items of software during development and maintenance, where the components and other items have been previously used as parts of other systems or are software components from vendors or suppliers. Reusing software items and components and especially COTS components is regarded as using available tested known components whose costs have already largely been charged elsewhere. For component reuse to happen, the specifications of the component must match the requirements of the system the component is to go into. This requires a rigorous and full definition of the functionality, expected and sought, and a mitigation or elimination of side effects. While this focus is consistent with the modularization practices noted earlier, the traditional practice has been to skip lightly in applying this focus. Mismatches between the specifications and the requirements, and the side effects, then are handled as software defects—i.e., faults to be corrected (hopefully before deployment!)

The term "spare parts" refers to an analogy between a hardware maintenance practice and one variety of software maintenance.[6] In maintaining electronic hardware, for example, often a quick way of doing it is to pull a faulty or obsolete circuit board and replace it with a circuit board believed to be free of faults or of a new design with changed capabilities. Functionality must still be matched or be deliberately changed as a part of the maintenance done.

Component Interactions

Advocates of software component usage also applaud the emphasis it encourages on the interactions of components and on the interfaces among components. This casts a search light on what often is only dimly visible about software—how the component parts interact. This can be especially helpful in software to support Mars missions, where maintenance has to be done quickly and accurately to meet fast-changing conditions. The components may be of many sizes, levels, and sources, yet all contribute by their interactions to the performance of the software system.

This visibility helps in software development by facilitating the work of different personnel and personnel working at different times or places because it helps show how the parts are integrated and fit together. This helps in setting requirements for software components and helps in locating components of particular specifications.

This visibility helps also in measuring accomplishment during software development and maintenance. The interfaces are between the components of the system, and between some components of the system and the environment in which the system operates, including interfaces with human users. The visibility comes from management having a complete catalog of the interfaces and then being able to note the status of each one during development and maintenance.

Encapsulation, Table-Drive, and Instrumentation

Using software components encourages three practices that usually improve software quality, and very rarely impair it. All three appear attractive as ways of improving the contribution of software systems in carrying out Mars missions. They are encapsulation, table-drive, and instrumentation.

- Encapsulation isolates aspects of the software to particular components, so that other components of the system can access or affect them only in a controlled manner. Object-oriented software typically makes use of encapsulation.[10] For the greatest effect, encapsulation has to be included in the architecture of the software system, and typically involves concentrating the "ownership" and use of data among a limited number of software components. This "information hiding" helps limit faults in software to specific components.

- Table-drive concentrates decision making in the software to data tables associated with particular software components. The system keeps the status of the system's operation updated continuously in data tables, and components of the system that make decisions about what to do and when to do it, access the tables. This availability of an overall view helps improve the robustness and reliability of the performance of the software system.

- Instrumentation helps in tracing to their causes, faults and defects in the performance of the software system. Instrumentation consists of inserting optional data capture instructions between the components to record the activity at the interfaces of the components and the functionality of the components. This is especially helpful with COTS software components since the information systems personnel typically have quite limited access to the interior workings of the COTS components.

THE CON SIDE OF USING COMPONENTS

Three Basic Factors

Opponents of using CBD and CBM list many arguments to support their position. Three factors of importance for software to support Mars missions underlie many of these arguments.

- The component needed cannot be found. This is a matter of availability in the COTS market place and availability from alternative sources. Some of these alternative sources include personal contacts, shareware listings, in-house repository of reusable software, existing software systems in the organization, on-line forums, computer hardware or software user groups, contacts from trade and industry associations, and contacts from professional associations. While sometimes no software components can be found that even vaguely meet the requirements for a component, usually the availability question comes down to ascertaining the closeness of the match between the specifications and the requirements. The quality of the documentation on both sides is critical in this matching process. The process of finding candidate components and of evaluating their match to the requirements incurs costs and takes time.

- The time and cost needed to make any of the available components match the requirements (i.e., tailor or customize them) may exceed the cost of an in-house development of a component. Most COTS components come with major restrictions on making modifications to them, while components from other sources are usually subject to fewer or even no restrictions.

- The quality of the available components is sometimes not known with confidence, a situation true for COTS components as well as components from other sources. The quality may have been assessed in an environment different from the one characterizing a specific Mars mission.

Customization of Components and Work Practices

Modifying software components incurs costs and takes time to tailor or customize their performance to meet user requirements. This raises the question of how much various modifications would cost and how much additional value the user would get from the software system incorporating the modified component. This question gets more significant the bigger the difference is between the requirements and the specifications. When the gap is large, as is sometimes the case with big COTS software components, the practical choice becomes:

- Modify the software component so that its specifications match the user requirements closely enough to permit the user to work the way it wants to, or
- Modify the way the user works enough to match the specifications of the available software component.

The first alternative (modify the component) incurs costs and takes time. The result is that the user personnel feel that they are in control in setting the way they do their work, and that the software system helps them do their work. Being able to set their own work practices helps the personnel define their group identify and turf, and helps keep up their motivation and morale. For example, a group focusing on the hydrological history of Mars will almost certainly believe the computer-implemented systems they use should do what their combined knowledge, skill, and experience dictate.

The second alternative (modify the work practices) also incurs costs and takes time, and because the user personnel have to walk a learning curve, they are likely to be less productive, make more errors, and produce work of a lower quality for awhile. They typically feel that the software system is dictating what they do, how they do it, and when they do it, and that they have lost control of their own destiny while they are walking the learning curve. They feel themselves to be less of a group and more like cogs in a wheel. For example, a group focusing on the hydrological history of Mars will almost certainly believe they are being forced to put up with time-wasting clerical work if they find themselves feeding data differently into a computer system because the system requires a data form and format different from what the personnel want to use.

Attempts to compromise and seek a middle ground at an acceptable cost and time by taking from each alternative are typically regarded as introducing some pollution, especially by the personnel most affected by changes in work practices. This discord within the group typically results in some reduction in motivation and morale.

Constraints on Modifications

COTS software components and components from other than in-house sources are rarely sold to the user in order to protect the vendors' or suppliers' intellectual property rights. Instead, the vendor or supplier usually licenses the components for use by the user with some constraints on modifications. These constraints are of special significance for software to support Mars missions because they affect the modifiability of the software that includes such components. These constraints usually take one or more of the following forms:

- Modifications are permitted if specifically approved in advance in writing by the COTS vendor or the component supplier,
- The vendor or supplier of the component licenses the component in a form inconvenient for modifications (such as object-code only) and with skimpy or no detailed documentation,
- Modifications not specifically approved in advance in writing, invalidate all or some of the COTS vendor's or the component supplier's warranties,

- Modifications not specifically approved in advance in writing, terminate or restrict the customer's (user's) right to receive vendor-made or supplier-made modifications, such as new releases, upgrades, or corrected versions of the software component,
- The vendor-made or supplier-made modifications counter or fail to work with the modifications previously made by the user, or
- Modifications not specifically approved in advance in writing, terminate the license, usually without recourse, and making any use by the (formerly) licensed user be a violation of the contract.

For non-COTS software components, the constraints are usually less onerous, but usually still present. A few software components are truly "freeware" in that the vendor or supplier makes the components available gratis, and asks no contract. Sometimes the vendor or supplier of freeware even provides the component in a convenient to modify form (such as source-code form), but providing gratis detailed documentation is very rare.

Slower Modifications

Making modifications to (i.e., doing maintenance on) COTS software components typically takes more time than does maintaining non-COTS components. In turn, maintaining non-COTS externally-acquired components is slower than maintaining in-house provided components. However, maintaining in-house provided components is typically somewhat faster than is maintaining non-component-based software.

Among the reasons for the slower maintenance process for software components are these:

- A configuration management system is needed to keep track of the status of all software components used in a system.
- Written permission in advance must be obtained from the vendor or supplier—typically a relatively slow paperwork process,
- Interactions among the software components have to be explored to confirm that a proposed modification in one component does not trigger unwanted consequences in one or more other components,
- Component vendors or suppliers may on a regular or irregular schedule issue modified versions of their components, so that proposed changes must be coordinated with the diverse release schedules of the vendors or suppliers,
- The user typically cannot directly make corrective modifications in a component without the express permission of the vendor or supplier, a permission that is not always granted ("That's a feature, not a bug!") and even if granted comes only after a delay,
- When faults or defects ("bugs") appear in the performance of component-using software, the interacts of the components usually results in each component vendor or supplier claiming that the difficulty is caused by some other vendor's or supplier's component (such "finger pointing" takes time to resolve),
- While the user may elect not to install and use the vendor-provided or supplier-provided modified components (new version releases), there may be license constraints to be satisfied that require advance planning by the user, and

- The user typically has very limited access to the inner workings of the component, and hence has to modify the component indirectly by modifying instead the data it uses or the data it produces or both.

Added Diversity

CBD and CBM can lead to a diversity in styles, conventions, and practices affecting maintenance and system security.[21] The richness of platform and language independence that component usage can yield, also leads to a richness of demands upon the personnel charged with making modifications to the software system. Everyone cannot be an expert in everything. The more differences in the software system on the hardware platforms used, the operating system environments used, and the computer programming languages used, mean that more personnel skills and knowledge are needed in doing the maintenance work. That typically translates to more personnel time, slower maintenance, and added cost—matters of importance for the support of Mars missions.

Added to that is the diversity of styles and practices embodied in the components themselves. Each component from a different source is nearly certain to be different in the design philosophy used and in the way it is used, in the way it has been implemented, and in the documentation provided, both in form and content. Underlying these differences are usually differences in architecture, because some components may fit much better in some software system architectures than others. Incompatibilities among the components and their respective fits to and in the overall system architecture result in slower and higher cost maintenance of the resulting software system—matters of importance for the support of Mars missions.

Difficulty in Functional Enhancement

Mars missions need to have built into them, like space missions generally, a flexibility and adaptability to handle with software maintenance, both planned and unexpected eventualities. Examples are different demands in different phases of a mission, and hardware failures such as an unexpectedly dead battery. In practice, this requires an ability to revise the functionality of a software system, easily, quickly, reliably, and hopefully inexpensively. Since usually functions have to be added to the software system, the variety of maintenance concerned with the revision of a system's functionality is usually termed functional enhancement or just enhancement.

For systems using software components, all of the con factors listed above in the three prior subsections of this paper add to the costs and slowness of enhancement. Some additional contributors are these:

- Usable COTS components may not be available, and usable non-COTS components cannot located,

- The component vendor or supplier of a component that could be modified to be usable may refuse permission to incorporate the needed enhancement,

- The component vendor or supplier of a component that could be modified to be usable may grant permission but either only after a significant delay, or insist on an unacceptable schedule, or insist on doing the enhancement itself on it own schedule in its own way,

- Preparing a request for permission to enhance a component takes about the same time and cost as preparing to make the enhancement in-house, and

- Multiple vendors or suppliers may have to be coordinated and brought to an agreement before an enhancement can be done involving a component.

Reduction in Corrective Responsiveness

While components whether COTS or non-COTS, can often claim to be pre-tested and to perform reliably, sometimes a software system fails in use because of a defect or fault in one or more of the components. Finding the cause of such bugs—failures to perform as specified—is difficult when one or more COTS or other components is involved. The interior workings of all but the in-house supplied components is usually protected; at best all that can be seen even with instrumentation are the messages going in and out of the component. While that can show that the component is involved in the failure to perform as specified, it rarely shows the cause of the failure, and does not normally enable any correction to be made to the component.

Instead, a pause ensues for documenting the involvement of the component, contacting the component vendor or supplier, asking it for correction, waiting for the vendor or supplier to analyze the situation and frame a response to the request and deliver the response. If the response is favorable, the vendor or supplier usually will schedule including some form of the correction in one of its future scheduled releases of a new version of the component. That rarely is less than a month away, and may be as much as a year. In addition, there is no assurance that the vendor's or supplier's correction when delivered will actually work in the user's environment.

Such delays can be serious on a Mars mission. Sometimes the window of opportunity is limited in which to get the software system working as specified, as when a human exploration team encounters something unexpected, or a space craft is to make a course correction. The consequences of missing the window are sometimes irreversible.

DISCUSSION

Significance for Mars Missions

For Mars missions in this era of cheaper, faster, and better, two of the pro factors are powerful—lower cost and faster development. But also powerful are the combined effect of the con factors—slower, less responsive, and sometimes more costly maintenance.[1] To that is added the usual experience: each dollar of development cost spent gets followed by about four dollars of spending for maintenance.

Partially offsetting some of the maintenance burden are the use of the pro factors of open standards, spare-parts maintenance, encapsulation, table-drive, software instrumentation, and component interaction visibility. Each of these depends not upon the use of components themselves but upon the management of the development and maintenance of the software systems used to support Mars missions. Only a rigorous, consistent, and persistent application of well-chosen local standard operating practices in software development and maintenance can achieve those offsets.

Clearly, having a well-defined software system architecture and being careful in the acceptance for use in the software system of the COTS and non-COTS software components to fit that architecture can greatly affect the success both CBD and CBM. For Mars missions, one requirement is always present—the ability to modify the software systems quickly and accurately to meeting changed conditions. That does not happen by itself during either development or maintenance. Maintainability has to be among the three top objectives during both development and maintenance, for otherwise as shown by experience, maintainability is lackluster or worse.

To help achieve a balance between the pro and con factors for Mars missions of component-based software systems, four techniques or practices can help. They are tailoring or customizing the code, using wrappers and middleware, encouraging in-house software reuse, and recognizing component certification.

Tailoring Code

Since most Mars mission will have some features and requirements in common, the opportunity exists to provide the software support for one mission by reusing portions of prior systems as components in new systems. Since all Mars missions will have some unique features and requirements and be able to draw upon a growing pool of experience, skill, and knowledge, tailoring or customizing the software systems will be essential.

Customizing or tailoring the system to do what the users want is nearly always needed for component-based software systems. Modifying a component-based software system is usually done in one of three ways to make it more closely act as and do what the systems' users want:

- Components can be included that alter what the system would otherwise do;
- In-house personnel can prepare and insert software specifically to modify the system, just as they could do with the organization's other software; and
- Modifications can sometimes be made directly to existing non-COTS components to change their functionality.

All of these add to costs and take time, but can yield benefits in making Mars mission software support be more effective.

Wrappers and Middleware

Wrappers are like masks and buffers—they make a component appear to act differently than it does. They may also act like translators, converting data on their way into or out of a component. Further, they may act to disguise the performance of a component so that it appears to be different. Wrappers are a major tool for customizing or tailoring a software system involving components and especially COTS components, and are usually prepared in-house by the organization's own personnel.

Wrappers get their name from the way they work. They act to intercept all data going to or coming from a specific component, and may under specific circumstances transform those data. The two main difficulties in using wrappers are handling the action of data interception, and in assessing which transformation to apply and when the conditions call for applying it. An example of a wrapper is software that permits a component written for a specific computer operating system environment to execute as if it had been written for a different operating system environment.

Middleware is software that acts as a buffer and communication link among or between components, and sometimes called "glue code." Its performance is similar to a wrapper, but is both more limited and more general. It is more general in that it usually applies to large components or groupings of smaller components. It is more limited in that its action usually focuses on data format or coding, and usually ignores the data values. While middleware is sometimes prepared in-house, some middleware is available as components, both COTS and non-COTS.

Using wrappers and middleware increases the pool of components that can be drawn upon in developing and maintaining software systems for the support of Mars missions. While both wrappers and middleware can be used in customizing or tailoring in CBD and CBM, wrappers can be more precisely targeted than can middleware, but are also more difficult to develop and maintain.

Reuse

Encouraging software reuse increases the pool of components that are potentially useful for incorporation in software for the support of Mars missions. The most fruitful places to look for possibly reusable software is in the software used with prior space missions that had to do a task or function similar to one expected on a Mars mission. A simple example is storing data transmitted from a remote probe.

The requirements to satisfy to make software reusable have been closely and extensively examined.[11] In brief summary, a few of these are:

- Clean and full documentation of functionality, interface, and design,
- Full specification of the operating environment assumed,
- Full disclosure of resource demands and side effects (such as of working storage needed), and
- Implementation faithful to the design and consistent with accepted standards.

Certification

Certification looks to the quality of components. It is a way of assuring potential users of a component that the component will reliably perform as specified. It adds the creditability of an independent certifying agency to the component vendor's or supplier's claims. To obtain certification, the vendor or supplier hires the certifying agency to examine the component and its performance, including usually independently testing it. If the certifying agency finds the component to live up to or exceed the vendor's or supplier's specification for it, the certifying agency then issues a certification that the vendor or supplier can use in its marketing literature.

The certifying agency accepts no responsibility for fitness for use, and no liability for any faults, defects, or other deficiencies. Software certification is still relatively uncommon but increasing as component usage increases, especially COTS usage. The value of certification depends partly on the thoroughness of the certification process used. That value can assist a potential user in selecting components when choices exist.

CONCLUSION

Software components **will** be used in developing and maintaining software supporting Mars missions. The questions are how much will they be used and for what? Management will see the pro factors as dominant, and treat mitigating the con factors as being primarily technology matters. Answering the "how much" and "for what" questions in practice will involve making tradeoffs, both on the management side and the technology side. This paper has reviewed the major pro and con factors affecting the tradeoffs.

In a larger context, the questions of "how much" and "for what" are actually particular forms of the classic "make or buy" decision that continually arises in organizations.[4] Resource availability, personnel skill and knowledge inventory, management expertise, mission requirements, and time schedule often are the critical factors in making the decision, and are not unique to Mars missions. When those factors are neutral or offsetting on net balance, then the CBD and CBM pro and con factors are decisive.

Management can do a lot to bolster the pro factors and to minimize the con factors. Raising the computer maturity of the software work to at least the equivalent of the CMM level 3, and extending the improvement to software maintenance are the two most significant steps.[8] Absent that, management can take some less comprehensive but still effective steps that also pull up the information technology, such as:

- Adopt and enforce consistently a set of local standards and standard practices, as for example, Cleanroom software,[13] covering all aspects of software development and maintenance, and embody them in a SOP (standard operating practices) manual,

- Include in the SOP a requirement that software be designed to be and be implemented as relatively small modules and/or as objects with relatively brief methods,[10] but in either case with data interfaces fully defined so that they can be put into a reuse repository,

- Include in the SOP a requirement that software reuse is supported and is the default choice as the source for components in development and maintenance,

- Include in the SOP that an integrated data dictionary is to be used for all data anywhere (including in objects) in the software system except those data encapsulated within a single module,

- Organize the software work so that each step is self-documenting and is done in a way that checks the quality of the work done previously done, and

- Encourage the use of strong SQA (software quality assurance) techniques such as inspections[5] and IV&V (independent verification and validation).[18]

The balance between the CBD and CBM pro and con factors is an ever shifting one that leaves a residual in the software system. As the system ages and accumulates the consequences of the compromises, the original system architecture degrades and maintenance becomes more costly and takes more time. Component-based software systems are not immune to this and systems to support the exploration or settlement of Mars can be expect to follow this usual pattern.

REFERENCES

1. D. J. Carney and P. A. Oberndorf, "The Commandments of COTS: Still in Search of the Promised Land," *CrossTalk*, Vol. 10, No. 5, May 1997, pp. 25-30.
2. N. Chapin, "Adaptable Software Needed for Mars Missions," a *Case for Mars V* paper pending publication in the American Astronautical Society, Science and Technology Series, Univelt, San Diego CA, 1993.
3. N. Chapin, "Software Maintenance Life Cycle," in *Proceedings Conference on Software Maintenance-1988*, IEEE Computer Society, Los Alamitos CA, pp. 6-13.
4. P. F. Drucker, *Management*, Harper & Row, Publishers, New York NY, 1974.
5. D. P. Freedman, and G. M. Weinberg, *Handbook of Walkthroughs, Inspections, and Technical Reviews*, Little Brown & Co., Boston MA, 1982.

6. T. Gilb, "Spare Parts Maintenance Strategy for Programs," *Computer Weekly*, Vol. 21, No. 502, May 16 1977, pp. 8-11.
7. D. N. Gray, J. Hotchkiss, S. LaForge, A. Shalit, and T. Weinberg, "Modern Languages and Microsoft's Component Object Model," *Communications of the ACM*, Vol. 41, No. 5, May 1998, pp. 55-65.
8. W. S. Humphrey, *Managing the Software Process*, Addison-Wesley Publishing Co., Reading MA, 1989.
9. IEEE, *IEEE Standard for Software Maintenance, IEEE Std 1219-1993*, Institute of Electrical and Electronic Engineers, Inc., New York NY, 1993.
10. I. Jacobsen, M. Christerson, P. Jonsson, and G. Övergaard, *Object-Oriented Software Engineering*, Addison-Wesley Publishing Co., Reading MA, 1992.
11. E.-A. Karlsson, Editor, *Software Reuse*, John Wiley & Sons, Inc., New York NY, 1995.
12. D. Kiely, "Are Components the Future of Software?" *Computer*, Vol. 31, No. 2, February 1998, pp. 10-11.
13. H. D. Mills, M. Dyer, and R. C. Linger, "Cleanroom software engineering," *IEEE Software*, Vol. 4, No. 5, May 1987, pp.19-25.
14. R. S. Pressman, *Software Engineering: A Practitioner's Approach*, McGraw-Hill Book Co., New York NY, 1996.
15. J. Ramanathan, and C. J. Shubra, "Use of Annotated Schemes for Developing Prototype Programs," *Software Engineering Notes*, Vol. 7, No. 5, December 1982, pp. 141-149.
16. D. B. Reiber, Editor, *The NASA Mars Conference*, Volume 71, *Science and Technology Series*, American Astronautical Society, Univelt, San Diego CA, 1988.
17. J. E. Sammet, *Programming Languages: History and Fundamentals*, Prentice-Hall, Inc., Englewood Cliffs NJ, 1969.
18. G. G. Schulmeyer and J. I. McManus, Editors, *Handbook of Software Quality Assurance*, Van Nostrand Reinhold, New York NY, 1987.
19. N. Talbert and J. McDermid, "The Cost of COTS," *Computer*, Vol. 31, No. 6, June 1998, pp. 46-52.
20. J. M. Voas, "Certifying Off-the-Shelf Software Components," *Computer*, Vol. 31, No. 6, June 1998, pp. 53-59.
21. Q. Zhong and N. Edwards, "Security Control for COTS Components," *Computer*, Vol. 31, No. 6, June 1998, pp. 67-73.

ADJUSTABLE AUTONOMY FOR HUMAN-CENTERED AUTONOMOUS SYSTEMS ON MARS

Gregory A. Dorais,[*] R. Peter Bonasso,[†] David Kortenkamp,[‡] Barney Pell[**] and Debra Schreckenghost[††]

We expect a variety of autonomous systems, from rovers to life-support systems, to play a critical role in the success of manned Mars missions. The crew and ground support personnel will want to control and be informed by these systems at varying levels of detail depending on the situation. Moreover, these systems will need to operate safely in the presence of people and cooperate with them effectively. We call such autonomous systems human-centered in contrast with traditional "black-box" autonomous systems. Our goal is to design a framework for human-centered autonomous systems that enables users to interact with these systems at whatever level of control is most appropriate whenever they so choose, but minimize the necessity for such interaction. This paper discusses on-going research at the NASA Ames Research Center and the Johnson Space Center in developing human-centered autonomous systems that can be used for a manned Mars mission.

1. INTRODUCTION AND MOTIVATION

Autonomous system operation at a remote Martian site provides the crew with more independence from ground operations support. Such autonomy is essential to reduce operations costs and to accommodate the ground communication delays and blackouts at such a site. Additionally, autonomous systems, e.g., automated control of life support systems and robots, can reduce crew workload [Schreckenghost et al., 1998] at the remote site. For long duration missions, however, the crew must be able to (partially or fully) disable the autonomous control of a system for routine maintenance (such as calibration or battery recharging) and occasional repair. The lack of in-line sensors that are sufficiently sensitive and reliable requires manual sampling and adjustment of control for some life support systems. Joint man-machine performance of tasks (referred to as traded control) can improve overall task performance by leveraging both human propensities and autonomous control software capabilities through appropriate task allocation [Kortenkamp, et al., 1997]. The crew may also desire to intervene opportunistically in the operation of both life support systems and robots to respond to novel situations, to configure for degraded mode operations, and to accommodate crew preferences. Such interaction with autonomous systems requires the ability to adjust the level of autonomy during system execution between manual operation and autonomous operation. The design of control systems for such adjustable autonomy is an important enabling technology for a manned mission to Mars.

---

[*] Caelum Research, NASA Ames Research Center, Mail Stop 269-2, Moffett Field, California 94035. E-mail: gadorais@ptolemy.arc.nasa.gov.
[†] TRACLabs, NASA Johnson Space Center, Houston Texas 77058. E-mail: bonasso@mickey.jsc.nasa.gov.
[‡] TRACLabs, NASA Johnson Space Center, Houston Texas 77058. E-mail: kortenkamp@jsc.nasa.gov.
[**] RIACS, NASA Ames Research Center, Mail Stop 269-2, Moffett Field, California 94035. E-mail: pell@ptolemy.arc.nasa.gov.
[††] TRACLabs, NASA Johnson Space Center, Houston Texas 77058. E-mail: schreck@mickey.jsc.nasa.gov.

1.1 Definition of Adjustable Autonomy and Human-Centered Autonomous Systems

A human-centered autonomous system is an autonomous system designed to interact with people intelligently. That is, the system recognizes people as intelligent agents it can (or must) inform and be informed by. These people may be in the environment of the autonomous system or remotely communicating with it. In contrast, a traditional "black box" autonomous system executes prewritten commands and generally treats people in its environment as objects if it recognizes them at all.

Typically, the level of autonomy of a system is determined when the system is designed. In many systems, the user can choose to run the system either autonomously or manually, but must choose one or the other. Adjustable autonomy describes the property of an autonomous system to change its level of autonomy to one of many levels while the system operates. A human user, another system, or the autonomous system itself may "adjust" the autonomous system's level of autonomy. A system's level of autonomy can refer to:

- how complex the commands it executes are
- how many of its sub-systems are being autonomously controlled
- under what circumstances will the system override manual control
- the duration of autonomous operation.

For example, consider an unmanned rover. The command, "find evidence of stratification in a rock" requires a higher level autonomy than, "go straight 10 meters." The rover is operating at a higher level of autonomy when it is controlling its motion as well as its science equipment than when it is just controlling one or the other. The rover overriding a user command that would cause harm or to automatically recover from a mechanical fault also requires a higher level of autonomy than when these safeguards are not in place. Finally, a higher level of autonomy is required for the rover to function autonomously for a year than for a day. Adjustable autonomy is discussed further in [Bonasso *et al.*, 97a][Pell *et al.*, 98].

1.2 Manned Trip to Mars Statistics

The fast-transit mission profile described in [Hoffman and Kaplan, 97] of a hypothetical manned Mars mission sets a launch date of 2/1/2014 arriving on Mars 150 days later on 7/1/2014. After a 619-day stay on Mars, the crew will depart Mars on 3/11/2016 and arrive on Earth after 110 days on 6/26/2016.

Crew time on a Mars mission will be a scarce resource. Figure 1 outlines a possible timeline of how a crew of 8 would allocate their time based on 600-day mission on the Mars surface. The figure is from the NASA Mars Reference mission [Hoffman and Kaplan, 97, p. 1-15] cited from [Cohen, 93]. In this figure, how the 24 (Earth) hours per day per person of the crew are expected to be used are averaged by task classes represented by each column. The production tasks, which differ throughout the mission, are separated into mission phases, represented by rows, of lengths indicated by the Mission Duration column whose total is 600 days.

This schedule is based on a crew of 8. However, we expect the crew size of the first manned Mars mission to be at least 4 but no more than 6. A smaller crew would probably result in a reduction in the total duration of manned surface excursions and an increase in the time devoted to system monitoring, inspection, calibration, maintenance, and repair per person. For example, during a recent 90-day, closed-system test at NASA Johnson Space Center, four crewmembers spent roughly 1.5 hours each day doing maintenance and repair. During this period, they had a continual (i.e., 24-hours per day) ground control presence and engineering support to help in these tasks [Lewis et al., 1998]. This 90-day test used only a small subset of the systems that would be involved in a Mars base. Also, the tasks described in Figure 1 do not include growing and processing food at the Mars base, but assume all food for the entire 600 days is pre-packaged and shipped from earth.

Personal: 14 hr/day					Overhead: 3 hr/day		Production: 7 hr/day	Mission Duration
Sleep, Sleep Prep, Dress, Undress: 8 hr	Hygiene, Cleaning, Personal Communication: 1hr	Recreation, Exercise, Relaxation: 1 hr	Eating, Meal Prep, Clean-up: 1 hr	One Day Off Per Week: 3 hr (per day)	General Planning, Reporting, Documentation, Earth Communication: 1 hr	Group Socialization, Meetings, Life Sciences Subject, Health Monitoring, Health Care: 1hr / System Monitoring, Inspection, Calibration, Maintenance, Repair: 1hr	Site Prep, Construction	90 days
							Verification	
							Week Off	7 days
							Local Excursions	50 days
							Analysis	
							Distant Excursion	10 days
							Analysis	40 days
							Week Off	7 days
							Distant Excursion	10 days
							Analysis	40 days
							Distant Excursion	10 days
							Analysis	40 days
							Week Off	7 days
							Distant Excursion	10 days
							Analysis	40 days
							Distant Excursion	10 days
							Analysis	40 days
							Flare Retreat	15 days
							Week Off	7 days
							Distant Excursion	10 days
							Analysis	40 days
							Distant Excursion	10 days
							Analysis	40 days
							Week Off	7 days
							Sys Shutdown	60 days

Figure 1 Possible time line for first Mars surface mission.

Although 600 days may sound like a long time, the time allocation schedule only has 7 distant excursions. In this schedule, a distant excursion is a multi-day trip on a manned rover to locations within a 5-day drive radius of the Mars base. Of course, the above schedule could be radically changed if a major malfunction were to occur. However, major changes may be necessary simply due to poorly estimating the time needed for the tasks.

Currently for a space shuttle mission, dozens of people are dedicated to system monitoring 24 hours per day. In order for the Mars crew members to achieve the low level of time for system monitoring, inspection, calibration, maintenance, and repair shown in the figure, the systems they use will need to reliably operate with minimal human assistance for nearly two years.

1.3 Motivation for Autonomy and Robots

Given the statistics described in the previous subsection, it is apparent that a successful mission that accomplishes scientific objectives (as opposed to a mission that just keeps the crew alive) will require advanced autonomy, both in terms of controlling the environment and in robots to replace manual labor. Automation and robots provide the following advantages:

- Increased safety of crew
- Increased safety of equipment
- Increased science capabilities
- Increased reliability of mission
- Decreased crew time for monitoring and maintenance
- Decreased mission operation costs
- Decreased demand on deep space ground antennas (DSN).

To illustrate the need for autonomy, consider the following example. Most of us trust the heating of our home to a simple autonomous system. It turns the heating unit on when the temperature sensed by a thermostat falls below an "on" temperature setting and turns the heating unit off when the temperature rises above an "off" temperature setting. We certainly wouldn't want to burden a crewmember with the need to continually monitor the temperature of each room in the Mars habitat turning on and off the heating unit as necessary if it can be avoided. Temperature control is just one of many tasks that must done continually to keep the crew alive and to enable them to accomplish their science objectives.

Making matters more complex, on Mars many systems will be interdependent making control more difficult than that described in the previous example. In most of our homes, the temperature control system is independent. We can turn the heating unit on and off as we please. One reason this is possible is because we are not constrained by our energy resources. For example, the water heater can be on at the same time as the furnace. We can use a large quantity of energy during the day and not have to worry about not having sufficient energy at night. On Mars, energy will be a major constraint. Using energy for one purpose may mean that energy will not be available for another important task at the same time or later. Crewmembers will not want to have to calculate all the ramifications of turning on a machine each time they do so to avoid undesirable effects. Sophisticated autonomous systems can continually manage constrained resources and plan the operation of various machines so there is not a conflict.

In summary, crew time is scarce on Mars and the system must be closed to calories (i.e., the crew cannot spend more calories producing food than the food provides for them!). The more computers and machines can free up crew time, the greater chance that the mission will meet all of its objectives.

1.4 Motivation for Adjustable Autonomy

It is impossible to predict what will be encountered on a Mars mission. Given the current state-of-the-art in autonomous systems, it is critical that human intervention be supported. Allowing for a an efficient mechanism for crew to give input to an autonomous control system will provide the following benefits to the mission:

- Increased autonomous system capabilities
- Decreased autonomous system development and testing cost
- Increased reliability of mission
- Increased user understanding, control, and trust of autonomous systems.

The goal of adjustable autonomy is to maximize the capability but minimize the necessity of human interaction with the controlled systems. Adjustable autonomy makes an autonomous system more versatile and easier for people to understand and to change the system's behavior.

Consider the following example of a Mars rover. On Mars, the round trip time for a radio signal to Earth can be up to 40 minutes. This is too long to tele-operate a rover from Earth for lengthy missions. During a manned Mars mission, a crewmember can tele-operate a rover. However, when the rover is travelling long distances over relatively uninteresting and benign terrain, a crewmember should not be needed to continually tele-operate the rover. Hence the need for autonomy. Nevertheless, complete autonomy does not solve the problem either. When the rover is traversing the terrain, we expect it will come to places that would be interesting to observe or treacherous to traverse. Unfortunately, commands like, "go out and find interesting images and soil samples for 10 days, then return safely," are beyond the scope of what autonomous rovers can do for the near future. Even if a rover could reliably execute such commands, sometimes people will want to more directly control the rover to make observations and perform tasks. When a crewmember does take more direct control of the rover, it is desirable to have the rover's autonomous systems make sure the operator doesn't do something by accident, such as drain the batteries so communication is lost or overturn the rover. In some cases, the crewmember may not wish to control the entire rover. Instead, the person will want the rover to continue its current mission and share control of only certain subsystems, such as a camera or scientific instrument. Finally, a high-performance completely autonomous system for a rover is difficult to design and might require too much power to operate the computer required on the rover. For such a system, a significant portion of the development time and the computer resources is required to handle the myriad of situations that a rover is unlikely to encounter. It is simpler to design an autonomous control system that can call on a human for help than one that is expected to handle any situation on its own. Adjustable autonomy supports these capabilities.

1.5 Overview of Paper

In the remainder of this paper, we describe how adjustable autonomy might affect the crew during a typical day at the Mars base. We then describe autonomous system requirements necessary to achieve the capabilities required. Following the system requirements, we present two systems developed by NASA to perform autonomous system control with adjustable autonomy: Remote Agent and 3T. We describe projects that have or are using these autonomous control systems as well as future NASA projects that may use enhanced versions of these systems. We then conclude.

2. A DAY IN THE LIFE OF A MARS BASE

To illustrate the range of operations required of adjustable autonomy systems, we describe a typical day in the life of the crew at a remote Martian site. The remote facility consists of enclosed chambers for growing crops, housing life support systems, conducting experiments, and providing quarters for the crew. There are six crewmembers, each with different specialized skills. These skills include operational knowledge of vehicle hardware, life support hardware, robotic hardware, and control software, as well as mission-specific knowledge required to perform scientific and medical experiments. Similarly, there are a variety of robots with different skills, such as the ability to maneuver inside or outside the remote facility and the ability to mount different manipulators or payload carriers. A joint activity plan specifies crew duties, robotic tasks, and control strategies for life support systems. This plan is based on mission objectives determined by Earth-based ground operations.

At the beginning of each day, the crew does a routine review of the performance overnight of autonomous life support systems. Next the crew discusses the previous day's activities and identifies any activities not completed. Additionally, the crew reviews the ground uplink for changes in mission objectives. Then the pre-planned activities for that day are reviewed and adjusted to accommodate these changes and crew preferences. For example, inspection of the water system may indicate the need to change a wick in the air evaporation water-processing module before the maintenance scheduled later in the week. Alternatively, the results of soil analysis performed by an autonomous rover yesterday may be sufficiently interesting to motivate taking additional samples today. This planning task is performed jointly by the crew and semi-autonomous planning software, where the crew specifies changes in mission objectives and crew preferences and the planning software constructs a modified plan.

In this example, the daily plan includes the following major tasks:

- harvest and replant wheat in the plant growth chamber
- sample and analyze soil at a specified site some distance from the Martian facility
- analyze crew medical information and samples
- repair a faulty circuit at one of the nuclear power plants.

Harvesting and replanting are traded control tasks, where the crew lifts trays of wheat onto a robotic transport mechanism that autonomously moves the trays to the harvesting area. Similarly, trays of wheat seedlings are moved by the robot into the plant chamber and placed in the growth bays by the crew. Sampling soil is an autonomous task performed by the tele-robotic rovers (TROVs). Crewmembers supervise this activity and only intervene when a sample merits closer inspection. This tele-operated sample inspection includes using cameras mounted on the TROV to view samples and manipulators on the TROV to take additional samples manually. Analyzing medical information and samples is a manual task assisted by knowledge-based analysis and visualization software. Repair of the power plants is an EVA task, requiring two crewmembers at the power plant and one crewmember monitoring from inside the surface laboratory. The autonomous power control system detected the fault the previous day and rerouted power distribution around the fault using model-based diagnostic software. The scheduled time required to perform this repair is minimal, because the autonomous diagnostic software has isolated the faulty portion of the circuit before the EVA. Humans are assisted by the tele-operated rovers, which carry tools and replacement parts to the plant and perform autonomous setup and configuration tasks in preparation for the repair.

This example of a daily plan also includes the following routine activities:

- sensor inspection of the plants
- change out of the wick in the air evaporation water processing module
- Human health activities such as exercise, eating, and sleeping.

Sensor inspection of the plant growth chamber is an autonomous robot task. A mobile robot with camera and sensor wand mounted on it traverses the plant chamber and records both visual data and sensor readings for each tray of plants in the chamber. This information is transmitted to the crew workstation and anomalous data are annunciated to the crew for manual inspection. A report summarizing the results of the inspection is generated automatically and reviewed later in the day by the crew. The change out of the wick is a manual task. The autonomous control software for the water system, however, operates concurrently with this operation and annunciates any omissions or misconfigurations in this manual operation (such as failing to turn off the air evap before removing the wick).

In addition to the tasks listed above, the adjustable autonomy system will also be performing its routine tasks of keeping the base operational. That means, monitoring all aspects of the base (water, gases, temperature, pressure, etc.) and operating valves, pumps, etc..., to maintain a healthy equilibrium. These activities happen without crew intervention or knowledge as long as all of the systems remain nominal. They run 24 hours a day, 7 days a week for the entire mission duration.

This simple scenario illustrates a rich variety of man-machine interactions. Humans perform tasks manually, perform tasks jointly with robots or automated software, supervise automated systems performing tasks, and respond to requests from automated systems for manual support. Automated software fulfills many roles including:

- activity planning and scheduling
- reactive control
- anomaly detection and alarm annunciation
- fault diagnosis and recovery.

Effectively supporting these different types of man-machine interaction requires that this automated software be designed for adjustable levels of autonomy. It also requires the design of multi-modal user interfaces for interacting with this software. We describe the requirements for adjustable autonomy software that robots and other systems in the next section.

3. REQUIREMENTS FOR A SYSTEM LIKE ABOVE

A life support system such as that described in the previous section will require intelligent planning, scheduling and control systems that can rigorously address several problems:

- Life support systems have difficult constraints including conservation/transformation of mass, crew availability, space availability and energy limitations.
- Life support systems are non-linear, respond indirectly to control signals and are characterized by long-term dynamics.

- Life support systems must react quickly to environmental changes that pose a danger to the crew.
- Life support systems are labor-bound, meaning that automation must provide for off-loading time intensive tasks from the human crew.
- Life support systems must be maintained and repaired by the crew, requiring that autonomous systems be able to maintain minimal life support during routine maintenance and be able to adjust the level of autonomy for crew intervention for repair or degraded mode operation.

Each of these five areas is mission-critical, in the sense that an autonomous system must deal with all of them to ensure success. Dealing with the first problem requires reasoning about time and other resources, and scheduling those resources to avoid conflicts while managing dynamic changes in resource availability (e.g., decreasing food stores). Dealing with the second problem requires the ability to plan for a set of distant goals and adapt the plan, on-the-fly, to new conditions. Dealing with the third problem requires tight sense-act loops that maintain system integrity in the face of environmental changes. Dealing with the fourth problem requires an integration of robotic control and scheduling with the overall monitoring and control of the life support system.

To support these goals, we have identified a core set of competencies that any autonomous system must have. These are described in the following subsections.

3.1 Sensing/Actuation/Reactivity

The autonomous system needs information about the environment so that it can take actions that will keep the system in equilibrium. Environmental sensors connected to control laws, which then connect to actuators, are critical to the success of an autonomous system. The control laws are written to keep a certain set point (for temperature, light, pH, etc.) and run at a high frequency, continually sampling the sensors and taking action. They form the lowest level of autonomous control. Of course, most control laws work only in very specific situations. As the environment approaches the boundaries of the control law's effectiveness, the system can become unstable. A key component of a reactive control law is *cognizant failure*, that is, the reactive control law should know when the environment has moved out of its control regime and alert a supervisory module that can make appropriate, higher-level adjustments [Gat, 1998].

3.2 Sequencing

Reactive control alone will not result in a stable system. Certain routine actions will need to be taken at regular intervals. These routine actions involve changing the reactive control system (for example, to switch from nighttime to daytime). This kind of control is the job of a *sequencer*. Of course, changes to the reactive control system may not have their intended effects. For this reason, the sequencer needs to *conditionally* execute sequences of actions. That is, the sequencer constantly checks the sensor values coming from the environment and decides its course of action based on that data. For example, the sequencer may begin heating up a chamber by turning a heater on. If the chamber temperature does not rise, the sequencer could respond by turning a second heater on. This second action is conditional on the result of the first action. A sequencer is also needed to insure that interacting subsystems are coordinated and do not conflict, for example, when they both require a common resource at the same time.

3.3 Planning/Scheduling

Intelligent planning software is used to construct dynamically task sequences to achieve mission objectives given a set of available agents (crew, robots) and a description of the facility environment (the initial situation). This can include both periodic activities to maintain consumables (like oxygen, food) and the facility (maintenance), and aperiodic activities (like experiments). Periodic activities would include planting and harvesting crops, processing and storing food, maintaining crew health (through exercise, meals, and free time), and maintaining robots, facility systems (like life support), and transportation systems. Aperiodic activities would include responding to system anomalies, robot and system repair, and scientific experimentation. Some of these activities must be prepared for weeks or even months ahead. Crop planning to maintain balanced food stores must be done well in advance due to lengthy crop growth time (e.g., 60-80 days for wheat). The usage of other consumables (like water, oxygen, carbon dioxide) must be monitored and control strategies altered to adjust for mass imbalances or product quality changes. Because of the complexity of such a plan, it should be generated iteratively, with interim steps for the crew to evaluate the plan "so far" and with support for changing goals and constraints to alter the plan at these interim points. The crew should be able to compare alternative plans resulting from these changes and from different optimization schemes, and to specify one of the alternatives for use or additional modification. They should be able to compare previous plans to current plans to identify similarities and highlight differences. The new plan will include activities affecting the control of robots and life support systems within the facility. Portions of the plan related to such control must be provided to the control software for execution.

Intelligent scheduling software is used to refine the task sequences generated by the planning software. Specifically, the scheduler will be used to adjust the time resolution between early and later versions of a plan. It will be used to generate alternative plans, optimized along different dimensions (e.g., optimized for minimum elapsed time, optimized for maximum usage of robots). A system, such as the one described in the previous section, often has several different dimensions that must be optimized. A scheduling system must allow users to understand the interactions and dependencies along the dimensions and to try different solutions with different optimization criteria.

Abstraction is also critical to the success of planning and scheduling activities. In our scenarios, the crew will often have to deal with planning and scheduling at a very high level (e.g., what crops do I need to plant now so they can be harvested in six months) and planning and scheduling at a detailed level (e.g., what is my next task). The autonomous system must be able to move between various time scales and levels of abstraction, presenting the correct level of information to the user at the correct time.

3.4 Model-Based Diagnosis and Recovery

When something goes wrong, a robust autonomous should figure out what went wrong and recover as best as it can. A model-based diagnosis and recovery system, such as Livingstone [Williams and Nayak, 96], does this. It is analogous to the autonomic and immune systems of a living creature. If the autonomous system has a model of the system it controls, it can use this to figure out what is the most likely cause that explains the observed symptoms as well as how can the system recover given this diagnosis so its mission can continue. For example, if the pressure of a tank is low, it could be because the tank has a leak, the pump blew a fuse, a valve is not open to fill the tank or not closed to keep the tank from draining. However, it could be that the tank pressure is not low and the pressure sensor is defective. By analyzing the system from other sensors, it may say the pressure is normal or suggest closing a valve, resetting the pump circuit breaker, or requesting a crewmember to check the tank for a leak.

3.5 Simulations

Both the crew and the intelligent planner use simulations of the physical systems. They allow for either the crew or the planner to play *what-if* games to test different strategies (i.e., what if I planted 10 trays of wheat and 5 trays of rice? What would my yield be? How much oxygen would be produced? How much water would it take?). Simulations can also allow for verification of the actual responses of the system to the predicted responses based on the simulation. Any inconsistencies could be flagged as a possible malfunction. For example, if turning on a heater in the simulation causes a rise in temperature, but turning on the same heater in the actual system does not, then the heater or temperature sensor may be faulty. There does not need to be one, complete, high-fidelity simulation of the entire system. Each subsystem may have its own simulation with varying levels of fidelity.

3.6 User Interfaces

Supervising autonomous software activities must be a low cost human task that can be performed remotely without vigilance monitoring. Providing the human supervisor with the right information to monitor autonomous software operations is central in designing effective user interfaces for supervisory control. The supervisor must be able to maintain with minimum effort an awareness of autonomous system operations and performance, and the conditions under which operations are conducted. She must be able to detect opportunities where human intervention can enhance the value of autonomous operations (such as human inspection of remote soil samples) as well as unusual or unexpected situations where human intervention is needed to maintain nominal operations. Under such conditions, it is important to provide status that can be quickly scanned about ongoing autonomous operations and the effects of these operations on the environment. Notification of important events, information requests, and anomalies should be highly salient to avoid vigilance monitoring. Easy access is needed to situation details (current states, recent activities, configuration changes) that help the supervisor become quickly oriented when opportunities or anomalies occur [Schreckenghost and Thronesbery, 1998].

When interaction is necessary, mixed initiative interaction provides an effective way for humans and autonomous software to interact. A system supports mixed initiative interaction if both the human and the software agent have explicit (and possibly distinct) goals to accomplish specific tasks and the ability to make decisions controlling how these goals will be achieved [Allen, 1994]. There are two types of tasks where mixed initiative interaction with autonomous control software is needed: during joint activity planning and during traded control with robots. For joint plan generation, the human and planning software interact to refine a plan iteratively. The human specifies goals and preferences, and the planner generates a plan to achieve these goals [Kortenkamp, *et al.*, 1997]. The crew evaluates the resulting plan and makes planning trade-offs at constraint violation and resource contention. The planner uses modifications from the crew to generate another plan. This process continues until an acceptable plan is generated. For traded control, the human and robot interact when task responsibility is handed over (when control is "traded"). Coordination of crew and robot activities through a joint activity plan reduces context registration problems by providing a shared view of ongoing activities. To accommodate the variability in complex environments, it is necessary to be able to dynamically change task assignment of agents. Finally, as crew and robot work together to accomplish a task, it is important to maintain a shared understanding of the ongoing situation. The robot must be able to monitor the effect of human activities. This may require enhanced sensing to monitor manual operations and situated memory updates. The crew must be able to track ongoing activities of the robot and to query the robot about its understanding of the state of the environment.

To enable such interaction, the user interface software must integrate with each process in the control architecture to provide the user with the capability to exchange information with and issue commands to the control software. Software provided for building the user interface includes:

- communication software to exchange data and commands with the control architecture
- software for manipulating data prior to display
- software for building a variety of display forms.

In addition to screen-based, direct manipulation graphical user interfaces, other modalities of interaction will be important in space exploration. Natural language and speech recognition are needed for tasks where the crews' hands are otherwise occupied or where interaction with a computer is not well-suited to the ongoing task (such as extra-vehicular activity (EVA) tasks or joint manipulation tasks with a robot). Alternative pointing mechanisms such as gestural interfaces also enable multi-modal interaction. Interaction with virtual environments to exercise control in the real environment can improve tele-operations such as remote soil sampling using a TROV or remote robotic maintenance and repair tasks. Visualization software can help manage and interpret data from experiments.

4. SOME CURRENT IMPLEMENTATIONS AND TESTS

NASA has developed and implemented two "state-of-the-art" autonomous robotic control systems of the type described in the last section: the Remote Agent and 3T. Each of these architectures is briefly described in this section along with the descriptions of their implementations.

4.1 Remote Agent

The Remote Agent (RA) is an agent architecture designed to control extraterrestrial systems autonomously for extended periods. The RA architecture is illustrated in Figure 2.

RA consists of four components: Mission Manager (MM), Planner/Scheduler (PS), Model-based Mode Identification and Recovery (MIR), and the Smart Executive (Exec). These components function as follows. Ground control sends the MM a *mission profile* to execute. A mission profile is a list of the goals the autonomous control system is directed to achieve over what may be a multi-year period. When Exec needs a new plan to execute, it requests a plan from MM. MM breaks the mission up into short periods, e.g., two weeks, and submits a plan request to the Planner/Scheduler to create a detailed plan for the next period as requested by Exec. MM insures that resources, such as fuel, needed later in the mission are not available for use in earlier plans. PS creates the requested plan and sends it to Exec to be executed. PS creates flexible, concurrent temporal plans. For example, a plan can instruct that multiple tasks be executed at the same time, but be flexible to when the tasks start and finish. Exec is designed to robustly execute such plans. Exec is a reactive, plan-execution system responsible for coordinating execution-time activities, including resource management, action definition, fault recovery, and configuration management.

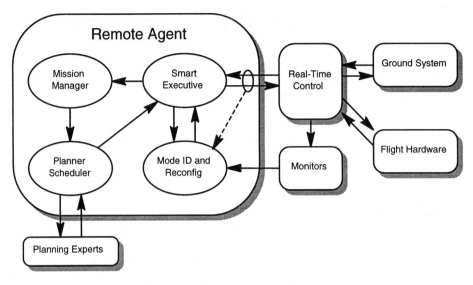

Figure 2   Remote Agent architecture.

In the real world, plan execution is complicated by the fact that subsystems fail and sensors are not always accurate. When something does go wrong, is can be difficult to quickly figure out what went wrong and what to do about it. MIR addresses this problem. MIR is a discrete, model-based controller that uses a declarative model of the system being controlled, e.g., a spacecraft. It provides an abstract level of the state of the spacecraft based on its model to Exec. For example, a spacecraft sensor may indicate that a valve is closed when it is, in fact, open. MIR reasons from its model and other sensor readings that the valve must be open. MIR reports to Exec the valve is open and Exec acts accordingly. Another function of MIR is to offer recovery procedures to Exec when Exec is unable to accomplish one of its tasks. Continuing the valve example, let us say that Exec requires the valve to be closed, but the command to close the valve does not work. Exec sends a request to MIR to figure out how to close the valve without disrupting any other tasks currently being executed. MIR would consult its model of the spacecraft and might instruct Exec to reset a valve solenoid circuit breaker, or to close another valve that will have the effect of closing the desired valve.

RA is discussed further in [Muscettola *et al.*, 98]. We continue by describing systems that have implemented a RA or some of its components.

4.1.1 Saturn Orbit Insertion Simulation

The Saturn Orbit Insertion (SOI) simulation pertains to the phase of a Cassini-like mission where the spacecraft must autonomously decelerate to enter an orbit around Saturn. In addition to handling the nominal procedures of taking images of Saturn's rings as it performs the SOI maneuver, it had to handle faults in real-time in a flight like manner. In the scenario, the main engine overheats when it starts its "burn." The autonomous system shuts down the engine immediately to prevent damage to the spacecraft. However, now the spacecraft is in danger of flying past Saturn. The spacecraft is too far from Earth to wait for a command. The autonomous system starts the backup engine that it previously prepared just in case it was needed.

The onboard planner quickly prepares an updated plan. The plan is executed as soon the spacecraft cools sufficiently. During the maneuver, a gyroscope fails to generate data. The system has also prepared for this problem because the backup gyroscope was already warmed up in case it was needed. Additional problems such as coordinating tasks so that there were no power overloads were also handled The SOI is described in more detail in [Pell *et al.*, 96]. This scenario demonstrates that an autonomous can prepare for and react to situations that a human might overlook or react to slowly to even if the person were onboard.

4.1.2 DS1

Deep Space One (DS1) is the first of a series of low-cost unmanned spacecraft missions whose mandate is to validate new spacecraft technologies. NASA scheduled the DS1 launch in October 1998. The science objective of DS1 is to approach then image an asteroid and a comet. An ion propulsion system and the Remote Agent autonomous control system are among the technologies being validated.

Figure 3   Drawing of Deep Space 1 Spacecraft (DS1) encountering a comet.

The DS1 Remote Agent will control DS1 for a 12-hour period and a 6-day period during the primary phase of the mission. During these periods, the Remote Agent will generate and execute plans on the spacecraft and recover from simulated spacecraft faults. These faults include a power bus status switch failure, a camera that cannot be turned off, and a thruster stuck closed. The DS1 Remote Agent will periodically generate plans as necessary based on the mission profile (goals & constraints) and the current state of the spacecraft. Model-based failure detection and recovery will be demonstrated. When the model-based recovery system cannot correct the problem, the planner will generate a new plan based on the diagnosed state of the spacecraft. [Bernard et al., 98] discusses the DS1 Remote Agent in depth. The DS1 Remote Agent supports adjustable autonomy by allowing the spacecraft to be controlled as follows:

- entirely from the ground using traditional command sequences

- partially from the ground by uplinking conditional command sequences to be executed with model-based recoveries performed as needed

- autonomously with various ground commands being executed while a plan is also being executed

- completely autonomous with no ground interaction - plans generated onboard as needed.

4.1.3 IpexT

IpexT (Integrated Planning and Execution for Telecommunications) is an autonomous control system prototype for managing satellite communications, particularly in crises. The autonomous control system prototype was developed using the RA Planner/Scheduler and Smart Executive. [Plaunt and Rajan, 98] discuss this system in more detail. This system supports adjustable autonomy by operating autonomously in normal circumstances, but permitting operators to take control at various levels, from satellite beam positioning to change the bandwidth of a region to the call priority and quality of a specific call.

4.1.4 Mars Rover Field Test

For this experiment, the Remote Agent Smart Executive and Model-based Diagnosis and Recovery components are being used to control a Mars rover prototype based on the Russian-built rover Marsokhod.

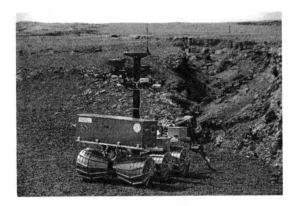

Figure 4   Marsokhod Rover.

The field test is scheduled for one week in November 1998 in rough desert terrain. The rover will be commanded by scientists to achieve various science goals by sending high-level commands to the Smart Executive (Exec). Exec will execute conditional plans and manage resources such as power so that the scientists do not have to. When performing sampling and science tests, scientists will share control of the rover with the Remote Agent.

4.2 3T

NASA Johnson Space Center and TRACLabs/Metrica Incorporated have, over the last several years [Bonasso *et al.* 97b], developed a multi-tiered cognitive architecture that consists of three interacting layers or tiers (and is thus known as 3T, see Figure 5).

- A set of hardware-specific control skills that represent the architecture's connection with the world. Control skills directly interact with the hardware to maintain a state in the environment. They take in sensory data and produce actions in a closed control loop. For example, a skill may adjust the flow of base or acid to maintain a pH level. A program called a *skill manager* schedules skills on the CPU, routes skill data and communicates with the sequencing layer of the architecture.

- A sequencing capability that differentially activates control skills using different input parameters to direct changes in the state of the world to accomplish specific tasks. For example, the sequencer may adjust the pH set point of a pH control skill based on overall environmental conditions. We are using the Reactive Action Packages (RAPs) system [Firby, 1989] for this portion of the architecture.

- A deliberative planning capability which reasons in depth about goals, resources and timing constraints. 3T currently uses a state-based non-linear hierarchical planner known as AP [Elsaesser and MacMillan, 1991]. AP determines which sequences are running to accomplish the overall system goals.

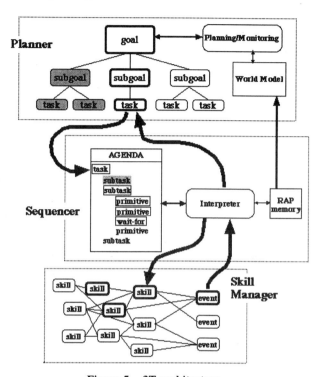

Figure 5   3T architecture.

The architecture works as follows. The deliberative layer begins by taking a high-level goal and synthesizes it into a partially ordered list of operators. Each of these operators corresponds to one or more RAPs in the sequencing layer. The RAP interpreter (sequencing layer) decomposes the selected RAP into other RAPs and finally activates a specific set of control skills in the skills layer. Also activated is a set of event monitors that notifies the sequencing layer of the occurrence of certain world conditions. The activated control skills will move the state of the world in a direction that should cause the desired events. The sequencing layer will terminate the actions, or replace them with new actions when the monitoring events are triggered, when a timeout occurs, or when a new message is received from the deliberative layer indicating a change of plan.

4.2.1 3T Autonomous Control System for Air Revitalization: Phase III Test

A series of manned tests that demonstrate advanced life support technology were conducted at the Johnson Space Center (JSC) under the Lunar/Mars Life Support Technology Project. The Phase III test, the fourth in this series of tests, was conducted in the fall of 1997. During the Phase III test, four crewmembers were isolated in an enclosed chamber for 91 days. Both water and air was regenerated for the crew using advanced life support systems. One of the innovative techniques demonstrated during this test is the use of plants for converting carbon dioxide (CO2) produced by crew respiration and solid waste incineration into oxygen (O2). Because the crew and the plants were located in different chambers, it was necessary to build the product gas transfer system to move gases among the crew chamber, the plant chamber, and the incinerator. Figure 6 shows the physical layout of the product gas transfer system.

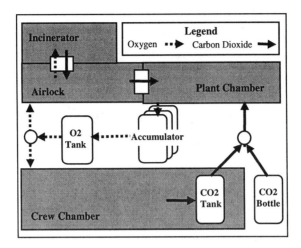

Figure 6   Physical Layout of the Product Gas Transfer System.

An objective for the Phase III test was to demonstrate that automated control software can reduce the workload of test article engineers (TAEs). We developed an automated control system for the product gas transfer system using the 3T control architecture [Schreckenghost et al., 98a]. The planning tier implements strategies for managing contended resources in the product gas transfer system. These strategies are used (1) to manage the storage and use of oxygen for the crew and for solid waste incineration, and (2) to schedule the airlock for crop germination, planting, and harvesting and for waste incineration. The reactive sequencing tier implements tactics to control the flow of gas. Gas is transferred to maintain O2 & CO2 concentrations in the plant chamber and to maintain O2 concentration in the airlock during incineration. The sequencer also detects caution & warning states and executes anomaly recovery procedures. The skill management tier interfaces the 3T control software to the product gas transfer hardware and archives data for analysis.

The 3T product gas transfer control system operated round-the-clock for 73 days. It typically operated with limited intervention by the TAEs and significant engineer workload reduction was demonstrated. During previous tests, TAEs spent at least 16 hours a day at the life support systems control workstation. Product gas transfer required 6-8 hours per week of shift work with 6 hours for each incineration (conducted every 4 days) and 3 hours for each harvest (conducted every 16-20 days).

Even with such minimal human intervention, it was important to design the product gas transfer control system for adjustable autonomy. We designed the system so that a human can replace each tier of automated control. Either the human or the planner can specify the control strategies for the top tier. For the middle tier, either the human or the sequencer can sequence the control actions. This design was useful during phased integration of the control system, allowing each tier to be integrated from the bottom up (starting with the skill manager). It also supported manual experimentation with novel control tactics during the test. A TAE could temporarily disable the automatic control of either oxygen or carbon dioxide transfer, manually reconfigure gas flow, then return to autonomous control. Manual control is executed through the autonomous software and this software does not overwrite human commands while manual control is active. Control setpoints and alarm thresholds are parameterized for manual fine-tuning of control strategies. For long duration tests like the Phase III test, it is important to design control systems to continue operating when control hardware is maintained and repaired. The product gas transfer control system adapts control automatically when sensors are taken out of the control loop manually for calibration or repair. Finally, the autonomous system can initiate a request for information that is only available from a human. For example, the autonomous software requests the TAEs for the results of lab analyses when such information is not available from in-line sensors.

4.2.2 Node 3

Node 3, to be launched in 2002, serves as a connecting module for the U.S. Habitation module, the U.S. crew return vehicle and future station additions. It contains two avionics racks and two life support racks. The crew and thermal systems division at JSC have been developing advanced water and air recovery systems which would be more efficient in terms of power and consumables than those life support systems originally included in node 3. A biological water processing system and a system to recover oxygen from CO2 via a water product are two such advanced systems.

This advanced life support comprises 8 subsystems that must be carefully coordinated to balance the gas and mass flows to be effective. Moreover, the human vigilance required must be minimal. To those ends, JSCs automation and avionics divisions are configuring the 3T control system to run this advanced life support. During ground testing, however, a variety of control techniques are being studied and this requires the ability to vary the autonomy of various subsystems. Such adjustable autonomy is integral to 3T.

4.2.3 Space shuttle Remote Manipulator System Assistant

3T [Bonasso et al., 97] is being used as the software framework for automating the job of NASA flight controllers as they track procedures executed by on-orbit astronauts. The RMS Assistant (RMSA) project focuses on automating the procedures relating to the shuttle's Remote Manipulator System (RMS) and is a pathfinder project for the automation of other shuttle operations. The RMSA system is designed to track the expected steps of the crew as they carry out RMS operations, detecting malfunctions in the RMS system from failures or improper configurations as well as improper or incomplete procedures by the crew. In this regard, it is a "flight controller in a box". Moving the flight controller functions on-orbit is part of a larger program of downsizing general space shuttle operations. The 3T architecture was designed for intelligent autonomous robots, but has an integral capability that allows adjustable autonomy to include full tele-operation. In that mode, the software acts as a monitoring system, and thus provides an "assistant" framework now and accommodates operations that are more autonomous in the future.

The RMSA was used to "flight follow" portions of the RMS checkout operations in shuttle flights STS80 (November 1996) and STS82 (February 97) as well as various RMS joint movements during payload deployment and retrieval [Bonasso et al., 98]. To flight follow, RMSA was run on the ground with access to the telemetry downlist, but without the crew's knowledge. Additional monitor RAPs were written to "watch" for certain telemetry cues that would indicate that the crew had begun each procedure. The RMSA showed that it could follow crew operations successfully even in the face of loss of data due to communications exclusions or procedures being skipped by the crew.

After the flight following, we held several demonstrations for various RMS-trained astronauts in the spring of 1997. These demonstrations used the RMS simulator to show RMSAs ability to guide checkout procedures, monitor payload deployment and retrieval and to handle off-nominal operations in either tele-operation, semi-autonomous and autonomous modes. The general reaction of the crew was positive, but since it would be sometime before the orbiter's avionics were upgraded to allow autonomous operations, the crew tasked us with developing the tele-operation interface. After a year of development, we began a series of training demonstrations with the crew this past spring that included displays of the RMSA task agenda, expected switch and mode settings, and an integrated VRML 3D synoptic display of the orbiter and RMS positions [3]. This combination was hailed by the crew as the best technology suite for use not only in shuttle RMS operations, but for space station assembly as well. We are currently exercising RMSA for the complete RMS Checkout procedures using mission control simulations in preparation for two Extended Mission Capability flights this winter.

5. FUTURE PROJECTS

The manned Mars missions require systems that can operate autonomously for extended periods as well as operate as in conjunction with people when necessary. Thus, these missions require autonomous software that has the strengths of both the Remote Agent and 3T. We are considering using adjustable autonomy on several projects that we briefly describe below. By enhancing our autonomous control software on the following projects, we are preparing the autonomous control technology needed for manned Mars missions. In doing so, the manned Mars mission will not have to incur the cost or suffer the delay of directly developing and validating these technologies.

5.1 Mars TransHab

The TransHab will provide the living and working space for the crew while in transit between the Earth and Mars. The design incorporates a central structural core with an inflatable outer shell. The central core is the structural backbone of the vehicle and provides a mounting surface for all required equipment and systems. The inflatable shell is packaged around the core so that the whole package can be launched in the shuttle. Once in orbit, the shell is inflated, providing most of the interior volume as well as micrometeoroid protection and thermal insulation. The TransHab is one vehicle for the overall architecture required for the Mars mission and has the largest impact on the overall mission mass since it must to go to Mars and back.

Several of the life support systems developed by JSCs crew and thermal system division for node 3 have even more advanced counterparts which will be light-weight, designed to fit in the TransHab core, and run in TransHab's energy-rich environment

In some Mars mission scenarios, a second TransHab will be placed in Mars orbit two years before it is occupied by the crew on the return trip to the Earth. During that time, the life support must be maintained in a stand-by configuration and then respond to the specific crew requirements that may be quite different from what was planned when the mission began.

3T is envisioned for the TransHab control system. Its adjustable autonomy capability will be a primary asset to allow the planned crew of six to adjust the control schemes when encountering unexpected regimes in the 200-day transit to Mars. As well, the adjustable autonomy will allow the crew to set changed, possibly novel activity schedules and profiles for the life support functions on the return trip.

5.2 BioPlex

The 3T control architecture has been selected for controlling computer-controlled machines (robotic and regenerative life support) in the BIOPlex facility to be completed at NASA JSC in 2000. The BIOPlex facility will be a ground-based, manned test facility for advanced life support technology destined for use in lunar and planetary bases, and planetary travel (such as Mars TransHab Project). It consists of five connected modules: two plant growth chambers, a crew habitation module, a life support module, and a laboratory. Regenerative life support systems include water recovery, air revitalization, solid waste management, and thermal/atmospheric control. Plant support systems include nutrient delivery, gas management, and thermal/humidity control. Robotic systems include transport, manipulation, and sensor/video scanning. Controlling these heterogeneous systems to maintain food supplies, water, and gas reservoirs, while minimizing solid waste reservoirs (inedible biomass and fecal matter), poses a challenging set of problems for planning and scheduling. The planner must balance conflicting system needs and account for cross system coupling, at time scales varying from hours to months. In this facility, human and robots will jointly execute tasks and must coordinate their efforts. A common/shared schedule for both crew and computer-controlled machines is needed to guarantee such coordination. This schedule must be sufficiently flexible to adapt to crew preferences while stable and robust for computer control. An integrated planning, scheduling, and control architecture that includes both fine time grain scheduling and optimization as well as long term crop planning will be required for BIOPlex. A more complete description of this on-going work is available in [Schreckenghost *et al.*, 1998].

5.3 Mars Rover

Although a manned Mars mission may be more than a decade away, NASA has already planned rover missions for launch in 2001, 2003 and 2005. The rover mission in 2001 is likely to be similar to the Sojourner rover, which operated on Mars in 1998 and had very limited autonomous capabilities. However, the goals for the 2003 and 2005 rovers are much more ambitious. Both will travel distances approaching 10km. To achieve this, the rovers will operate completely autonomous while traversing much of Mars. In case of failures, the rovers are expected to autonomously recover. When the rovers reach points selected by scientists, commands from scientists will be uplinked to the rovers to be executed robustly. For example, a high-level command sequence from a scientist might be: if you have the time and the energy to meet your other mission goals, go to this rock, remove the dust, and put the Alpha Proton Xray Spectrometer at this point (as selected by pointing to an image of the rock). If you are collecting valid data, continue collecting data for 2 hours, otherwise continue your mission.

During any day, the rover will communicate with ground for only a couple of hours. To maximize the ability for scientists to perform science, the rover should be at "interesting spots" during the periods and must plan its day accordingly. The scientists would like to be able see the telemetry sent by the rover and respond that same day rather than have to wait for the next day in order to get the commands exactly right to accomplish their science goals and safeguard the rover. By supporting high-level commanding with model-based recoveries, resource management, and autonomous planning, adjustable autonomy addresses this need.

5.4 Mars In-situ Propellant Production

A novel approach for reducing the mass of a mission that involves returning from Mars is in situ propellant production (ISPP) on Mars. That is, an ISPP reactor uses CO2 from the Martian atmosphere, hydrogen brought from Earth, and energy (solar or nuclear) to produce the methane and oxygen that will fuel the return vehicle as well as oxygen for life support. In the Mars Reference Mission [Hoffman and Kaplan, 97], an ISPP system runs autonomously for nearly two years before the arrival of the first crewmember on Mars. Its purpose is to produce the fuel for the primary manned-return vehicle. The ISPP system then runs for two more years to produce the fuel for the backup manned-return vehicle. In order to test the reasonableness of this approach, a Mars Sample Return mission that uses an ISPP to fuel the unmanned return vehicle is under study. The proposed launch date is in 2005. In this unmanned mission, an ISPP system, a rover, and a Mars Ascent Vehicle (MAV) will be sent to Mars. While the ISPP system is fueling the MAV, the rover is collecting rock and soil samples that the MAV will return to Earth (or take into Mars orbit to be returned to Earth by another spacecraft).

One of the autonomy challenges ISPP presents is dealing with slowly degrading performance and long-term goals. In addition to failures that may happen suddenly, the performance of the ISPP is expected to degrade over time for several reasons (primarily contamination). Therefore, the rate at which propellant accumulates will decrease over time in a way that may be difficult to predict. The autonomous system must make decisions about whether to operate less efficiently to keep propellant production high, decrease the production rate to conserve energy, or request that parts be cleaned or replaced. Throughout the mission, crewmember or ground personnel may adjust how the autonomous system makes these decisions.

5.5 DS3

Deep Space Three (DS3) is a set of three spacecraft to be launched in a single vehicle in 2001. The mission goal is to test a large optical interferometer formed by the three spacecraft. The purpose of developing large spaceborne interferometer is to discover earth-sized planets around other stars. In space, the three spacecraft will form an equilateral triangle with each side from 100m to 1 km depending on the accuracy desired. The spacecraft must maintain their positions relative to each other with accuracy of ± 1 cm. The spacecraft are expected to target 50 stars during its 6-month mission.

This mission is interesting from an autonomous system viewpoint because three spacecraft must be precisely coordinated. In addition, the propellant on each spacecraft limits the life of the mission. The quantity on each spacecraft may vary considerably as the mission progresses. Once the propellant of one spacecraft is exhausted, they will no longer be able to form an equilateral triangle. To address this problem, the resource manager of the autonomous system will attempt to compensate. For example, if one spacecraft is low on propellant, the autonomous system may instruct the other two to use more propellant to reduce the propellant consumed by the one that is low. Moreover, scientists will want to aim these spacecraft like they would a large telescope. The plans must be made so that the most important images can be taken while minimizing the propellant consumed. This combination of autonomous operation and human interaction makes it an excellent candidate for an autonomous system that supports adjustable autonomy.

5.6 DS4

Deep Space Four (DS4) is a spacecraft with a 200kg lander that will land on the comet Tempel 1. The lander will take images of the comet, collect and analyze samples up to one meter below the surface. The lander will return to Earth with up to 100 cubic centimeters of comet material.

Figure 7   Drawing of Deep Space 4 Lander Prototype.

Although this science goal is laudable, the primary mission goal is to test advanced technologies, including autonomous control systems, necessary for landing spacecraft on small bodies in space, e.g., asteroids. NASA scheduled the DS4 launch for 2003, the comet rendezvous in 2005, and return to Earth in 2010.

This mission presents an interesting challenge for autonomy. Scientists and spacecraft designers have only a vague idea of what to expect when the lander attempts to land on a comet. During the landing sequence, the lander must operate autonomously due to the delay of the radio signal to Earth and back. The lander must appropriately react to whatever state it finds itself in, including hardware failures due to flying into the comet's tail and impacting the comet. Once on the comet, scientists would like to give specific commands to the lander based on data transmitted from the lander. However, the lander must complete its mission even if it loses its communication link to Earth and return with a comet sample.

5.7 Reusable Launch Vehicles (Shuttle, X33, VentureStar)

The projects mentioned above have focused on autonomous systems in space and on Mars. However, autonomous systems will play an essential role launching people and material from Earth to space and will reducing the cost and risk of Mars missions.

Figure 8   Drawing of X-33.

Autonomous systems are being considered to manage ground operations for reusable launch vehicles, in particular, the Space shuttle, X-33, and VentureStar. The autonomous system will collect data from the spacecraft while it is in flight and determine what needs to be maintained. Unlike, the previous autonomous systems, this autonomous system will rely on many people to execute its maintenance plan. When people or the autonomous system discovers additional problems or if maintenance and repairs are not made according to schedule, the autonomous system creates a recovery plan to stay on schedule and in budget as guided by the constraints human managers place on it.

6. SUMMARY

NASA is engaged in several projects that "push the envelope" on the design of autonomous systems that support adjustable autonomy. Current NASA autonomous systems include the Remote Agent and 3T. The Remote Agent was designed for controlling unmanned spacecraft and rover for extended periods of time while managing consumable resources throughout a mission and using model-based diagnosis and recovery to handle failures. 3T was designed to interact with humans. This includes control by humans at various levels and controlling systems in environments with humans, e.g., life-support system and a robot working with a human. Manned missions to Mars will require autonomous systems with features from both of these autonomous systems. Crewmembers will operate and maintain several systems, from life support, to fuel production, to rovers. Without autonomy, they would be spending most of their time just trying to stay alive. However, even with completely autonomous systems, crewmembers would be frustrated by how to repair them, how to get them to do exactly what they want, and even to understand why the systems are behaving as they do or predicting how the systems will behave under certain conditions. Adjustable autonomy is essential to address these problems and vital for autonomous systems that interact with people on Mars.

7. ACKNOWLEDGEMENTS

The authors acknowledge the efforts of Chuck Fry, Bob Kanefsky, Ron Keesing, Jim Kurien, Bill Millar, Nicola Muscettola, Pandu Nayak, Chris Plaunt, Kanna Rajan, Brian Williams of NASA Ames; Douglas Bernard, Ed Gamble, Erann Gat, Nicolas Rouquette, and Ben Smith of JPL for their efforts in developing the Remote Agent. The authors also acknowledge Cliff Farmer, Mary Beth Edeen , and Karen Meyers for their support of the work using 3T at JSC. Images of the DS1, DS4, X33 spacecraft and Marsokhod rover were provided by NASA.

8. REFERENCES

[Allen, 1994] Allen, J. "Mixed initiative planning: Position paper". *ARPA/Rome Labs Planning Initiative Workshop*, Feb. 1994.

[Bonasso et al., 97a] Bonasso, R. P., D. Kortenkamp, and T. Whitney. "Using a robot control architecture to automate space station shuttle operations". *Proceedings of the Fourteenth National Conference on Artificial Intelligence and Ninth Conference on Innovative Applications of Artificial Intelligence*, Cambridge, Mass., AAAI Press, 1997.

[Bonasso et al. 97b] Bonasso, R. P., R. J. Firby, E. Gat, D. Kortenkamp, D. Miller, M. Slack. "Experiences with an Architecture for Intelligent, Reactive Agents". *Journal of Experimental and Theoretical Artificial Intelligence*, 9(2), 1997.

[Bonasso et al., 98] Bonasso, R. P., R. Kerr, K. Jenks, and G. Johnson. "Using the 3T Architecture for Tracking Shuttle RMS Procedures", *IEEE International Joint Symposia on Intelligence and Systems (SIS)*, Washington D.C., May 1998.

[Bernard et al., 98] Bernard, D. E., G. A. Dorais, C. Fry, E. B. Gamble Jr., B. Kanefsky, J. Kurien, W. Millar, N. Muscettola, P. P. Nayak, B. Pell, K. Rajan, N. Rouquette, B. Smith, and B. C. Williams. "Design of the Remote Agent experiment for spacecraft autonomy". In *Proceedings of the IEEE Aerospace Conference*, Snowmass, CO, 1998.

[Cohen, 93] Cohen, M. "Mars surface habitation study". Presentation at the Mars Exploration Study Workshop II, NASA Conference Publication 3243. Conference held at NASA Ames Research Center, May 24-25, 1993.

[Elsaesser and MacMillan, 1991] Elsaesser, C., and E. MacMillan. "Representations and Algorithms for Multi Agent Adversarial Planning", MITRE Technical Report MTR-199191W000207, 1991.

[Firby, 1989] Firby, R. J. *Adaptive Execution in Complex Dynamic Worlds*. Ph.D. Thesis, Yale University, 1989.

[Gat, 1998] Gat, E. Three-Layer Architectures, *Artificial Intelligence and Mobile Robots*, D. Kortenkamp, R. P. Bonasso, and R. Murphy, eds., AAAI/MIT Press, Cambridge, MA, 1998.

[Hoffman and Kaplan, 97] Hoffman, S. J., and D. I. Kaplan, Eds. *Human Exploration of Mars: The Reference Mission of the NASA Mars Exploration Study Team*. NASA Special Publication 6107, Johnson Space Center: Houston, TX, July 1997.

[Kortenkamp et al., 97] Kortenkamp, D., P. Bonasso, D. Ryan, and D. Schreckenghost. "Traded control with autonomous robots as mixed initiative interaction". AAAI-97 Spring Symposium. *Workshop on Mixed Initiative Interaction*, Mar. 1997.

[Lewis et al., 1998] Lewis, J. F., N. J. C. Packham, V. L. Kloeris, and L. N. Supra. "The Lunar-Mars Life Support Test Project Phase III 90-day Test: The Crew Perspective". *28th International Conference on Environmental Systems*, July 1998.

[Muscettola et al., 98] Muscettola, N., P. P. Nayak, B. Pell, and B. C. Williams. "Remote Agent: To boldly go where no AI system has gone before". *Artificial Intelligence*, 103(1/2), August 1998. To appear.

[Pell et al., 96] Pell, B., D. Bernard, S. A. Chien, E. Gat, N. Muscettola, P. P. Nayak, M. D. Wagner, and B. C. Williams. "A remote agent prototype for spacecraft autonomy". In *Proceedings of the SPIE Conference on Optical Science, Engineering, and Instrumentation*, 1996.

[Pell et al., 98] Pell, B., S. Sawyer, D. E. Bernard, N. Muscettola, and B. Smith. "Mission operations with an autonomous agent". In *Proceedings of the IEEE Aerospace Conference*, Snowmass, CO, 1998.

[Plaunt and Rajan, 98] Plaunt, C., and K. Rajan. "IpexT: integrated planning and execution for military satellite tele-communications". In *Working Notes of the AIPS Workshop*, 1998.

[Schreckenghost and Thronesbery, 1998] Schreckenghost, D. and C. Thronesbery. "Integrated Display for Supervisory Control of Space Operations". *Human Factors and Ergonomics Society*. 42nd Annual Meeting, Chicago, IL, October 1998.

[Schreckenghost et al., 98] Schreckenghost, D., D. Ryan, C. Thronesbery, P. Bonasso, and D. Poirot. "Intelligent control of life support systems for space habitats". In *Proceedings of the Fifteenth National Conference on Artificial Intelligence and Tenth Conference on Innovative Applications of Artificial Intelligence*. Madison, WI, AAAI Press, pp. 1140-1145, July 1998.

[Schreckenghost et al., 98a] Schreckenghost, D., M. Edeen, P. Bonasso, and J. Erickson. "Intelligent control of product gas transfer for air revitalization". In *Proceedings of 28th International Conference on Environmental Systems*. Danvers, MA, July 1998.

[Williams and Nayak, 96] Williams, B. C., and P. P. Nayak. "A model-based approach to reactive self-configuring systems". In *Proceedings of the Thirteenth National Conference on Artificial Intelligence*. Cambridge, Mass., AAAI Press, pp. 971-978, 1996.

MODEL-BASED AUTONOMY FOR ROBUST MARS OPERATIONS

James A. Kurien,[*] P. Pandurang Nayak[†] and Brian C. Williams
NASA Ames Research Center, MS 269-2, Moffett Field, California 94035.
E-mail: {kurien, nayak, williams}@ptolemy.arc.nasa.gov.

Space missions have historically relied upon a large ground staff, numbering in the hundreds for complex missions, to maintain routine operations. When an anomaly occurs, this small army of engineers attempts to identify and work around the problem. A piloted Mars mission, with its multiyear duration, cost pressures, half-hour communication delays and two-week blackouts cannot be closely controlled by a battalion of engineers on Earth. Flight crew involvement in routine system operations must also be minimized to maximize science return. It also may be unrealistic to require the crew have the expertise in each mission subsystem needed to diagnose a system failure and effect a timely repair, as engineers did for Apollo 13.

Enter model-based autonomy, which allows complex systems to autonomously maintain operation despite failures or anomalous conditions, contributing to safe, robust, and minimally supervised operation of spacecraft, life support, ISRU and power systems. Autonomous reasoning is central to the approach. A reasoning algorithm uses a logical or mathematical model of a system to infer how to operate the system, diagnose failures and generate appropriate behavior to repair or reconfigure the system in response.

The "plug-and-play" nature of the models enables low cost development of autonomy for multiple platforms. Declarative, reusable models capture relevant aspects of the behavior of simple devices (e.g. valves or thrusters). Reasoning algorithms combine device models to create a model of the system-wide interactions and behavior of a complex, unique artifact such as a spacecraft. Rather than requiring engineers to envision all possible interactions and failures at design time or perform analysis during the mission, the reasoning engine generates the appropriate response to the current situation, taking into account its system-wide knowledge, the current state, and even sensor failures or unexpected behavior.

1. INTRODUCTION

Exploring and ultimately settling Mars will be a milestone in the development of our civilization and an uncompromising measure of our courage, cleverness and resolve. Accordingly, it will also be an unprecedented technical challenge, involving multiple interdependent mission elements, multiyear duration, incredible budgetary pressure and the duty to protect human lives in a harsh environment millions of miles from Earth. Evidence of the utility of highly capable, robust and coordinated autonomous systems in meeting this challenge pervades mission scenarios such as Mars Direct [1] and the NASA Mars Reference Mission [2].

Model-based autonomy involves the use of automated reasoning engines and high level models of the system being controlled to generate correct system behavior on the fly, even in the face of failures or anomalous situations. This approach is proving to be a robust and cost

---

[*] Caelum Research Corporation.
[†] Recom Technologies.

effective method for developing more highly capable autonomous systems than have been deployed in the past and might prove invaluable to the development of piloted missions to Mars.

The next section of this paper describes why autonomous systems are needed to explore Mars. Section 3 briefly discusses the varieties of model-based autonomy research going on at NASA Ames Research Center. Section 4 discusses how this work can contribute to cheap, safe, robust, and minimally supervised systems on Mars. Section 5 describes Livingstone, one of the reasoning engines developed at Ames that will be tested onboard a spacecraft next year. Section 6 describes a number of Mars-related testbeds that are making use of model-based autonomy technology.

This paper is meant to serve as an introduction to the concepts behind model-based autonomy for those who are not computer scientists and as a rough position paper regarding how those concepts might assist in a journey to Mars. The References section contains pointers to a number of papers on model-based autonomy with more technical detail and concrete explication.

2. THE UTILITY OF AUTONOMY ON MARS

The need for robust, inexpensive and productive operation of remote assets on Mars appears throughout both the Mars Direct scenario and the Mars Reference Mission. In both of these mission designs, initial mission elements such as in-situ propellant production (ISPP) plants and the crew return vehicle must be able to operate for a period years in a harsh environment with limited downlink capabilities and a reduced set of ground control personnel. Such systems must maintain efficient operation in spite of unexpected failures, novel environmental phenomena and degraded system capabilities. Safety places high demands on system robustness: the crew cannot depart Earth if propellant plant down time results in inadequate production or if the return vehicle cannot verify nominal operation.

Once the crew does depart Earth, they will be travelling two orders of magnitude farther from home than the Apollo crews. They will be separated from mission control by thirty-minute communication delays and potentially multi-day communication blackouts imposed by the relative positions of Mars and the Earth. There will be a number of systems upon which the crew's ability to reach Mars or survive an abort to Earth will depend: life support, attitude control, propulsion, communications and power generation are examples. While only life support might seem to require immediate response to anomalies, many other situations require on board response as well: losing attitude control during an aerobreaking maneuver, failures which need to be quickly safed, and loss of communications with Earth are all cases in point.

Once on the Martian surface, maximization of exploration becomes a focus in addition to safety. We do not yet have the resources to send crews of fifteen to Mars to run a Martian science outpost and support system. Hence crew involvement in routine operations such as controlling the life support system or maintaining rovers must be minimized and minor anomalies must be resolved locally rather than awaiting ground analysis. In addition, to maximize science return in this unknown environment, operations on Mars must be able to rapidly adapt to take advantage of new science opportunities or make the best of degraded capabilities.

These challenges to maintaining safety and productivity on Mars from Earth for several years are daunting when one considers the current state of mission operations. Current piloted missions rely upon near-instantaneous contact with hundreds of engineers and operators on the ground. In addition, recent attempts to teleoperate relatively simple systems for ninety days on

Mars resulted in a considerable fraction of the mission being used to determine the state of the remote system and return it to productive operation, often over the course of a day or more [3].

The Reference Mission therefore explicitly calls for autonomous systems on Mars to allow unmanned systems to robustly prepare for human arrival, to protect crew and resources by rapidly responding to critical failures, to free explorers from routine operations and to control operations costs for this complex, multi-year mission. In this context, autonomy means the ability to correctly react to a wide range of circumstances, both usual and anomalous, without the need for direct human supervision. If available, a robust onboard autonomy capability would enable safer, more affordable missions to Mars by allowing complex systems such as life support systems or spacecraft attitude control systems to operate for extended periods of time without supervision over a wide range of nominal and anomalous operating conditions. The benefits would be increased safety and reduced downtime for mission critical systems, leverage of scarce human skills by automation of routine tasks, and reduced ground operations due to unattended recovery from anomalies and less detailed commanding requirements.

Currently, NASA's operational experience with the type of high capability, failure-tolerant autonomy described in the Reference Mission is low. To date, no fully automated power plants, life support, or cryogen plants have been deployed. Some automated planning and scheduling has been used to pre-compute command sequences for spacecraft and to schedule space shuttle refurbishment, but no deployed system has autonomously replanned its mission activities in the field. In addition, the robotic systems that have been deployed in space have been almost entirely dependent upon pre-computed command sequences relayed from Earth controllers, and have not been highly autonomous in the sense conveyed above.

Of course, every unmanned system sent into space has required some level of autonomy: if a spacecraft cannot at least point its antenna at Earth and wait for help after the expected kinds of failures, it is likely to be lost. Currently programmers and mission control operators use their commonsense understanding of hardware and mission goals to produce code and control sequences that will allow a spacecraft or other system to achieve some goal while allowing for some (usually very small) amount of uncertainty in the environment. This has the disadvantages of being relatively time intensive, error prone, and not particularly reusable. Because of the amount of analysis involved, such systems usually allow for uncertainty by being extremely conservative and provide the minimal amount of adaptability necessary to raise the likelihood of survival of the spacecraft. If an anomaly occurs the spacecraft or other system typically halts all activity, achieves a safe mode, and awaits further instructions. One notable exception is the attitude and articulation control system on the Cassini spacecraft, which represents the state of the art in deployed spacecraft autonomy [4] and which has not been replicated on the "faster, better, cheaper" missions which have followed.

The cost to develop highly robust autonomous control software and the ability of such systems to improve safety and productivity of assets deployed on Mars (or deep space or Europa for that matter) are significant risk factors that impact NASA's ability to accurately plan and scope future missions. One intent of the work described in the paper, model-based autonomy, is to demonstrate that highly robust autonomous systems can be developed more easily and more cheaply than the more modest systems which have been deployed to date.

What is Model-Based Autonomy?

Model-based autonomy refers to the achievement of robust, autonomous operation through a growing set of reusable artificial intelligence (AI) reasoning algorithms that reason about first principles models of physical systems (e.g. spacecraft). In this context, a *model* is a logical or mathematical representation of a physical object or piece of software. A *first principles* model captures what is true about behavior or structure of the object (e.g. fluid flows through an open valve unless it is clogged). This is as opposed to traditional programs or rule-based expert systems, which capture what to do (e.g. turn on valves A, B, & C to start fuel flow) but unfortunately work only in certain implicit contexts (e.g. valves A, B, & C are working and A, B & C happen to control the fuel flow).

Since model-based autonomous systems do not contain an encoding of what to do in each situation, they must reason about the appropriate action to take or conclusion to draw based upon their models and the currently available information about the environment. The past few decades of AI research have produced reasoning engines that can plan a course of action to achieve a goal, identify the current state of a physical system, reconfigure that system to enable some function (e.g. make the engine thrust) and so on from a first principles model.

Reasoning directly from the model, the current observations of the world and the task at hand provides many large advantages over traditional software development. Not least among these are that the system is robust in uncertain environments since it was not hard coded to respond to certain situations, the models and inference engines can be reused, and the models explicitly capture the assumptions about the system that are being relied upon to control it.

3. MODEL-BASED AUTONOMY AT NASA AMES

The Autonomous Systems Group within the Computational Sciences Division at NASA Ames Research Center focuses on pursuing basic computer science research to increase the competence of autonomous systems and on providing basic autonomy technologies to NASA mission centers.

Many of the approximately twenty researchers within the group focus on model-based approaches to autonomy. As described above, a model-based autonomy architecture uses a declarative specification of a system's components and their interconnections to reason about the system as a whole and to provide general capabilities such as resource planning, execution monitoring, fault diagnosis and automated recovery. These compositional (plug-and-play) models adapt the generic architecture to a specific platform, enabling rapid development of autonomous control and health maintenance software for new systems.

Briefly, here is a small sampling of the many autonomy technologies being investigated within the Autonomous Systems Group that could be considered model-based. In Section 5, we will focus on one technology in slightly more detail.

- Model-based discrete controllers
  This line of research seeks to create a general engine for providing discrete control of a hardware and software system using only a declarative model of that system's components. This reasoning engine must infer the most likely state (referred to as a *mode*) of each component of the controlled system (including failures), accept a high level configuration goal (e.g. make the spacecraft produce thrust), and return a set of commands for the system's components that will achieve the configuration goal. The engine must use all available data to correctly handle sensor failure, multiple faults and novel recoveries.

  The Livingstone system for state identification and reconfiguration is such an engine, which has been developed at NASA Ames [5] and is being deployed in a number of NASA projects described below and in Section 6.

- Model-Based Decompositional Learning (MDL)
  This research uses a model of a system's structure to decompose complex parameter estimation problems into largely independent and solvable estimation subproblems. The Moriarty system is being developed at Ames to provide this capability and has been applied to developing a highly efficient office building environment controller [6].

- Planning & Scheduling
  Planning and scheduling research develops algorithms that accept a goal some system must achieve and emit a plan for causing the system to achieve the goal within the bounds of the resources allotted. The plan might be a simple sequence of commands or more generally it might be a partially ordered set of commands, a recurring schedule, or a plan with contingent branches which are executed based upon the outcome of previous actions. Unlike the simple configuration achievement of the discrete controller mentioned above, a planning and scheduling system must deliberate more thoroughly to achieve goals that involve oversubscribed resources, irreversible actions, or time-based constraints such as "take pictures of the following ten asteroids in the next hour".

  There are several planning and scheduling systems being pursued at NASA Ames, including the PS planner/scheduler [7]. Previous applications of NASA Ames planning technology include scheduling of observation requests onto an array of automated telescopes and scheduling of ground processing for the space shuttle.

  The Livingstone engine, the PS planning and scheduling system and the Smart Executive plan execution system [8] have been combined to form the Remote Agent autonomy architecture. NASA Ames and the New Millennium Program at the Jet Propulsion Laboratory (JPL) developed the Remote Agent architecture and will flight-test it during a weeklong autonomous operations experiment on the New Millennium Deep Space 1 spacecraft. During this week, control of the spacecraft will be turned over to the Remote Agent system. PS will generate plans that achieve high-level mission goals specified by the ground controllers. The Smart Executive will decompose those partially ordered plans into a sequence of commands to the spacecraft and execute them. Livingstone will determine the state of the spacecraft at each command and notify the Smart Executive if any failures or unexpected results have occurred. If any such failures occur, Livingstone will be used to find a repair or workaround that allows the plan to continue execution. Simulated failures will also be injected for the purpose of testing, as one cannot rely on a spacecraft failing during a particular week. If the plan cannot be executed, the Smart Executive will inform the planner of the degraded capabilities of the spacecraft and ask for a new plan that still achieves the mission goals. Much more detail on the Remote Agent and its experiment on Deep Space 1 can be found in [9] and [10].

  In addition to developing the core automated reasoning engines that enable model-based autonomy, NASA Ames is also pursing a number of research directions intended to further reduce the cost and effort required to develop and operate a model based autonomous system. Below two representative examples are described.

- Model-Based Programming Environments
  This task provides engineer-friendly tools to visually develop the models used by the above reasoning engines, generate test suites from the model definitions, and enable collaboration during model development.

- Human Centered Autonomous Agents

    This task draws upon the lessons learned during development of Remote Agent. It seeks to develop methods to smooth collaboration between humans and autonomous systems, providing variable levels of autonomy and enabling cooperation between humans and autonomous systems. The intent is to minimize the need for human involvement in routine operations, but to avoid interfering with it when it is needed or simply desired. The goal is to allow a small team to interact with and direct one or more autonomous systems and to allow humans to quickly assimilate the situation should an autonomous system realize it is out of scope and request human intervention.

4. BENEFITS OF MODEL-BASED AUTONOMY

Our initial experience has been that model-based autonomy can drastically reduce cost compared to traditional software development while increasing robustness, safety and maintainability of the autonomous system. We believe this experience will carry through to additional NASA missions that use this type of technology to achieve high capability autonomy. We discuss a number of areas where model-based autonomy can provide benefits below.

Safety and Reliability

Model-based autonomous systems increase safety and efficiency primarily via timely and correct response to anomalies. They can detect failures as soon as or very shortly after they occur and automatically reason through system-wide ramifications. They can then immediately notify personnel or autonomously attempt to mitigate the effects of the problem. A rapid, well-reasoned response minimizes the impact of failures or unforeseen scenarios.

Reliability is also increased by raising the visibility of the mapping between the control system and the apparatus it is meant to control. For example, with hard-coded fault protection designs knowledge about the controlled system is implicit rather than explicit. This means that we require the fault protection software developers to understand the system, understand the system-wide ramifications of various symptoms that might arise and actions that might be taken in response (taking into account the system might have experienced previous failures by the time the current problem occurs). The results of that understanding must then be encoded in a low-level procedural programming language. If the system is modified, as is often the case with one-of-a-kind hardware, that mapping from understanding to procedural code must be reconstructed and the necessary modifications to the code made, a likely time for the insertion of errors. Under these conditions even developing a fault protection system which has been scaled back to the minimal essentials can be a challenging task.

In contrast, with model-based fault diagnosis the fault protection software engineers explicitly model how the system behaves in nominal and (if known) failure cases. Relevant assumptions about the behavior and structure of the controlled system are explicitly stated in a declarative model, which is easily inspected and understood. Appropriate behavior is generated by operation of the heavily tested inference algorithms on the explicit models, not by low level statements of standard procedural software codes, which have an implicit and tenuous correspondence to written requirements or the software developer's evolving mental model of the system.

Finally, an approach which features reuse of reasoning engines and component models over multiple mission elements benefits reliability, cost and capability in the same manner that standard engines for database or graphics functionality benefit other software developers. Cost is reduced as development of the autonomy capability is amortized over several applications. Reliability is increased as the architecture's inference engines are invoked thousands of times before flight and as testing aggregates over multiple applications of the system. Capability is increased as reasoning engine improvements made and verified for one application are available for use in subsequent applications.

Development Effort

Model-based systems have proven extremely easy to design, implement and maintain. Applications consist of explicit, easily acquired "common-sense" level models of the controlled system. Gone is the need to develop procedural code that mingles implicit system models with implementation details, complicating development and maintenance. There is also no need to constantly reason through the interactions of the system's many components in order maintain code or rules for every possible scenario or spacecraft configuration.

Declarative, high level models of the structure and behavior of relevant device in the system can be quickly developed, unit tested and plugged together. Appropriate system-wide behavior is then generated. Device models capture only information about the local device, and not procedures for controlling or recovering the device within the context of some specific spacecraft or system. This makes them both reusable to model different systems and easily modified or replaced as a specific system being modeled is revised.

As the models are declarative and contain only first principles knowledge rather than context-dependent procedures or rules, the model of a single system can also be reused in a number of contexts. In the Remote Agent experiment, a single set of device models is used to monitor execution of commands, diagnose failures, provide recoveries, generate discrete event simulators and automatically produce descriptions of the spacecraft. Reuse could be taken much farther than time allowed during the Remote Agent experiment. One could use the same model to compile a static fault tree, compile interface definitions for communicating with the Remote Agent, and so on. In addition by historical accident the high level planning and scheduling component of Remote Agent did not accept the same model as was used for all of the above tasks, but an effort is underway to unify the two languages before Remote Agent is for additional mission operations.

Operations Cost

Cost goals demand that mission planning be streamlined and mission control intervention in routine spacecraft and Mars surface operations are minimized. Systems that can autonomously monitor, adjust and repair themselves, even in the face of novel situations, obviously decrease the time the ground controllers or crew must devote to such tasks. In addition, when human involvement is required or desired, our goal is to drastically reduce operator effort, allowing an operator to control more systems or a crewmember to operate a system more quickly or with less detailed training. Model-based autonomy allows very high level interaction. The current state of the system can be presented in hierarchical diagrams generated from the models. Commands can be goals to be achieved rather than detailed instructions. Significantly, when a user wants to interact with an model-based system, its explicit internal representation allows it to explain why it has made certain inferences, why it decided to take certain actions or what it would do in a hypothetical situation. These features combine to make operating a model-based system potentially far more intuitive, cooperative and efficient than current spacecraft operations.

5. LIVINGSTONE

As mentioned in Section 3, Livingstone is a model-based discrete controller. Its function is to infer the current state (mode) of each relevant device making up the system being controlled and to recommend actions that can reconfigure the system so that it achieves the currently desired configuration goals, if possible. In practice, these configuration goals could be provided by a human operating some apparatus by issuing high level configuration commands, or by some automated system such as the Smart Executive (Exec) mentioned above, which decomposes a high level plan into a series of configuration goals to be achieved. Purely for the sake of the discussion below, we will assume the Exec is providing the configuration goals and that the system being controlled is a spacecraft.

To track the modes of system devices, Livingstone eavesdrops on commands that are sent to the spacecraft hardware by the Exec. As each command is executed, Livingstone receives observations from spacecraft's sensors, abstracted by monitors in the real time control software for the Attitude Control Subsystem (ACS), communications bus, or whatever hardware is present. Livingstone combines these commands and observations with declarative models of the spacecraft components to determine the current state of the system and report it to the Exec. A pathologically simple example is shown schematically in Figure 1. In the nominal case, Livingstone merely confirms that the commands had the expected effect on spacecraft state. In case of failure, Livingstone diagnoses the failure and the current state of the spacecraft and provides a recovery recommendation. A single set of models and algorithms are exploited for command confirmation, diagnosis and recovery.

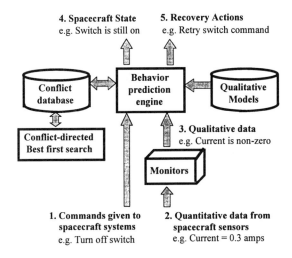

Figure 1   Information Flow in Livingstone.

The capabilities of the Livingstone inference engine can be divided into two parts: mode identification (MI) and mode reconfiguration (MR). MI is responsible for identifying the current operating or failure mode of each component in the spacecraft. Following a component failure, MR is responsible for suggesting reconfiguration actions that restore the spacecraft to a configuration that achieves all current configuration goals required by the Exec. Livingstone can be viewed as a discrete model-based controller in which MI provides the sensing component and MR provides the actuation component. MI's mode inference allows the Exec to reason about the state of the spacecraft in terms of component modes or even high level capabilities such as "able to produce thrust" rather than in terms of low level sensor values. MR supports the run-time generation of novel reconfiguration actions to return components to the desired mode or to re-enable high level capabilities such as "able to produce thrust".

Livingstone uses algorithms adapted from model-based diagnosis [11, 12] to provide the above functions. The key idea underlying model-based diagnosis is that a combination of component modes is a possible description of the current state of the spacecraft only if the set of models associated with these modes is consistent with the observed sensor values. Following de Kleer and Williams [13], MI uses a conflict directed best-first search to find the most likely combination of component modes consistent with the observations. Analogously, MR uses the same search to find the least-cost combination of commands that achieve the desired goals in the next state. Furthermore, both MI and MR use the same system model to perform their function. The combination of a single search algorithm with a single model, and the process of exercising these through multiple uses, contributes significantly to the robustness of the complete system. Note that this methodology is independent of the actual set of available sensors and commands. Furthermore, it does not require that all aspects of the spacecraft state are directly observable, providing an elegant solution to the problem of limited observability.

The use of model-based diagnosis algorithms immediately provides Livingstone with a number of additional features. First, the search algorithms are sound and complete, providing a guarantee of coverage with respect to the models used. Second, the model building methodology is modular, which simplifies model construction and maintenance, and supports reuse. Third, the algorithms extend smoothly to handling multiple faults and recoveries that involve multiple commands. Fourth, while the algorithms do not require explicit fault models for each component, they can easily exploit available fault models to find likely failures and possible recoveries.

Livingstone extends the basic ideas of model-based diagnosis by modeling each component as a finite state machine, and the whole spacecraft as a set of concurrent, synchronous state machines. Modeling the spacecraft as a concurrent machine allows Livingstone to effectively track concurrent state changes caused either by deliberate command or by component failures. An important feature is that the behavior of each component state or mode is captured using abstract, or qualitative, models [14]. These models describe qualities of the spacecraft's structure or behavior without the detail needed for precise numerical prediction, making abstract models much easier to acquire and verify than quantitative engineering models. Examples of qualities captured are the power, data and hydraulic connectivity of spacecraft components and the directions in which each thruster provides torque. While such models cannot quantify how the spacecraft would perform with a failed thruster for example, they can be used to infer which thrusters are failed given only the signs of the errors in spacecraft orientation. Such inferences are robust since small changes in the underlying parameters do not affect the abstract behavior of the spacecraft. In addition, abstract models can be reduced to a set of clauses in propositional logic. This form allows behavior prediction to take place via unit propagation, a restricted and very efficient inference procedure.

It is important to note that the Livingstone models are not required to be explicit or complete with respect to the actual physical components. Often models do not explicitly represent the cause for a given behavior in terms of a component's physical structure. For example, there are numerous causes for a stuck switch: the driver has failed, excessive current has welded it shut, and so on. If the observable behavior and recovery for all causes of a stuck switch are the same, Livingstone need not closely model the physical structure responsible for these fine distinctions. Models are always incomplete in that they have an explicit unknown failure mode.

Any component behavior that is inconsistent with all known nominal and failure modes is consistent with the unknown failure mode. In this way, Livingstone can still infer that a component has failed, though the failure was not foreseen or was simply left unmodeled because no recovery is possible. By modeling only to the level of detail required to make relevant distinctions in diagnosis (distinctions that prescribe different recoveries or different operation of the system) we can describe a system with qualitative "common-sense" models which are compact and quite easily written.

6. MARS RELATED APPLICATIONS

The intent behind model-based autonomy is to create generic, high capability reasoning systems that can be adapted to a wide range of applications simply by writing the appropriate models. As such, model-based autonomy might be able to contribute to the control of a variety of elements of a piloted Mars mission. In this early stage of Mars mission definition, model-based autonomy is involved in the prototyping of a number of specific mission elements.

Closed-Loop Ecological Life Support Systems (CELSS)

In order to transport and support humans for Mars expeditions, NASA's Human Exploration and Development of Space (HEDS) requirements state a need for autonomous operation of life support, ISRU and transport equipment. During a Mars expedition, autonomous plant operations would allow unmanned systems to prepare for human arrival, protect crew and resources by rapidly responding to critical failures, and free humans from routine operations, allowing greater exploration.

At NASA's Johnson Space Center (JSC), a closed loop life support testbed called Bioplex has been constructed. The Bioplex consists of three sections: a three story cylindrical living quarters similar to the Mars habitats discussed in various mission proposals; a plant chamber where wheat is grown to provide food and exchange CO_2 for O_2; and an incinerator chamber used to eliminate solid waste and produce CO_2. The most recent Bioplex testing is referred to as the Product Gas Transfer phase as it concentrates on generation and distribution of product gases (CO_2 from the crew and incinerator and O_2 from the plants) and does not yet address issues such as waste water recycling or power management.

A JSC advanced development group has developed an autonomous control system to operate the product gas transfer phase of Bioplex [15]. This system, based upon the 3T autonomy architecture [16], maintains the appropriate atmosphere in each chamber by extracting and storing product gases and coordinating activities such as firing the incinerator or opening the plant chamber for human access. The system successfully controlled gas transfer during test in which a human crew inhabited the Bioplex for ninety days. It was not expected, however, to maintain operation in the face of failures, though many would likely occur over a 4-year mission.

We are currently working to integrate the Livingstone mode identification and reconfiguration engine with JSC's 3T architecture, adding to it the ability to determine the current state of the testbed and respond to anomalous situations or failures by performing high level, system-wide reasoning. This will result in a single, reusable architecture which maintains the best possible operation of a regenerative life support system and other complex physical plants during both nominal operation and failures, somewhat analogous to the autonomic and immune functions of a living organism.

We intend to demonstrate the combined system by maintaining operation of the testbed over an extended test period and providing both fully autonomous and human-centered operation. To test the system, an outside examiner will be employed to introduce failures into the testbed as desired which the system will diagnose and attempt to mitigate.

The second goal is to demonstrate and extend the ability of model-based systems to reduce analysis, development and operations costs. The testbed application will be rapidly developed with tools that could be used to develop mission applications. Users will develop and operate the testbed by manipulating explicit models with visual tools. If previous experience is to be believed, far less effort will be required to develop, understand and revise the system than in an approach where system model is implicit but still must be maintained.

If successful, this demonstration will increase the likelihood that autonomy technologies being developed by NASA are appropriate and sufficiently mature when they are required for HEDS missions to Mars and other destinations. It will also ensure that the necessary technologies can be integrated and will identify needed extensions before such shortcomings could impact the critical path of a mission. In addition, JSC will have a prototype of a reusable, fault-tolerant, high-capability autonomous control system and the expertise to apply this system to a flight experiment or mission. This could be applied to any complex physical system that must be controlled and maintained over an extended period of time such as spacecraft, power plants, ISRU machinery, and autonomous or semiautonomous surface vehicles.

In-Situ Resource Utilization

In-situ resource utilization, or "living off the land", is critical to making a piloted Mars mission robust and affordable [1]. More specifically, it is envisioned that in-situ propellant production (ISPP) plants will arrive on Mars years before humans and begin combining hydrogen brought from Earth with CO_2 from the Martian atmosphere to create methane. This fuel will power the ascent vehicle that will lift the crew off Mars to begin their trip home in addition to powering any methane-fueled surface vehicles the astronauts might possess.

Though the chemical reactions involved are conceptually quite simple, on Mars they are somewhat complicated by issues such as the low atmospheric pressure and slow contamination of the ISPP catalysts by trace elements in the Martian atmosphere. To ensure that adequate ISPP capability is available for future Mars missions, NASA has begun to explore ISPP designs and build prototype hardware for operation in Mars-like test chambers. Both JSC and NASA Kennedy Space Center (KSC) are involved in early ISPP development, and the KSC team is integrating Livingstone into their ISPP prototyping efforts.

The short-term focus of this collaboration is to integrate Livingstone's ability to diagnose and mitigate failures with existing KSC model-based technology to gain experience with a model-based monitoring, diagnosis and recovery system for ISPP. A secondary short-term goal is to determine if any other autonomy technology previously invested in by NASA, for example the Smart Executive, can be reused on the ISPP testbed, thus increasing capability without greatly increasing cost.

A longer term goal is to continue research into control of physical systems which must continuously adjust their operation to unforeseen degradation in capability (for example an ISPP unit where Martian dust covers solar panels or slowly clogs air filters) rather than taking a discrete recovery action as Livingstone does. Related issues include reasoning about hybrid discrete/continuous systems, predictive diagnosis and relearning models of the continuous dynamic behavior of the system. This research should contribute to development of ISPP and other systems that are robust and yet run at the ragged edge of optimality throughout their lifetimes, neither being overly conservative nor exceeding their remaining degraded capabilities.

Autonomous Rovers

The Remote Agent system, described above and consisting of a planner, a smart execution system and Livingstone, is being adapted for use on the NASA Ames Marsokhod rover as part of an effort to demonstrate increased rover autonomy. That effort is described in [17].

7. ACKNOWLEDGEMENTS

This paper touches on the work of a great many people too numerous to name here. The Autonomous Systems Group at NASA Ames Research Center consists of about twenty computer science researchers pursuing all manners of autonomy research, much of which was not mentioned here. Members of the JPL New Millennium Program and AI Group contributed to the Remote Agent architecture and to making it work on a flight platform. Advanced development groups at NASA JSC (3T and Bioplex PGT), NASA KSC (ISPP and KATE) and JPL (space based interferometry) have shared their expertise with us and are helping to push the model-based autonomy technologies described here forward.

REFERENCES

Many of the following papers may be found on the World Wide Web at
http://ic-www.arc.nasa.gov/ic/projects/mba/

1. R. Zubrin and R. Wagner. *The Case for Mars: The plan to settle the Red Planet and why we must*. The Free Press, 1996.
2. S. J. Hoffman and D. I. Kaplan, editors. *Human Exploration of Mars: The Reference Mission of the NASA Mars Exploration Study Team*. NASA Special Publication 6107. July 1997.
3. A. H. Mishkin, J. C. Morrison, T. T. Nguyen, H. W. Stone, B. K. Cooper and B. H. Wilcox. "Experiences with operations and autonomy of the Mars Pathfinder microrover". In *Proceedings of the IEEE Aerospace Conference*, Snowmass, CO 1998.
4. G. M. Brown, D. E. Bernard and R. D. Rasmussen. "Attitude and articulation control for the Cassini Spacecraft: A fault tolerance overview". In *14th AIAA/IEEE Digital Avionics Systems Conference*, Cambridge, MA, November 1995.
5. B. C. Williams and P. Nayak, "A Model-based Approach to Reactive Self-Configuring Systems", *Proceedings of AAAI-96*, 1996.
6. B. C. Williams and B. Millar. 1996. "Automated Decomposition of Model-based Learning Problems". *In Proceedings of QR-96*.
7. N. Muscettola, B. Smith, C. Fry, S. Chien, K. Rajan, G Rabideau and D. Yan, "Onboard Planning for New Millennium Deep Space One Autonomy", *Proceedings of IEEE Aerospace Conference*, 1997.
8. B. Pell, E. Gat, R. Keesing, N. Muscettola, and B. Smith. Robust periodic planning and execution for autonomous spacecraft.
9. B. Pell, D. E. Bernard, S. A. Chien, E. Gat, N. Muscettola, P. P. Nayak, M. D. Wagner, and B. C. Williams, "An Autonomous Spacecraft Agent Prototype", *Proceedings of the First International Conference on Autonomous Agents*, 1997.
10. D. E. Bernard *et al.* "Design of the Remote Agent Experiment for Spacecraft Autonomy". *Proceedings of IEEE Aero-98*.
11. J. de Kleer and B. C. Williams, "Diagnosing Multiple Faults", *Artificial Intelligence*, Vol. 32, Number 1, 1987.
12. J. de Kleer and B. C. Williams, "Diagnosis With Behavioral Modes", *Proceedings of IJCAI-89*, 1989.
13. J. de Kleer and B. C. Williams, *Artificial Intelligence*, Volume 51, Elsevier, 1991.
14. S. Weld and J. de Kleer, *Readings in Qualitative Reasoning About Physical Systems*, Morgan Kaufmann Publishers, Inc., San Mateo, California, 1990.
15. D. Schreckenghost, M. Edeen, R. P. Bonasso, and J. Erickson. "Intelligent control of product gas transfer for air revitalization". Abstract submitted for 28th International Conference on Environmental Systems (ICES), July 1998.

16. R. P. Bonasso, R. J. Firby, E. Gat, D. Kortenkamp, D. Miller and M. Slack. "Experiences with an architecture for intelligent, reactive agents". In *Journal of Experimental and Theoretical AI*, 1997.
17. J. Bresina, G. A. Dorais, K. Golden, D. E. Smith, R. Washington, "Autonomous Rovers for Human Exploration of Mars". *Proceedings of the First Annual Mars Society Conference*. Boulder, CO, August 1998. See this volume.
18. B. C. Williams and P. P. Nayak. "Immobile Robots: AI in the New Millennium". In *AI Magazine*, Fall 1996.
19. B. C. Williams and P. P. Nayak. "A Reactive Planner for a Model-based Executive". In *Proceedings of IJCAI-97*.
20. N. Muscettola. HSTS: Integrating planning and scheduling. In Mark Fox and Monte Zweben, editors, *Intelligent Scheduling*. Morgan Kaufmann, 1994.
21. V. Gupta, R. Jagadeesan, V. Saraswat. "Computing with Continuous Change". *Science of Computer Programming*, 1997.

Chapter 10
THE QUESTION OF LIFE

Mars meteorite investigator Everett Gibson makes the case for relic life.

LIFE ON MARS:
EVIDENCE WITHIN MARTIAN METEORITES*

Everett K. Gibson, Jr.,† David S. McKay‡ and Kathie Thomas-Keprta**

The Viking Mission to Mars in 1976 was the first attempt to search for live *in situ* on another planet with experiments designed by scientists on Earth. There were five experiments on the Viking spacecrafts which were designed to search for evidence of life. They were the cameras which photographed the surrounding areas of the landing site in hopes of seeing evidence of movement associated with a living system. The gas chromatograph/mass spectrometer instrument was to seek the presence of reduced carbon compounds, at the parts per billion levels, in surface soil samples. The labelled release, the gas exchange and the metabolic release experiments were a series of three biology experiments which were to seek evidence of biological activity within the soil samples. At the conclusion of the Viking mission, the consensus of the scientific community was that the results were all negative. However, Levin, a member of the original Viking team noted that one of the life detection experiments (the labelled release experiment) did in fact detect evidence of life (Levin and Straat, 1981). He has assembled all of the arguments used against this interpretation, and then proceeded to give a rebuttal for each. Whether one agrees or disagrees with the viewpoint of Levin, his paper makes very interesting reading and is available on the web at: **www.biospherics.com**.

Bogard and Johnson (1983) showed that a meteorite (EETA79001) of the Shergottite-Nakhlite-Chassignite (SNC) class contained trapped noble gases which matched the Martian atmospheric gases measured by Viking. For the first time the scientific community had samples of another planet (besides the moon) available for study and those materials were from Mars. Additional recent examination (Bogard and Garrison, 1998) of the group of SNC meteorites has shown that six of the thirteen Martian meteorites contain samples of the atmosphere. In addition to the trapped gases, the unique composition of the oxygen isotopes within the silicate minerals of the SNC meteorites shows they were from a unique oxygen reservoir within our solar system (Clayton and Mayeda, 1983; Romanek *et al.*, 1998).

In 1996, we suggested that features within the Martian meteorite ALH84001 could be interpreted as evidence of past biogenic activity (McKay *et al.*, 1996). The four lines of evidence was based upon the presence of carbonate globules which had been formed at temperatures fa-

---

* Everett K. Gibson, Jr., David S. McKay and Kathie Thomas-Keprta were the co-leaders of the team that first reported evidence of possible past biological activity within the ALH84001 meteorite. They all work at NASA Johnson Space Center. Gibson is a geochemist and meteorite specialist, McKay is a geologist and electron microscopist and Thomas-Keprta is a biologist who applies electron microscopy techniques to the study of terrestrial and extraterrestrial materials. Gibson and McKay work for NASA and Thomas-Keprta is a senior scientist with Lockheed-Martin Inc.

† SN2, Planetary Sciences, NASA Johnson Space Center, Houston, Texas 77058.
E-mail: everett.k.gibson@jsc.nasa.gov.

‡ SN2, Planetary Sciences, NASA Johnson Space Center, Houston, Texas 77058.

** C23, Lockheed Martin Corp., NASA Road 1, Houston, Texas 77058.

vorable for life, the presence of biominerals (i.e. magnetites) with the characteristics of those formed by magnetotactic bacteria, the presence of indigenous reduced carbon within Martian materials, and the presence of unique shapes and sizes of morphological features such as segmented structures and elongated tubular features. Each of features could be interpreted as having origins different from biogenic but the unique spatial relationships suggested to our team that they were evidence of past biogenic activity recorded within the meteorite. Both criticism and support has been directed toward our hypothesis (Anders, 1996; Bradley *et al.* 1996, 1997, 1998; Bada *et al.*, 1998; Oro, 1998; Valley *et al.*, 1997; Krischvank *et al.*, 1997; Friedmann *et al.*, 1998; Hoover *et al.*, 1997).

An overview of the status of our original hypothesis will be presented and we will examine those critical points of our hypothesis in light of recent criticisms. In our view, the specific criticisms of many of the critics reflect a selective and possibly one-sided sampling of the available peer-reviewed scientific literature.

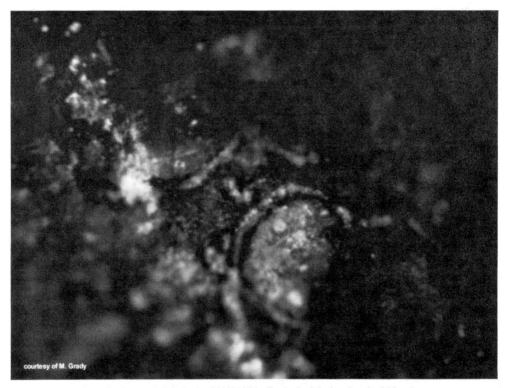

Figure 1 Carbonate globule in ALH84001. Typical globule size is 200 microns with a black-white-black rim.

TEMPERATURE OF FORMATION OF THE CARBONATE GLOBULES

The carbonate globules within ALH84001 are the component within which evidence of past biogenic activity is found (Figure 1). None of the other thirteen Martian meteorites contain carbonate globules. We believe that the signatures of past biogenic activity are clearly evident within the carbonate globules. Understanding the formation conditions and temperatures associated with carbonates holds the key to possible evidence of past Martian life. We (Romanek *et al.*, 1994; McKay *et al.*, 1996) and others (i.e. Valley *et al.* 1997: Warren, 1998; Romenek and

Treiman, 1998; McSween and Harvey, 1998) have proposed that these globules formed at low temperatures by precipitation or evaporated from a carbon dioxide saturated fluid. Borg *et al.* (1998) have reported that the formation age of the carbonate globules was 3.9 b.y. as determined by Rb-Sr measurements. This formation age is during a period of time (4.1 to 3.9 b.y. ago) when Mars had just undergone extensive bombardment and the crust would have been extensively fractured. The movement of ground waters through these cracks and fractures surrounding the craters would have been facilitated. The planet was experiencing a period of extensive surface waters with elevated temperatures, and greater atmospheric pressures. The existence of a Martian ocean and standing waters which could have evaporated was possible (Head *et al.*, 1998).

Romanek *et al.* (1994) initially showed the oxygen and carbon isotopic compositions of the carbonate globules indicated temperatures of formation below 110°C. On the other hand, Harvey and McSween (1996) along with Bradley *et al.* (1996, 1997, 1998) argued that the carbonates formed at temperatures above 600°C. A number of papers were subsequently published examining the formation temperatures of the carbonates, some advocating high temperature and others low temperature. Warren (1998) reviewed all of the high temperature models and concluded that none of them could be supported by the available analytical data. He went on to propose a low temperature origin such as carbonate precipitation or evaporation in cracked bedrock or from drying Martian lakes. Subsequently, McSween and Harvey (1998) have proposed a similar model.

Barrat *et al.* (1998) showed that small carbonate globules grew within cracks of the Tatahouine meteorite which fell in Tunesia in 1931; these globules are rather similar to the ones in Allen Hills 84001 and they contain small calcite rods and spheres which may be mineral precipitates or may be fossilized bacteria from earth. While no one disputes that the carbonates in ALH84001 formed on Mars, the Tatahouine meteorite demonstrates that carbonate globules can form in a very arid environment near the ground surface and at relatively low temperatures (within the range of liquid water). This provides a clear mechanism for low temperature formation of the carbonates on Mars.

The main proponents of the high temperature origin (Harvey and McSween, 1996) have modified their position with a new model for the formation of ALH84001 carbonates from Martian salt-rich water (brines) at low temperatures (McSween and Harvey, 1998). It is not clear whether they have abandoned their advocacy of the high temperature model, but the fact that they have proposed a low-temperature model shows that they may no longer be very certain about their high temperature model.

In summary, in spite of the critic's claims, it seems clear to us and to objective observers that a low-temperature origin for the carbonates has become the most reasonable model in the scientific community, and the Tatahouine meteorite provides us with an actual example of similar carbonates formed at relatively low-temperatures on the earth. At the 61st annual meeting of the Meteoritical Society in Dublin, Ireland in July 1998, there was only a single paper presented in support of the high temperature formation model for the carbonates in ALH84001 and that paper by Scott *et al.* (1998) showed evidence that elevated temperature processes may have been associated with the impact processes operating after carbonate formation and which fractured the carbonate globules present in ALH84001. For a summary of these papers, their full references, objective comments, see the section written by Dr. Allen Treiman on the Lunar and Planetary Institute (LPI) web page: **cass.jsc.nasa.gov/lpi/meteorites/alhnpap.html**.

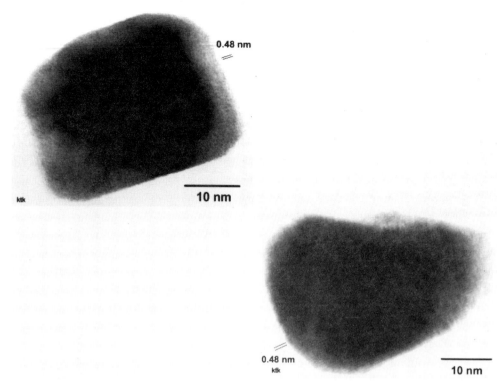

Figure 2 Magnetite (Fe$_3$O$_4$) grains in ALH84001 carbonate globule rim. Typical size is 40 to 60 microns.

ORIGIN OF THE MAGNETITES IN THE CARBONATE GLOBULES

We originally proposed that a portion of the tiny magnetite (Fe$_3$O$_4$) grains in the rims of the carbonate globules were very similar to magnetites produced by magnetotactic bacteria (McKay *et al.* 1996; Thomas-Keprta *et al.* 1997, 1998) (Figure 2). Bradley *et al.* (1996, 1997, 1998) suggest that the magnetite are formed at high temperatures from either condensation from a vapor or shock heating of iron-rich carbonate which might precipitate magnetite (These papers are summarized on the LPI web page). At the 29th Lunar and Planetary Science Conference, Thomas-Keprta *et al.* (1998) presented very strong evidence that the sizes and shapes of more than 80 percent of the magnetites within the carbonates were identical in size (40 to 60 nanometer), in shape parameters (parallelopiped or teardrop shaped), and in chemical compositions (essentially pure Fe and oxygen with no minor or trace elements of significant concentrations) to magnetites produced by magnetotactic bacteria on earth. Yet these magnetites were found within carbonates which were formed on Mars. They showed that the remaining magnetite morphologies may possibly result from non-biologic processes such as inorganic precipitation from a solution, although biologic processes cannot be ruled out for them as well. We find it difficult to believe that the carbonates formed at low-temperatures (as most of the evidence now indicates), but that the tens of thousands of magnetite grains within each carbonate formed at high temperatures. We have proposed (Thomas-Keprta *et al.*, 1998a, b; Gibson *et al.*, 1998) that the magnetites of high temperature origin which Bradley and coworkers (1996, 1997, 1998) have recognized were produced during the early bombardment of Mars when elevated temperatures were present and metasomatism processes were in operation. These magnetites were produced prior to the production of the magnetites in the rims of the carbonate globules and have simply become trapped within the globules at the time of their formation from fluids on the Martian surface. Recent studies (Thomas-Keprta et al., 1998, 1999) have shown that the high temperature magnetite whiskers of the type discussed by Bradley et al., (1997, 1998) account for less than 5 to 7 percent of all of the magnetites present within the carbonate globules.

SULFUR ISOTOPES

Shaerer et al. (1996) and Greenwood et al. (1997) suggested that the sulfur isotopes in the ALH84001 sulfides do not reflect any biologic role in their formation. As Gibson et al. (1997, 1998) noted both groups had not analyzed the sulfides associated with the biogenic activity. No evidence has been offered that the sulfur isotopic compositions of the greigite (Fe_3O_4) have been analyzed. The biogenic sulfide phase is in the 40 to 60 nanometer size range and the ion microprobe beam size (typically 10 to 20 microns in size) is simply too large to analyze the small biogenic sulfides. The only sulfides which have been accurately analyzed are the relatively large micron sized pyrites in the meteorite's silicate groundmass, and indeed, they do not show obvious signs of biologic processing which can change their isotopic ratios. McKay et al. (1996) never proposed that these larger sulfides were biogenic, and it is unclear whether they are associated in any way with the carbonates. Finally, the majority of bacteria on earth which assist with sulfide formation do not cause isotopic changes in the sulfur components. Therefore, the ALH84001 sulfur isotope data currently available are not relevant to the question of bacteria activity; they neither support it or undermine it.

THE SIGNIFICANCE OF THE POLYCYCLIC AROMATIC HYDROCARBONS (PAHs)

Several critics (i.e. Oro, 1998; Becker and Bada, 1997) argues that (1) the PAHs we reported in ALH84001 are not diagnostic of life since PAHs form at high temperatures, that (2) the PAHs in this Martian meteorite are almost identical to those in the carbonaceous chondrite Murchison, presumed to have come from the asteroid belt, not Mars. In addition, Bada et al. (1998) have suggested that the PAHs are probably contamination products from Antarctic ice. In the original paper, McKay et al. (1996) did not claim that PAHs are biologic markers. Indeed, we certainly were aware that PAHs are not typically found in living systems or bacteria. What McKay et al. (1996) proposed was that the PAHs might be by-products from decay and fossilization of bacteria. It has been well documented that PAHs do form as living systems die and decay, and they form at relatively low temperatures. The PAHs in ALH84001 are not identical to those in Murchison or any other carbonaceous chondrite (Clemett et al., 1998). They differ in the presence or absence of some of the specific compounds and in the relative abundance of each major compound.

Clemett et al. (1998) have presented definitive evidence on the indigenous polycyclic aromatic hydrocarbons (PAHs) present in ALH84001. They showed that the Antarctic environment does not contribute to the introduction of PAHs into ALH84001. Concentrations of PAHs were measured in numerous ordinary chondrite meteorites which had longer residence periods in the Antarctic ice than ALH84001. In no case did the concentrations of PAHs increase as a function of exposure or residence time in the Antarctic environment. Clemett et al. (1998) showed that interplanetary dust particles (IDPs) from the Antarctic ice field near where ALH84001 was collected did not have enrichments in PAH concentrations. The study shows clearly that a distinction exists between ALH84001 and all other analyzed carbonaceous chondrites in their reduced carbon abundances. Clemett et al. (1998) also shows that the Antarctic ice collected from Allen Hills under very clean conditions, contains virtually undetectable levels of PAHs. They also showed that Antarctic micrometeorites each contain their own distinctive PAH fingerprint which is not only unlike any other but is also unlike ALH84001. These micrometeorites are known to have unusually large surface areas to which organics within the ice or melt water could have adsorbed. If all the PAHs in these micrometeorites came from contamination from the ice, they should all be identical in fingerprint each other reflecting the PAHs in the ice, but they are all clearly distinctive. Clemett et al. (1998) also analyzed more PAH profiles from the exterior fusion crust of ALH84001 to the interior and found again that the abundance always increases inward from the heated surface of the meteorite. Contamination from the ice would provide the opposite profile.

Another point on PAHs, Stephan et al. (1998) suggests that PAHs are everywhere in the ALH84001 meteorite including in the center of the igneous mineral grains, and if anything, PAHs are depleted in the carbonate rather than concentrated. It must be noted that these analyses were made on polished thin sections of the meteorite whereas the analyses of Clemett et al. (1998) were made on freshly broken surfaces. Most polishing procedures use organic solvents or organic diamond paste and require considerable handling in less than clean conditions. Could the thin section making process have introduced PAH contamination? Or could it have smeared out existing PAHs over the polished surface so that they no longer reflect their original location and abundance? Stephan et al. (1998) noted identical PAH concentrations in the pyroxenes and in the carbonates. Additional analysis by Flynn et al. (1999) showed variations of organic component concentrations within the carbonate interior and rims. Several isolated regions of PAH enrichments were observed. It is our opinion (Gibson et al., 1998) that the results reported by Stephan et al. (1998) represent contamination and do not provide meaningful information about the true state of reduced carbon components within ALH84001.

In another recent study, Clemett and colleagues (in preparation) have shown that PAHs of similar abundances and compositions to those present in ALH84001 are associated with fossilized bacteria in subsurface basalt samples of the Columbia River Basalts Group. If further studies on PAHs within samples containing known evidence of biogenic activity show specific abundance patterns or compositions of PAHs, they have the opportunity to offer another biomarker for studying fossil evidence of biological activity.

In summary, we and other objective observers believe that the new Clemett et al. (1998) studies show definitively that indigenous reduced carbon components are present within ALH84001 meteorite and confirm that a biogenic origin on Mars is still a reasonable hypothesis. Thus, for the first time the presence of reduced carbon compounds has been identified in Martian materials. This accomplishes one of the goals of the Viking experiments some 20 years after the mission.

AMINO ACIDS

At no time has our research team suggested the presence of amino acids within ALH84001. We did not analyze this meteorite for amino acids. Previous studies of amino acids within meteorites from Antarctica have shown that the Antarctic environment (i.e. melt waters and ice) (McDonald et al. 1995; Bada et al., 1998) contribute to contamination of the sample with amino acids. Therefore, any study of amino acids which is carried out on meteorites from Antarctica must be viewed with considerable skepticism.

While amino acids are clearly produced by living systems, they move around easily in water solutions because of their high solubility as opposed to PAHs which are essentially insoluble in water. Therefore, we question the use of amino acids as a reliable biomarker for Martian meteorites collected in Antarctica (Gibson *et al.*, 1998), and have never advocated such use. Amino acids may make good biomarkers in other settings, for example on actual samples returned from Mars or possibly other Martian meteorites which have not resided in the Antarctic.

TERRESTRIAL CARBON CONTAMINATION

Carbon-14 measurements such as those of Jull *et al.* (1998) show modern-day atmospheric carbon-14 is incorporated into all meteorites which have resided on the Antarctic ice fields. Typically, the secondary weathering products such as carbonates have incorporated modern-day carbon-14. All components with the carbon-14 signature should be viewed as contaminants introduced by the terrestrial environment. Jull *et al.* (1998) measurements showed that most (at least 80%) of the organic carbon in ALH84001 contains significant modern-day carbon-14 and is therefore terrestrial. This terrestrial carbon may include amino acids and organic material from the air. It may even include terrestrial organisms. However, Jull *et al.* (1998) also showed one component with a carbon isotopic composition of -18 per mil which did not contain modern-day carbon-14. This component resisted the acid dissolution which dissolved the carbonates. Yet, it persisted in the meteorite to temperatures above 400°C. They interpreted this component to be pre-terrestrial (that is Martian) and either organic carbon (i.e. kerogen or possibly PAHs) from Mars on a very unusual carbonate entirely unlike the bulk of the Martian carbonate. This component makes up 20% of the organic carbon in the meteorite. This carbon component is potentially one of the major discoveries in understanding the nature of carbon within ALH84001. Oro (1998) failed to point out the discovery by Jull *et al.* (1998) of this mystery carbon bearing component which is definitely not from earth (because it has no terrestrial carbon-14) and is almost certainly from Mars. This component may include the PAHs found by McKay *et al.* (1996) and Clemett *et al.* (1998) and would be additional proof that they are from Mars.

Grady *et al.* (1994, 1998) have shown that ALH84001 has a carbon isotopic composition for selected components suggestive of both terrestrial contaminants along with indigenous Martian carbon phases. The Martian carbon components are both magmatic carbon along with carbonate globule carbon. Within the globules there may be both carbon associated with the carbonate and an organic carbon phase which has an isotopic carbon composition similar to the carbon-14 free component identified by Jull *et al.* (1998). Flynn *et al.* (1997, 1998, 1999) have shown that that there is an indigenous organic carbon component within the carbonate globules which is distributed unevenly throughout the globules.

In summary, yes, this meteorite contains significant Antarctic contamination in the form of terrestrial organic carbon. This carbon may include terrestrial amino acids. However, data published this year by Jull *et al.* (1998), Grady *et al.* (1994, 1998) and Clemett *et al.* (1998) have shown that in addition to the terrestrial organic carbon, the meteorite contains indigenous Martian organic carbon.

MICROBIOLOGICAL EXAMINATION OF METEORITES

Oro (1998) notes that careful microbiological examination of any meteorite must be carried out if it is to be shown to contain evidence of past biogenic activity. Detailed studies of the microbiological activity within Antarctic meteorites is currently underway. Any investigator must show the differences between organic components introduced into the meteorite on Earth (i.e. Antarctica) from those which might be indigenous to the meteorite (i.e. Martian). Steele *et al.* (1999) has recently examined chondritic meteorites collected in Antarctica for evidence of microbial activity. Their results show some meteorite samples are contaminated with Antarctic microorganisms but other meteorites are not contaminated with terrestrial organisms. Detailed studies of ALH84001 for Antarctic microorganisms are underway by our group.

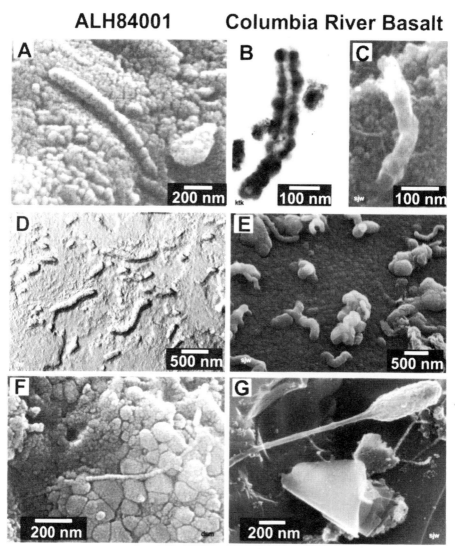

Figure 3 Comparison of features in ALH84001 (left half of image: A, D, F) to similar-size mineralized filaments in Columbia River Basalt samples (right half of image; B, C, E, G). Transmission electron microscopy (TEM) image (B) is not same feature shown in scanning electron microscopy (SEM) image (C). SEM image of freshly fractured surfaces from ALH84001 are shown in A and F; D shows replicate of small region of a carbonate globule surface from ALH84001 (TEM image by H. Vali). Features in Martian meteorite ALH84001 are similar in size and morphology to those from Columbia River Basalt. Martian features may be mineralized remains of extremely tiny cells but also may be mineralized filaments and appendages from larger cells. (Photo from Thomas *et al.*, 1988a. *Geology* 26, 1031-1035.

MORPHOLOGICAL- AND MICROFOSSIL-LIKE STRUCTURES IN ALH84001

The appearance of the segmented structure within ALH84001 clearly attracted the attention of the media along with the man-on-the-street because it resembled a possible microorganism. J. William Schopf of University of California at Los Angeles was one of the earliest critics who stated the feature was too small to be a microfossil or nanobacteria (Schopf, 1999). He based the criticism on his extensive studies of the oldest rocks found on the Earth and on volume calculations which show that bacteria smaller than about 0.1 micrometers are too small to contain even a minimum amount of the enzyme molecules used by bacteria to survive. Schopf had primarily used optical microscopy techniques for his investigations and he had never observed such small structures within the terrestrial fossil record. However, visible optical microscopy with a light microscope is not capable of examining features typically below about 1/4 of a micrometer (250 nanometers). It is understandable that Schopf and other critics noted that fossilized cells or filaments much smaller than this did not exist within ancient earth rocks. It is interesting to note that the recognized size of bacteria or nanobacteria had decreased by a factor of four since our 1996 report (Kojander *et al.*, 1998). Research by Kojander *et al.* (1998) and colleagues have demonstrated that bacteria in the size range of only 70 nanometers exist within selected fluids including blood. Kojander's recent work on nanobacteria contribution to kidney stone growth also demonstrates that bacteria in the 200 nanometer size range have the necessary components present survive.

On the other hand, modern electron microscopes are capable of revealing features at least 100 times smaller. Using these electron microscopes, various researchers have identified cells and cell products smaller than 100 nanometers in rocks and soils. Thomas-Keprta *et al.* (1998) documents bacterial appendages (20-200 nanometers in size) from cells which were grown from ground water taken from 0.5 to 1.0 km or more depth within the Columbia Basalt in Washington (Stevens and McKinley, 1995). The cell appendages are similar in size and morphology to unattached, mineralized filaments recognized in Martian meteorite ALH84001. They are associated with mineralized bacteria cells, and are clearly products and appendages of the bacteria. They are not present on control chips immersed in sterilized water. The close resemblance in size, morphology, and composition (iron oxide-rich) supports our interpretation that the mineralized, unattached filaments from the Martian meteorite ALH84001 may be biogenic.

NEW BACTERIA-LIKE FEATURES IN ALH84001

We are also finding new features in ALH84001 which greatly resemble bacteria (Figure 3). These features occur on and in carbonate phases in this chip. They consist of a few spheres but are mostly of elongated forms. Many of the elongated forms are segmented or consist of the spheres joined together in a row. They show a restricted size range, of individual units or width of segmented elongated units. They form abundant clumps or colonies in some areas and are absent in other areas. None show obvious mineral crystal faces. They are seemingly in random orientation. Features which can be interpreted as cell division are common. They are quite different in morphology from the magnetite grains typically found within carbonate dark rims. We considered whether they were artifacts of sample preparation, mineral precipitates, etch artifacts from natural weathering, tiny bacteria, or bacteria fragments. They do not appear on most areas of the same chip, particularly on the pyroxene, and are therefore unlikely to be artifacts of preparation. None show crystal faces, and many are curved with differing degrees of curvature. They do not resemble the magnetites known to be common in the rims of the carbonates. The seemingly random orientation of the elongated features would not be consistent with a weathering artifact of the carbonate; such weathering artifacts should show alignment with the carbonate crystal structure, and these forms do not. If these features were larger, most microbiologist would instantly identify them as bacteria. However, their small size, roughly 40 nanometers in diameter, makes them smaller than any known terrestrial bacteria. The smallest terrestrial bacteria cell described in the literature is approximately 60 nanometers in diameter, so the difference is not great. Alternatively, they may be fragments of bacteria (filaments, appendages, etc.). The occurrence of these features in cracks and projecting from the carbonate in the third dimension strongly suggests that they are in integral part of the Martian carbonates and not something added later in Antarctica. We are now attempting to make chemical analyses of these features and to cross section them to transmission electron microscope analysis.

We propose that the new data produced by our research group and other workers have shown strong additional evidence that fossilized bacteria exist in this meteorite. We are also working on documenting changes which may occur during the fossilization process. Thomas-Keprta et al. (1998) describe the fossilization process which occurs with bacteria and cells membranes grown from Columbia River Basalt groundwater maintained in a sealed container with basalt chips. The organic and cell-wall membrane undergo change during the fossilization process. For these samples, the cell-wall begins to acquire silicon and iron while the overall morphology of the original cell is retained. Other workers who have studied fossilization have shown that the original organic cell wall completely disappears in some cases, but the shape of the cells is retained.

SUMMARY

Our findings have stimulated new research on the study for evidence of past Martian life in indigenous Martian materials. Whether we are right or wrong about the discover of past life on Mars, the results will be a win-win situation for the scientific community because we will generate new data on Martian materials. These studies will better prepare us for the time when automated missions will deliver documented samples from known sites on Mars. With the proper sample site selections, the probability exists for answering definitively the ultimate question, "Did life ever exist or does it presently exist on Mars?" At the present time, we believe that the scientific data obtained from the study of Martian meteorites is beginning to show evidence of past biological activity on Mars.

REFERENCES

E. Anders (1996) *Science* 274, 2119-2121.

G. Bada et al. (1998) *Science* 279, 362-365.

Barrat et al. (1998) *Science* 280, 412-414.

L. Becker et al. (1997) *Geochimica Cosmochimica Acta* 62, 475-481.

D. Bogard and P. Johnson (1983) *Science* 221, 651-655.

D. Bogard and D. Garrison (1998) *Meteoritics and Planetary Science* 33, A19.

L. Borg et al., (1999) *Science* (Submitted)

J. Bradley et al. (1996) *Geochemica Cosmochimica Acta* 60, 5149-5155.

J. Bradley et al. (1997) *Meteoritics and Planetary Science* 32, A20.

J. Bradley et al. (1998) *Meteoritics and Planetary Science* 33, 765-773.

R. N. Clayton and T. Mayeda (1983) *Earth Planetary Science Letters* 62, 1-6.

S. Clemett et al. (1998) *Faraday Discussions* 109, 417-436.

P. Cloud and K. Morrison (1979) *Precambian Research* 9, 81-91.

G. Flynn et al. (1997) *Lunar Planetary Science XXVIII*, 367-368.

G. Flynn *et al.* (1998) *Meteoritics and Planetary Sciences* 33, A50-A51.

G. Flynn et. al. (1999) *Lunar Planetary Science XXX*, abst. 1087, LPI, Houston, TX (CD-ROM).

I. Friedmann *et al.* (1998) Workshop on Issues of Martian Meteorites, LPI Contrib. 956, pp. 5-6. Lunar Planetary Institute, Houston, TX.

E. K. Gibson *et al.* (1996) *Science* 274, 2119-2124.

E. K. Gibson *et al.* (1997) *Scientific American* 277, 58-65.

E. K. Gibson *et al.* (1998) *BioAstronomy News* 10, No. 3, 1-6.

M. Grady *et al.* (1994) *Meteoritics* 29, 469.

M. Grady *et al.* (1998) *GEOSCIENCE 98* (abstract)

J. Greenwood *et al.* (1997) *Geochimica Cosmochimica Acta* 61, 4449-4453.

R. Harvey and H. McSween (1996) *Nature* 382, 49-51.

J. Head *et al.*, (1998) *Meteoritics and Planetary Science* 33, A66.

R. Hoover (1997) *Proc. SPIE*, 3111, 115-135.

T. Jull *et al.* (1998) *Science* 279, 366-369.

E. O. Kojander *et al.* (1998), *Proc. SPIE 3441*, 86-94.

J. L. Kirschvink *et al.* (1997) *Science* 275, 1629-1633.

G. Levin and P. Straat (1977) *J. Geophysical Research* 82, 4663-4667..

G. D. McDonald and J.L. Bada (1995) *Geochimica Cosmochimica Acta* 55, 1179-1184.

D. S. McKay *et al.* (1996) *Science* 273, 924-930.

D. S. McKay *et al.* (1997) *Lunar and Planetary Science XXVIII*, 919-920.

H. McSween and R. Harvey (1998) *Meteoritics and Planetary Science* 33, A103.

J. Oro (1998) *BioAstronomy News* 10, no. 2, 1-6.

C. S. Romanek and A. Treiman (1998) *Meteoritics and Planetary Science* 33, 775-784.

C. S. Romanek *et al.* (1994) *Nature* 372, 655-657.

J. W. Schopf (1999) *The Cradle of Life*, Princeton Press, Princeton, NJ. 367 pgs.

J. W. Schopf and M. Walker (1983) *In Earth's Earliest Biosphere: Its Origin and Evolution*, Ed. J. W. Schopf, 214-239, Princeton Press.

C. Shearer *et al.* (1996) *Geochemistry Cosmochemica Acta* 60, 2921-2926.

A. Steele *et al.* (1999) *Lunar Planetary Science XXX*, Abst. 1321. LPI, Houston, TX, (CD-ROM).

T. Stephan *et al.* (1998) *Meteoritics and Planetary Sciences* 33, A149-A150.

T. O. Stevens and J.P. McKinley (1995) *Science* 270, 450-455.

K. Thomas-Keprta *et al.* (1997) *Lunar and Planetary Sciences XXVIII*, 1433-1434.

K. Thomas-Keprta *et al.* (1998a) *Geology* 26, 1031-1035.

K. Thomas-Keprta *et al.* (1999) *Science* (in preparation)

J. Valley *et al.* (1997) *Science* 275, 1633-1638.

P. Warren (1998) *J. of Geophysical Research* 103, E7, 16759-16773.

MAR 98-047

ESA EXOBIOLOGY ACTIVITIES

Gerhard Kminek*

The *Microgravity and Manned Spaceflight Directorate* of the *European Space Agency* (ESA) has initiated two studies, one in 1996 and one in 1997, in order to assess the interest and the capabilities for a European effort in exobiology research. Both science teams were composed of senior scientists in the fields of microbiology, geology, cosmochemistry and related disciplines. The first report focused on reviewing the places in our solar system where life or pre-biotic evolution might have occurred sometime in the past or at present. The second report focused on an exobiology package for a Mars lander. The scientific objectives to achieve were to identify and characterize the oxidants, to find morphological and chemical signatures of extinct life, and to determine the chirality of organic compounds if present. In order to minimize ambiguities, an assembly of instruments was proposed to carry out the in-situ investigation. Forward contamination and sterilization issues shall be addressed from the very first stage of the program (Phase A).

The whole exobiology system, including the sample acquisition, preparation and handling system has a mass of about 26 kg.

The science team report is the baseline for a nine month phase A study with industry and universities that will start mid- to third quarter 1998. In addition, exobiology will be a major cornerstone of the next microgravity program of ESA.

INTRODUCTION

The European Space Agency (ESA) has carried out studies and experiments related to exobiology in Low Earth Orbit (LEO) in the past. The Exobiology and Radiation Assembly (ERA) was part of the European Retrievable Carrier (EURECA) which was operational in 1992/93. ERA experiments were designed to study the impact of the LEO environment on cellular (e.g. spores), sub-cellular (e.g. membranes) and molecular (e.g. amino acids) systems that have been exposed to radiation and/or vacuum.

As a continuation of these experiments, ESA will put a Space Exposure Biology Assembly (SEBA) for a period of at least three years on the International Space Station (ISS), starting in 2002. SEBA is composed of two parts:

1. EXPOSE, which is a multiple sample container for radiation/vacuum exposure of small samples, including dark (protected from UV) reference samples. EXPOSE will be part of an EXPRESS pallet on the ISS truss.

2. MATROSHKA, which is a commercially available RANDO® phantom that is used in radiation therapy, consisting of natural bones that are embedded in tissue-equivalent plastic to simulate human tissue and organs. Information on the biological impact of ionizing radiation on dedicated tissue and organ equivalents during EVA's is the main goal of the

---

* ESA/ESTEC, TOS-E, Postbus 299, NL-2200 AG Noordwijk, The Netherlands.

experiment. MATROSHKA will most likely be put on the exterior of one of the Russian ISS modules.

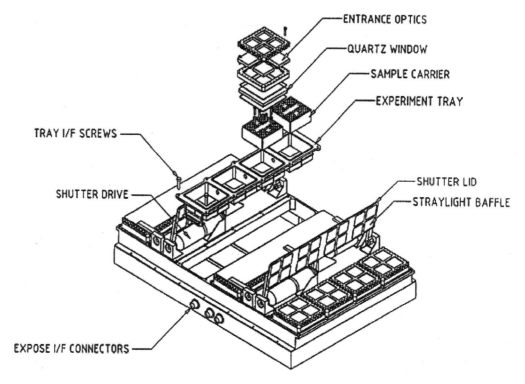

Figure 1   Expose exploded view (Contract Report SEBA-KT-RP-007, ESA).

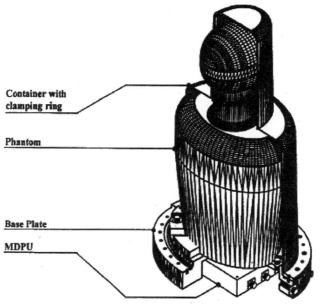

Figure 2   Matroshka configuration. MDPU is the Matroshka Data Processing Unit (Contract Report SEBA-KT-RP-007, ESA).

After the announcement of possible biogenic activity in the Martian meteorite ALH 84001 in the summer of 1996 (D. S. McKay *et al.*; 1996), more attention has been given to the possibility of pre-biotic and biotic evolution outside the Earth. Whether the crucial step from pre-biotic chemical evolution to biochemistry really happened somewhere else in our solar system is of secondary importance in this respect.

In response to that, the Microgravity and Manned Spaceflight Directorate (MGS) of ESA initiated a science team study with the title: *'The Search for Life in the Solar System'*. The members of the science team were senior scientists in the disciplines of geochemistry, physics, biochemistry, cosmochemistry, molecular biology and some other related disciplines. The main objective of this study has been to identify environments within the solar system with appropriate conditions for pre-biotic evolution or the occurrence of biochemistry.

The recommendation of the report issued at the end of the study was that Mars had the most likely appropriate conditions in the past. Further, it will be frequently visited by robotic spacecraft in the near future. Therefore, Mars represents a reasonable target for further studies related to exobiology.

Following the advice from the science team, ESA initiated a second study with a different group of senior scientists to focus on environments, bio-signatures and instrumentation that would be required for an exobiology package that goes to Mars. The second science team finished their work and issued a report in June 1998 entitled: *'The Search for Life on Mars'*. Some of the results of this second report will be presented in this paper.

The science team split up in three groups according to their background and interest with different objectives.

TEAM	OBJECTIVE
Team 1	To define those sub-surface and near surface environments of Mars where evidence of past life might best be sought and to establish a shortlist of preferred landing sites for a future exobiology mission.
Team 2	To study specifically the chemical analysis aspect of an exobiology multi-user facility.
Team 3	To define a set of imaging and spectroscopic systems which will allow a search for evidence of extinct microbial life at all scales.

Figure 3 Task teams.

The individual teams came up with the following conclusions:

Team 1: The major goal was to define environments with exobiological potential and, related to that, a set of landing sites. The presence of water and energy, either chemical or photosynthetic, in any period of time is the only requirement that was set forth for a specific location. Taking terrestrial analogies of environments where life can flourish, several potential environments have been identified:

- Lacustrine environments: these are areas where standing bodies of water existed for prolonged period of time. Remote sensing signatures for this kind of environment are channels entering craters, erosion features like terraces and a higher mineral concentration on the floor originating from evaporitic deposits.

- Sebkha's: these are areas where water was occasionally present but was dry most of the time. In Sebkha's, the evaporitic deposits are exposed to weathering effects during periods of dryness. Layering of deposits reflect the dry/wet cycle. Elevated mineral concentration, associated with these areas, provide a good remote sensing signature.
  A special form of deposit related sometimes to Sebkha's is duricrust. Duricrust forms very hard layers of cement when mineral loaded water evaporates. Biological activity is usually found in these environments and the preservation of extinct life is known to be good. Nevertheless, because of its inherent hardness it is difficult to extract samples.

- Thermal spring deposits: related to volcanic activity and ground water. Biological activity is usually high but the preservation of bio-signatures is somewhat difficult. Racemization processes, e.g. for amino acids, is accelerated by the presence of water and higher temperatures so that only racemic mixtures are present at the end, which degrade to other compounds (e.g. amines, hydrocarbons and kerogen) (J. L. Bada, G. D. McDonald; 1995, A. Kanavaroti, R. Mancinelli; 1990).

- Permafrost: although viable biological life has been identified in permafrost layer on Earth (e.g. D. A. Gilichinsky *et al.*; 1993), limited accessibility of the Martian permafrost will restrict the search in these areas.

Although the team pointed out several landing areas, the consensus was to wait for the Mars Global Surveyor (MGS) data before selecting primary landing sites.

Team 2: In order to search for chemical indicators of life, a thorough characterization of the environment has to be carried out beforehand to avoid ambiguous conclusions. Chemical, mineralogical and petrologic investigations are therefore part of the exobiology package. At the same time, minerals can be used as a bio-signature as well because certain life-forms produce minerals during their metabolic process. These mineral usually have a different crystallography, morphology and/or isotope fractionation.

Isotope fractionation of different elements is a major bio-marker on its own. Carbon fractionation (C^{13}/C^{12}) in favor of C^{12} is well established for photosynthetic processes and stable over geological timescales (M. Schidlowsky; 1988). Other fractionation processes include hydrogen/deuterium fractionation in methanogenic processes as well as nitrogen (N^{15}/N^{14}) and sulfur (S^{34}/S^{32}) fractionation. However, identifying sulfur fractionation is kind of tricky because it is very difficult to get rid of the sulfur as soon as it is injected in a GCMS system.

Although the non-biogenic fractionation processes on Mars are not well understood (B. M. Jakosky, J. H. Jones; 1997), the relative direction of the shift due to biological processes may give enough information to distinguish between inorganic and organic origin of the fractionation process.

Another useful bio-marker is molecular analysis. There are different kinds of organic and inorganic molecules that can be taken as indicators for biogenic processes. Related to that are different means to identify them. The most important inorganic molecules are water and oxidants such as hydrogen peroxide. The organic molecules can be classified in three groups:

- Volatile light organics like methane,
- Medium molecular weight organics like hydrocarbons and amino acids,
- Macromolecular components like kerogen and oligo's.

After identifying organic molecules, determination of their chirality can set limits for terrestrial contamination and distinguish between non-biogenic and biogenic origin. In order to be able to identify all these bio-markers, a set of scientific instruments has been proposed by the team:

INSTRUMENT	PERFORMANCE	ANALYSIS
Alpha-Proton-X-Ray spectrometer (APXS)	1 wt %	Elemental analysis
Moessbauer Spectrometer	1 % of total Fe 1-50 % of bulk (G. Klingelhoefer et al.; 1995)	Iron bearing mineral analysis, oxidation state, elemental analysis
RAMAN Spectrometer	Spectral range: 200-3500 wn Spectral resolution: 8 wn Spatial resolution: ~ 1 μm	Molecular analysis of organics and minerals
IR Spectrometer	Spectral range: 0.8-10 μm Spectral resolution: >100 ($\lambda/\Delta\lambda$) Spatial resolution: 200 μm	Molecular analysis of organics and minerals
Pyrolytic Gas Chromatograph & Mass Spectrometer	TBD	Analysis of inorganic/organic compounds, isotopic ratio and chirality determination.

Figure 4   Payload for analytical analysis.

The APXS and the Moessbauer spectrometers provide an absolute abundance of elements and identify the mineralogical composition of iron-bearing minerals, as well as the oxidation state of the sample. Optical methods for sample selection, which will enhance the capabilities of mineral identification, will be discussed later. The complementary RAMAN- and the IR spectrometers are able to identify certain minerals and organic molecules (e.g. clays, hydrates, carbonates, C-H bonds of organic molecules, etc.). The wavelength of the IR spectrometer and the laser for the RAMAN has been chosen to cover compounds of relevance to exobiology. The PYR-GCMS unit is used for multiple measurements: for volatile compounds, the GC separates the different species according to their chemical reactivity and molecular weight and inject the particle beam for detailed mass analysis into the MS. The pyrolyser can be used to release non-volatile products to the GCMS unit. It can also be used for stepped combustion processes to differentiate between organics and carbonates in samples. Not included in the list is a dedicated instrument to measure the chirality of organic compounds. New developments in pharmacological instrumentation could help to find a small but capable instrument for that purpose (e.g. K. Bodenhoefer et al.; 1987).

Team 3: Identification of macroscopic and microscopic bio-markers and characterization of the environment with optical methods requires a set of instruments with different resolution. Apart from the identification of different mineralogies and sample selection, bio-markers of interest are: biofilm layers (biolamenae) in the range of hundreds of microns to several millimeter, individual biofilms in the range of less than 100 microns, and microfossils in the range of 5 microns or less. In order to find and focus on the individual markers, a set of instruments was proposed:

INSTRUMENT	PERFORMANCE	EXOBIOLOGICAL OBJECTIVES	OTHER OBJECTIVES
Stereo Panoramic Camera	1000x1000 px 0.3 mrad/px 14 filters	Study rock types/gross bio-features	Landing site survey
Low Resolution Microscope	0.1 mm/px 5 filters	Examine samples prior to high res. and chemical studies	Examine samples prior to high res. and chemical studies
Optical Microscope	Res.: < 3 micron DoF: < 20 micron	Fossil identification	Mineral studies
Atomic Force Microscope (AFM)	Res.: 1nm FoV: 1x1 micron to 50x50 micron	High res. structural analysis	High res. structural analysis

Figure 5   Payload for imaging.

The main concern for optical investigation of samples is the sample preparation. Cutting and polishing is usually required in order to get reasonable surfaces for microscopic investigations.

IMPLEMENTATION

With the whole instrument assembly, identification of all major bio-markers is possible. Most of the instruments complement each other in order to reduce ambiguities in interpreting the results.

System	Mass (kg)
Sub-surface drill system	6.5
Sample handling/distribution	4.0
Sample sectioning	0.3
Sample grinding	0.4
Low res. Microscope	0.2
Optical microscope	0.3
Atomic Force Microscope (AFM)	1.5
Microscopy transfer stage	1.0
APXS	0.5
Moessbauer spectrometer	0.5
RAMAN spectrometer	1.5
IR spectrometer	1.0
PYR-GCMS unit	5.5
Oxidant detector	0.4
(Laser ablation ICP mass spectrometer)	2.5
total	**26.1**

Figure 6   Preliminary payload mass breakdown.

Sample preparation is crucial, in both the chemical and observational methods. Sample grinding (PYR-GCMS unit, chirality-unit), cutting and polishing (optical methods, APXS and Moessbauer spectrometers) will be required. Using small machinery that is rigid and stable enough to deliver the precision needed, as well as the material selected for that purpose, have to be evaluated. In earth-bound labs, diamond saws are used for cutting petrologic samples. The requirement to determine carbon at the ppm level may conflict with using diamond saws (the same is true for using carbide drill bits).

The consensus was to go for a drill system consisting of several drill stems adding up to an effective drill length of approximately 1.5 meters. Although it is not clear how deep the oxidants penetrate the Martian subsurface, measurements of the oxidant concentration can at least solve this problem. The whole drill system with the instrument package shall be mounted on a lander. A rover-based system cannot give the required stability and space to accommodate all the instruments and mechanical systems. In this case, mobility of the system must be sacrificed for practicality.

CONTAMINATION

Forward contamination and sterilization must be considered from the very beginning of the program. Article IX of *the Outer Space Treaty* of 1967 states that one has to 'avoid harmful contamination' of outer space. NPG 8020.12 of the *NASA Planetary Protection Document* sets the requirements for all NASA robotic extraterrestrial spacecraft and NPG 5340.1 defines the basic procedure for assessing spacecraft contamination. Because the exobiology package is likely to fly on a NASA mission to Mars, all these regulations (and others) must be followed. In the USA, the regulations are implemented by NASA. They are accepted by COSPAR (Committee on Space Research) of the International Council of Scientific Unions. Reviews and periodic revisions are the result of newly obtained knowledge about planetary exploration and terrestrial ecosystems (J. Rummel,[*] private communication; SSB Biological Contamination of Mars, 1992).

Although the general believe of the Space Science Board (SSB) of the National Research Council (NRC) is that no terrestrial organism can survive on the surface of Mars today (SSB Mars Sample Return; 1997), forward contamination of carbon bearing compounds and microorganisms has to be minimized in order to achieve the required sensitivity for the scientific instruments. The first step is to select the SSB category of the mission without impacting the planet and the mission. Next is the twofold issue of carbon contamination and sterilization. Carbon contamination can be avoided by using materials that are carbon-free and/or have appropriate encapsulation. Sterilization is partly a design criteria and partly dependent on the bioload assessment. All spacecraft parts have to be accessible after integration for the appropriate sterilization procedure and a bioload assessment has to be done after that. More capable analytical techniques are available today that can detect single microorganisms (e.g. PCR (Polymerase Chain Reaction)). Including non-cultivable microorganism in the bioload assessment is of crucial importance since only a small fraction of all microorganisms is cultivable. The detection limit of the analytical instruments and the category of the mission profile determines the level of contamination. Naturally, there is the impact on the cost related to spacecraft design considerations and changes in order to comply with sterilization procedures.

Because the exobiology package is only part of a bigger system, considerations of other contaminants or common protocols are required.

CONCLUSIONS

The science team report is the baseline for a nine month phase A study with industry and universities that will start mid- to third quarter 1998. In addition, ESA will propose a new microgravity program (following EMIR-2 and the MFC) that will include elements of the exobiology package. Discussions with ESA Member States started already (as of July 1998) and are assumed to last until mid 1999. If approved, the five-year program will start in 2000.

---

[*]   J. Rummel, NASA-HQ, Planetary Protection Officer - NASA.

ACKNOWLEDGMENTS

This work has been initiated, supported and directed by Dr. Paul Clancy, Manned Spaceflight and Microgravity Directorate, European Space Agency. Members of the science team:

Chairman: André Brack, Centre de Biophysique Moleculaire, CNRS, Orléans, France.

Beda Hofmann, Naturhistorisches Museum, Bern, Switzerland.

Gerda Horneck, Institute of Aerospace Medicine, DLR, Porz-Wahn, Germany.

Gero Kurat, Naturhistorisches Museum, Wien, Austria.

James Maxwell, Chemistry Dept. Bristol University, Bristol, England.

Gian G. Ori, Dipartimento di Scienze, Universita d'Annunzio, Pescara, Italy.

Colin Pillinger, Planetary Sciences Research Institute, Open University, Milton Keynes, England.

François Raulin, LISA, CNRS, Universités de Paris 7 et 12, Créteil, France.

Nicolas Thomas, MPI fur Aeronomie, Lindau, Germany.

Frances Westall, Planetary Sciences Branch, NASA-Johnson Space Centre, Houston, USA.

Secretary: Brian Fitton, ESC, Noordwijk, Holland.

REFERENCES

Bada, J. L., McDonald, G. D., "Amino Acid Racemization on Mars: Implications for the Preservation of Biomolecules from an Extinct Martian Biota", *Icarus* 114, 1995.

Bodenhoefer, K. *et al.*, "Chiral Discrimination using piezoelectric and optical gas sensors", *Nature*, 387, 1997.

Gilichinsky, D. A. *et al.*, "Cryoprotective Properties of Water in the Earth Cryolithosphere and its Role in Exobiology", *Origins of Life and Evolution of the Biosphere*, 23, 1993.

Jakosky, B. M., Jones, J. H., "The History of Martian Volatiles", *Review of Geophysics*, 35,1 February 1997.

Kanavaroti, A., Mancinelli, R., "Could Organic Matter Have Been Preserved on Mars for 3.5 Billion Years", *Icarus* 84, 1990.

Klingelhoefer, G. *et al.*, "Moessbauer spectroscopy in space", *Hyperfine Interact.*, 95, 1995.

McKay, D. S. *et al.*, "Search for Past Life on Mars: Possible Relic Biogenic Activity in Martian Meteorite ALH84001", *Science*, Vol. 273, 16 Aug. 1996.

Schidlowsky, M., "A 3,800-million-year isotopic record of life from carbon in sedimentary rocks", *Nature*, Vol. 333, 26 May 1988.

SSB "Mars Sample Return, Issues and Recommendations", Task Group on Issues in Sample Return, Space Studies Board, Commission on Physical Sciences, Mathematics, and Applications, National research Council; National Academic Press, Washington D.C. 1997.

SSB "Biological Contamination of Mars, Issues and Recommendations", Task Group on Planetary Protection, Space Studies Board, Commission on Physical Sciences, Mathematics, and Applications, National research Council; National Academic Press, Washington D.C. 1992.

PRESERVING POSSIBLE MARTIAN LIFE

Mark Lupisella[*]

As we expand our presence in the solar system, novel and challenging scientific and policy issues will face us. A relatively near-term issue requiring attention involves questions regarding the *in situ* human search for and discovery of primitive extraterrestrial life—Mars being an obvious candidate. Such a search and potential discovery is clearly of paramount importance for science and will pose unique and complex mission planning and policy questions regarding how we should search for and interact with that life. This paper will explore the scientific, mission planning, and policy issues associated with the search for and interaction with possible primitive extraterrestrial life, with an emphasis on issues regarding the preservation of such life.

Some of the questions to be considered are: To what extent could effects of human presence compromise possible indigenous life forms? To what extent can we control those effects (e.g. will biological contamination be local or global?) What are the criteria for assessing the biological status of designated locales as well as the entire planet (e.g. can we extrapolate from a few strategic missions?) What should our policies be regarding our interaction with primitive forms of extraterrestrial life?

Central to the science and mission planning issues is the role and feasibility of applying decision theory, risk analysis, and modeling techniques. Central to many of the policy aspects are issues of value. Exploring this overall issue responsibly requires a holistic understanding of how both of these dimensions of the issue interrelate.

1. INTRODUCTION

The primary emphasis of this paper will be on forward adverse effects with respect to the first human presence on Mars since such effects (in the form of contamination concerns) regarding robotic exploration have been reasonably well addressed. (However, contamination issues regarding robotic exploration may still require attention and are inevitably intertwined with an overall human mission planning strategy as indicated in Figure 2-1, Mission Planning Decision Tree for Preserving Possible Martian Life From a Human Presence.) This paper will also briefly address the policy and underlying philosophical aspects regarding the preservation of possible extraterrestrial life.

Before proceeding, however, it will be useful to first establish, in a general way, the need for addressing this issue. Such a need can be generally derived from three sources: (1) international law, (2) scientific value, and (3) public interest.

1.1 International Law

Regarding international law, Article IX of the United Nations Outer Space Treaty of 1967 states:

---

[*] University of Maryland and NASA Goddard Space Flight Center, Greenbelt Road, Code 584, Greenbelt Maryland 20770. E-mail: Mark.Lupisella@gsfc.nasa.gov.

States Parties to the Treaty shall pursue studies of outer space, including the moon and other celestial bodies, and conduct exploration of them so as to avoid their harmful contamination...and where necessary, shall adopt appropriate measures for this purpose.[1]

In addition, Article VII of the Moon Treaty of 1984 clarifies and improves upon the more general obligations of the Outer Space Treaty by stating:

In exploring and using the moon, States Parties shall take measures to prevent the disruption of the existing balance of its environment, whether by introducing adverse changes in that environment, by its harmful contamination through the introduction of extra-environmental matter or otherwise.[2]

Interestingly, the United States, while a signatory to the Outer Space Treaty, is not a signatory to the Moon Treaty. However, the Outer Space Treaty and Moon Treaty have been interpreted by some to apply to non-signatories.[3] Unfortunately, both treaties are quite general, lacking specific standards as well as formal international mechanisms for enforcement.[4]

In 1983, after having gathered more data about Mars and the solar system in general, NASA moved away from the previous probabilistic standards for planetary protection procedures and adopted a less rigid policy involving five categories of missions and associated planetary protection requirements. There is no category addressing human missions. This may be because it is thought that once a human mission is underway, forward planetary protection will not be relevant.[5] Also, human missions to the planets are not a near-term concern. However, this paper will indicate that such a category may be required sooner than later partly because an important requirement of such a category will likely involve obtaining some understanding of the biological status of a given locale or the entire planet via a complex series of strategic robotic precursor missions—possibly numbering in the hundreds or more—for which long-term planning might be required sooner than later. There will also be issues of back-contamination that would most certainly warrant a planetary protection category for human missions that return astronauts and samples—similar to what was done with Apollo.

1.2 Science

Although the preservation of extraterrestrial environments is important for scientific knowledge in general, a primary concern of planetary protection is to ensure the integrity of life-detection experiments by minimizing the chance of a false-positive result.[6] The NASA Viking mission of 1976 is the paradigm example. Each lander had three life-detection experiments and was heat sterilized and encapsulated in a bioshield that was released upon arrival at Mars.

Underlying these concerns, of course, is the widely acknowledged importance of discovering the second data point that biology is so desperate for. This places a high value on preserving and avoiding masking possible indigenous extraterrestrial life due to terrestrial contamination.[7] The National Academy of Sciences Space Science Board (SSB) writes: "...Forward contamination...is a significant threat to interpretation of results of *in situ* experiments specifically designed to search for evidence of extant or fossil Martian microorganisms", and that protecting Mars from terrestrial contamination so as to not jeopardize future life-detection experiments is "profoundly important".[8] In a report on back contamination, the SSB also writes:

It will be important to stringently avoid the possibility that terrestrial organisms, their remains, or organic matter in general could inadvertently be incorporated into sample material returned from Mars. Contamination with terrestrial material would compromise the integrity of the sample by adding confusing background to potential discoveries related to extinct or extant life on Mars. DNA and proteins of terrestrial origin could likely be unambiguously identified, but other organic material might not be so easily distinguished. The search for candidate martian organic biomarkers would be confounded by the presence of terrestrial material. Because the detection of life or evidence of prebiotic chemistry is a key objective of Mars exploration, considerable effort to avoid such contamination is justified.[9]

However, some suggest that contamination concerns are unfounded. In his recent book, *The Case For Mars*, Robert Zubrin suggests primarily three reasons for why back contamination should not be an issue—and presumably the essence of these arguments apply to forward contamination as well. One, Mars and Earth have exchanged much material already. Two, life almost certainly does not exist on the Martian surface. Three, the co-evolutionary dependence of pathogens and hosts makes it impossible for Martian and terrestrial organisms to adversely affect each other.[10] These are reasonable suggestions, but it may not be that simple. The fact that material has been exchanged between our planets does not mean that contamination has occurred in the way it could with a significantly more intrusive mission (i.e. a human mission) to Mars. Also, if panspermia has occurred, then Martian organisms could be genetically compatible with new organisms that arrive via contamination, hence calling into question Zubrin's third claim that a lack of co-evolutionary dependence should mitigate contamination concerns. Secondly, the lack of existence of life on the surface cannot be known with confidence until we conduct more missions to explore the planet. The recent announcement of water on the moon, tantalizing evidence of sub-surface liquid environments (perhaps water) on Europa, and possible evidence of biological remnants in a Martian meteorite indicate how unpredictable our solar system can be. Even if we were to confirm that no life exists on the surface of Mars, there is, as Zubrin himself acknowledges, the possibility of sub-surface life—which should still be of great concern since our intrusive missions could contaminate the sub-surface of Mars via drilling and other activities.[11] Surface or subsurface life could also be adversely affected by toxic substances, predation, competition, and general environmental modifications, making infection with its co-evolutionary dependence only one consideration among many.[12] Lastly, we only understand one kind of biology. How confident are we, or can we be, that life on Mars will be consistent with our present understanding of life when we really only have one data point?

1.3 Public Concern

There is also the issue of anticipating and addressing public concern. As there have been in the past, there will be public interest groups attempting to ensure that NASA and other space agencies are not only doing what is perceived to be environmentally/politically correct, but perhaps morally correct as well. Species preservations groups will have a new cause to champion, and it should be assumed that they will not hesitate to act as an obstacle if they have any reason to believe that the proper precautions are not being implemented. Environmentalists opposing the use of nuclear power sources have been able to delay launches in the past. In this light, planning now to address the questions posed in this paper could help mitigate future opposition to sending humans to Mars.[13]

2. STRATEGIC MISSION PLANNING TO PRESERVE POSSIBLE EXTRATERRESTRIAL LIFE

Figure 2-1, Mission Planning Decision Tree for Preserving Possible Martian Life From Human Presence, is a preliminary attempt to frame the issues regarding long-term planning for a human mission to Mars—primarily with respect to the issue of forward contamination. An underlying assumption of the decision tree is that the scientific value of preserving extraterrestrial life is high enough that many in the science community would agree that these questions are worth pursuing, and possibly high enough to justify a fairly conservative mission planning approach as suggested by the decision tree. This section will step through the decision tree and make a first order assessment of the questions posed.

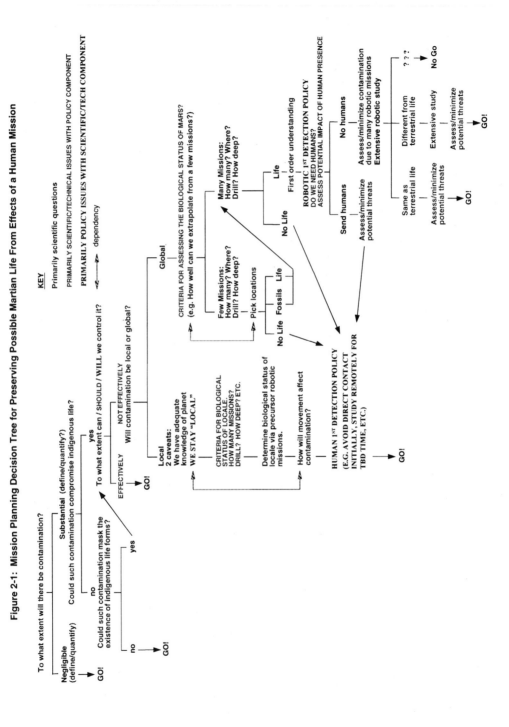

Figure 2-1: Mission Planning Decision Tree for Preserving Possible Martian Life From Effects of a Human Mission

The decision tree poses a series of questions with respect to the issue of minimizing adverse affects on potential indigenous extraterrestrial life due to a human presence on Mars. Before touching on those questions, it is important to establish what it is meant by contamination. The Task Group on Planetary Protection of Space Studies Board, in their report, *Biological Contamination of Mars: Issues and Recommendations*, concurred with a previous NASA report[14] that "forward contamination more broadly defined to include contamination by terrestrial organic matter associated with intact cells or cell components" since such material is a "significant threat to interpretation of results of *in situ* experiments specifically designed to search for evidence of extant or fossil Martian microorganisms."[15] An even broader relevant definition of contamination includes non-biological elements such as new compounds from rocket exhaust or airborne pollution from industrial chemicals.[16] With this broad definition of contamination in mind, we can now consider some important mission planning questions as shown in Figure 2-1.

To what extent will there be contamination? This question has been addressed in a preliminary manner by Chris McKay and Wanda Davis. The thinking expressed in their paper is that contamination is inevitable if humans are present.[17] To pursue this with rigor, however, we should try to establish *the extent* to which there will be contamination since the amount and kind will likely be critical to decision making. If contamination possibilities are thought to be negligible, a human mission will not, and should not, be prevented from occurring as soon as it is feasible—politically and technically.

If it is thought that there could be contamination to levels that are deemed significant, we must then ask: *Could such contamination compromise indigenous life-forms?* A conservative answer to the question is yes, it is possible. But again, this requires substantial analysis. What is the probability? Is it even feasible to establish such probabilities with any confidence? What kinds of effects could there be and to what degree? While co-evolutionary dependence seems important, and may even be essential for organisms to effect each other, how confident can we be that this "principle" of terrestrial life applies universally given that we base this idea on one biological data point? So we might want to assess the relative probabilities of direct adverse effects given panspermia vs. a separate origin. Is the latter a probability of zero? The Space Studies Board says no.[18] What are the chances for indirect adverse effects via toxin production or competition for resources? For example, perhaps various life-processes could be more efficient with different chiralities characteristic of other life forms. What is the probability that non-biological elements such as rocket exhaust or industrial chemicals could compromise indigenous life-forms? Given that a single kind of life-form on earth might have caused the extinction of all others early on in the evolution of life, could such a scenario occur if foreign organisms are brought to Mars?[19] If we obtain an appropriate level of confidence that contamination will not adversely effect possible indigenous life, then GO!

Otherwise, we should ask: *Could such contamination mask the existence of indigenous life-forms?* A masking effect, if possible, will presumably have a dependence on whether or not the contaminating organisms are dead or viable, either as dormant or active organisms. Dead organisms will probably not have a significant masking effect for life-detection experiments based on life processes such as metabolism—as were the Viking mission experiments. However, dead organisms might have a masking effect for simple observation based detection devices such as microscopes and robotic life-detection devices—although with humans present, detailed analysis could be done that might mitigate this problem. While perhaps not the most likely scenario, we might consider that dead terrestrial organisms, after having been on Mars for some time, will not be recognizable as terrestrial organisms. For example, there might only remain fragments of organisms or the organisms might undergo some sort of physical modification, making it difficult, if not impossible, to rule out an indigenous source. It may also be very difficult to tell whether or not the resident organisms were deposited by the mission or whether they arrived via panspermia—an important scientific question in its own right. If we're confident that masking effects are not significant, then GO!

Otherwise, if we determine there is an unacceptable chance of masking possible indigenous life, we should ask: *To what extent can/should/will we control contamination?* The "should" and "will" part of this question are both important for a realistic assessment of the outcome of this decision point. That is, we may determine that we *can* control contamination effectively, but that perhaps, for various reasons, we shouldn't; and even if we think we should, an honest assessment should prompt us to consider that ultimately, other forces could prevail, resulting in an absence of contamination control. Whether or not we *will* actually control contamination is a legitimate and interesting question. It is legitimate because we often don't do what we think we should do. It is reasonable to suspect that many people who think we should control contamination, will also think that, ultimately, because of our exploitive, destructive, and selfish nature, we will not.[20] This leads to why the question is so interesting because it goes right to the heart of humanity's power to control its destiny. These are obviously complex issues that cannot be addressed here with the attention they require, but they will be touched on briefly during mention of the policy challenges represented on the decision tree, as well as during later discussions of the larger philosophical context of this issue. For now, we will concern ourselves with the more practical issue of the feasibility of contamination control.

At least one person has suggested that absolute containment of all terrestrial biology is, in principle, possible and even desirable over the less certain method of obtaining all the relevant planetary data to determine that contamination will not cause adverse effects. Joseph Sharp rightly points out that an entire technology has been developed to contain dangerous biological agents, and that while such an effort for the first human Mars mission would be quite expensive, in the long run, it may be the only sure approach as long as no failures occur.[21] The back contamination containment procedures for the Apollo Astronauts could be a useful starting point for addressing forward contamination containment issues of the kind suggested by Sharp. However, given the expense and absolute requirements of such an approach, it is worthwhile to consider the implications of the more realistic suggestion made by McKay and Davis that contamination is likely if humans establish a presence on Mars.

If, however, there is a reasonable chance for controlling contamination, the difficult problem is to assess *the extent* to which we can control it. While Apollo did not have rigorous containment procedures for preventing the contamination of the lunar environment, some steps were taken to reduce and inventory such contamination. For example, a bacterial filter system on the lunar module was used to prevent contamination of the lunar surface when the cabin atmosphere was released.[22] NASA also adopted as official policy, aseptic subsurface drilling, decontamination and contained storage of waste materials, and biological and organic material inventory requirements.[23] Understanding the amounts and kinds of contamination that are released into the Martian environment will be important for dealing with this overall issue. Will we be able to completely isolate a given locale, in which contamination controls could be quite loose? Or will we want or be able to rigorously contain contamination for all areas and activities? If we're confident about contamination control, then GO!

If not, *will contamination be local or global?* McKay and Davis have briefly touched on this question by suggesting that biological contaminants such as human bacteria may not survive Martian oxidizing surface conditions and ultraviolet radiation exposure. However, we should consider the possibility that dead or viable organisms could potentially be distributed over a significant distance, perhaps globally, since large, sometimes global, dust storms are known to occur.[24] Indeed, McKay and Davis acknowledge this possibility when they say that "regions distant from the base may receive a lower bioburden."[25] The likely non-viability, and hence insignificant spread of contaminant organisms on the surface, while reasonable as a first order assessment, should be analyzed with as much scientific rigor as possible, paying close attention to the *continuous* source of contamination due to a human presence, possibilities of subsurface contamination, and other sources of contamination.

If it is thought that contamination will be local, *what are the criteria for determining the biological status of a designated locale?* It may be prudent to assume that contaminants will at least be present and possibly viable over a designated locale where humans first land; so it will be important to understand what will be required to obtain confidence about the biological status of the locale in question, since contamination could compromise possible indigenous ecosystems. Robotic precursor missions and possibly tele-robotic missions from an on-orbit station or moon to a potential landing site are obvious ways to remotely obtain knowledge about the biological status of the locale. The interesting challenge is to determine what level of confidence we require and what kinds and numbers of missions will be needed to establish that confidence. Understanding subsurface possibilities will be critical since a human landing site will likely result in contamination of the top few meters of the soil.[26] If this is the case, drilling missions, or at least subsurface penetrating missions, seem to be obvious candidate precursor missions. These missions could be similar to the penetrator missions being planned for execution in the next few years with the crucial difference being that life-detection experiments need to be present and, of course, functional after experiencing high impact forces. It should also be acknowledged that a human base will probably have the ability to drill to considerable depths below the surface (possibly to or below the permafrost level) for both exploratory and resource prospecting reasons (e.g. searching for water), which could possibly result in contamination of an otherwise protected subsurface environmental niche. It seems this is a reasonable activity to expect the first human mission to engage in, and so sterilized robotic precursor missions may also be required to drill to comparable depths in order to assess the biological status of the subsurface environment for a certain designated locale where humans will first land.

Another important element for sending humans to Mars, as shown on the decision tree, is to establish a Human First Detection Policy for the search for and potential interaction with an extraterrestrial life-form. Search and post-discovery protocols will need to be established, presumably by the international community under a body such as the United Nations Committee on the Peaceful Uses of Outer Space (COPOUS)—perhaps via the Committee on Space Research (COSPAR). Should a remote reconnaissance of some locale from an established base be done before sending humans out into that targeted area? To what extent should a continuous and rigorous search for signs of life be conducted while humans expand their presence on the surface? What will be the de-contamination requirements and procedures for humans interacting with the Martian environment? If drilling is to occur, to what depth, for what purposes, and how will the associated contamination risks be mitigated? We have had some experience dealing with this kind of isolated environment on earth, and presumably, similar requirements and procedures would be applicable to Mars, as well. Additional potential threats to indigenous life such as mechanical disturbances from rovers and moving equipment that might disturb and expose previously protected niches and additional seemingly benign environmental factors such as water, heat and light should be studied and incorporated into policies for how humans will interact with the new environment.

Regarding post-discovery policy, what will we do when we first discover a possible sign of extraterrestrial life? Should we immediately take a sample for lab analysis, or study it remotely first? Some might contend that we should leave the area completely until we obtain more knowledge of that potential life-form via remote sensing and tele-robotic vehicles. Still others, as extreme and impractical as it may sound, might suggest that we leave the planet entirely, perhaps for ethical reasons,[27] or at least until more is known about the nature of that life and its distribution on the planet. How rigid will such a policy be? All these questions will, of course, be examined as they relate to Astronaut safety—which, along with the more subtle forward contamination concerns, will likely require many years of scientific research and international policy formulation before a healthy consensus can be reached. Once this Human First Detection Policy is established, then GO!

If it is thought that contamination could be global, we must try to *establish the criteria for assessing with an appropriate level of confidence the biological status of the entire planet*. This is, of course, a very tricky question—partly because we only have one data point, the earth, on which to base any criteria. However, it still may be possible to establish criteria that should be satisfied before having some appropriate level confidence about the biological status of Mars. If a few strategic missions are adequate, how many, and of what kind? If many missions are required, the same set of questions hold, with a key long-term question being the total number of missions required since this will drive the overall timetable for getting humans to Mars. If many missions turn out to be required, we should try to address the associated issues now to ensure that all preliminary steps are taken in an efficient manner as we plan our first presence on another planet.[28]

If only a few strategic missions are required, and precursor robotic exploration doesn't find any signs of life, then establish Human First Detection Policy, and GO! If signs of extinct or extant life are found, that could imply that the determination that only a few strategic missions would be adequate to assess the biological status of Mars should be called into question, suggesting that many robotic precursor missions may be needed to assess the biological status of Mars. If many missions appear to be required, assess how many and of what kind. If no life is found, establish first Human First Detection Policy, then GO!

If life is found, try to understand what the data suggests regarding its nature. Establish and consult the Robotic First Detection Policy (presumably an international effort which should help address questions associated with what should be done prior and post a robotic detection of life. For example, what are the criteria for assessing whether humans should go immediately? What kind adverse effects are possible—mutual or otherwise?) This kind policy, which will be discussed further in a subsequent section, should be informed by research regarding possible impacts to indigenous Martian ecosystems—another key question that will be explored in more detail in the following section. If humans are needed, or if more generally, it is determined that humans should go immediately regardless, then get clear on the Human First Contact Policy, and GO!

If it is decided that humans should not go immediately, we will want to conduct extensive robotic study to understand that life, eliminating, as much as possible, the contamination effects due to many such missions. When the threshold for obtaining as much understanding as is reasonable via robotic exploration is reached, then GO!

For this decision tree, a "no-go" decision would be considered final because the decision tree allows for an extended period of time during which a no-go decision would essentially be in effect until there was enough confidence to send a human mission. Such a no-go conclusion would be extreme and would require an extremely compelling justification. Indeed, it should be noted that there may be circumstances which some would see as justification for such an absolute no-go decision. For example, it is conceivable that if Mars is teeming with a very dangerous form of life, a decision could be made to "quarantine" the planet for an indefinite period of time. However, as indicated above, the more likely scenario under such circumstances is that since humans will want to study those life forms *in situ* as we do dangerous organisms on earth, we will likely simply take whatever time and action necessary to have confidence about the first mission. There is, however, another possibility that could lead to a no-go decision. Political and ethical reasons for keeping humans away from Mars could prevail. For example, there will be those who will suggest that Mars is its own environment, its own world, that deserves to exist unaltered by human interference, especially given our propensity for facilitating undesirable environmental degradation.

3. KEY POLICY ISSUES

There appear to be at least two major areas requiring comprehensive rigorous policy analysis. One is what can be thought of as a "Robotic First Detection Policy". This policy would have something to say about what steps should be taken regarding robotic missions to minimize/avoid compromising possible extraterrestrial life before and after possible signs of life are discovered by a robotic vehicle. The second can be thought of as a "Human First Detection Policy" which would be concerned with pre and post-detection issues involving a human *in situ* search for and discovery of Martian life. The former is not intended to be the focus of this paper, although some brief thoughts will be offered since robotic exploration issues relate to human exploration.

An analog for the policy work being suggested here exists in the Search for Extraterrestrial Intelligence, or SETI, community. There exists a Declaration of Principles Concerning Activities Following the Detection of Extraterrestrial Intelligence which provides guidelines regarding how organizations should react in response to evidence of a detection. Although such a possibility may be legitimately perceived by many to be remote, it is nonetheless, wise to be prepared for such a possibility. The same should apply to what may be a more likely possibility of discovering primitive forms of extraterrestrial life.[29] Policy should be driven not only the likelihood of an event, but its significance, as well.

Pursuit of the policy and science considered in this paper could also help in choosing among different policy directions raised by Bruce Murray who has suggested three kinds of objectives that need to be decided regarding Mars exploration: (1) Open-ended exploration leading to human mission vs. accomplishing focused scientific objectives. (2) Priority for early detection of decisive evidence of life, past or present, vs. determination of key unifying global processes. (3) Technological evolution for long range exploration vs. expedient approach to near-term objectives.[30] Although by no means definitive, the concerns raised in this paper can help make choices from Murray's list by considering life-detection as the centerpiece for Mars exploration, hence suggesting we might choose the following: (1) accomplishing focused scientific objectives, (2) early detection of decisive evidence of life, and (3) expedient approach to near-term objectives.

Also, addressing these questions now will not be wasted if we were to indeed find a lifeless Mars. This kind of planning can only help prepare us as we move out into the rest of the solar system in search of life.

3.1 Robotic First Detection Policy

3.1.1 Pre-detection

Pre-detection planning regarding robotic missions has been addressed broadly in the form of the 1967 Outer Space Treaty and more specifically in the form of contamination prevention measures implemented by space faring nations. However, the absence of rigorous international enforcement mechanisms may have allowed unacceptable contamination to occur in the past and may allow it to continue in the future. As more nations become space-faring, effective international mechanisms for enforcing contamination regulations might be necessary. Also, the new NASA policy doesn't require sterilization unless there are life-detection experiments involved. Cleaning is certainly required, but it is unclear whether this is adequate, especially when we consider the accumulation of many robotic precursor missions. Contamination could be more of a concern after a large number of missions are launched.[31]

This policy work may also include exploring guidelines regarding how robotic vehicles could best operate so as to reduce possible adverse affects on indigenous ecosystems. For example, some form of remote surveillance (either from orbit or from other vehicles/stations on the ground) of a potentially life-bearing environment may be prudent before sending rovers to the specific locale in question. This might be particularly relevant if rovers have not been adequately sterilized, which, as suggested above, the present policy could allow for if life-detection experiments were not involved.

3.1.2 Post-detection

A discovery of indigenous life by a robotic vehicle may not present any severe difficulties if we take the proper contamination precautions, and if we are willing to take the time needed after the discovery to make policy decisions about how to proceed—which will be driven largely by the circumstances. However, it may be prudent to consider some of these robotic post-detection issues now in order to prepare ourselves to whatever extent is appropriate.

For example, upon the discovery of the first sign of life, should a robotic vehicle leave the immediate locale for remote study so as to minimize impacts to that environment? Will it depend on the kind of vehicle that makes the detection? Will we opt for an immediate sample return of those life forms? Perhaps we will want to send humans immediately to the site which has evidence of extant life? Or perhaps we should take a very conservative approach and study that life via robotic explorers for an extended period of time so as to not disturb the immediate discovery site. If we choose robotic exploration, will it be of a remote nature, say from a low orbit or nearby moon, or will we land one or many vehicles at the immediate site as soon as possible?

These are not easy questions, especially considering the amount of speculation that's involved. However, it is no different, in principle, than what goes on with most contingency planning—something NASA knows how to do.

3.2 Human First Detection Policy

Although it may be prudent to address the above questions now to whatever extent we are able, we are likely to at least have time to do so after a robotic discovery is made. We may not have that luxury if *in situ* humans make the first discovery. Significant contamination leakage is likely. There will be momentum, political and otherwise, much of which is emerging now, which could be hard to curtail, especially once humans are there. Most importantly, with humans on the scene, it will be prudent to at least establish in advance some decision making mechanisms, presumably of an international nature, to deal with post-detection activities. Preferably, an international forum should establish in advance at least general, if not specific, guidelines for pre and post-detection protocols and follow-on activities.

This policy, in addition to addressing direct detection scenarios and associated issues, can, and perhaps should, be broadened to include an assessment of the overall kind of approach we will take in preparing a human mission to Mars. Here, we might capture the general issue of whether we should or will take a conservative overall planning approach with respect to the preservation of potential extraterrestrial life.

3.2.1 Pre-detection

Guidelines should be established for activities that could jeopardize indigenous ecosystems while humans are present. Contamination measures are a part of this, but there are also issues such as establishing surveillance procedures before entering an area, guidelines for movement in an area, procedures for digging and drilling, procedures for releasing waste and dealing with rocket exhaust, etc. Such guidelines for pre-detection activities of human activity may help preserve key environments where life could exist, undetected. Emphasizing minimally intrusive procedures may be one such guideline. If we are prepared to send humans, and we are not confident about possible contamination effects, perhaps we might want to define a restricted area that human activity would be confined to, especially if we think contamination effects could be global. This is also related to our understanding of how movement will affect the spread of contamination. We may also wish to consider various forms of search/detection protocols to guide astronauts' activities as they relate specifically to the search for life on Mars. Criteria for determining that any given locale is devoid of life might also be useful to help have confidence regarding the relaxation of procedures for activities in that area.

As indicated previously, construed broadly, human pre-detection policy issues can address whether or not the issues on the mission planning decision tree, taken individually and collectively, are worthy of rigorous pursuit. How important is the preservation of extraterrestrial life? How much confidence do we want to have regarding the biological status of any given locale or of the entire planet before possibly jeopardizing indigenous ecosystems with a potentially intrusive human mission? Even if we think we should and can exercise rigorous contamination controls for a human presence, *will* we? It may be that many could agree a conservative approach is warranted and even feasible, but that may not be enough for it to be realized since many forces could conspire to relax such a cautious exploratory approach. If we have some sense for this ahead of time, perhaps we will want to consider planning for many robotic precursor missions to obtain a significant degree of confidence that Mars is dead before going with the first human mission.[32]

3.2.2 Post-detection

If and when human explorers first discover life on Mars, should the astronauts leave the immediate site and do remote analysis before disturbing the site and possibly the life form any further? Or should an astronaut take a sample immediately? If so, should the sample be sterilized immediately? Will we require a quarantine facility on the surface to study possible life forms, or will it be safer to send a sample to an orbiting laboratory so as to contain any possible adverse effects? More generally, will we be prepared, technically and politically, to deal with such a discovery *in situ*? For the first mission, it may not be feasible to send and build the appropriate technology and facilities to cope with discovering extraterrestrial life. As extreme as it may sound, some might suggest that we should leave the planet entirely until we are more certain about possible mutual effects. Some may go further and suggest we leave and never return so that life can be allowed to evolve and flourish without human interference.

But we humans will not likely be able to resist the temptation of studying such a discovery. We will send more missions and probably establish a robust scientific outpost to study the new life form. Might this eventually lead to a small community as we become more efficient at utilizing the Martian resources? Should potential population growth, either by immigration or reproduction, be controlled so as to avoid jeopardizing the indigenous biota of Mars?

Clearly these are difficult questions—partly because we have so little relevant data, and partly because they are very long-term issues. Nevertheless, as mentioned previously, exploring these issues now as part of long term contingency planning is probably wise since there is time to collect the relevant data and seek a healthy international consensus.

The nature of the life that is discovered will clearly be of critical importance in exploring these issues. A bit more specifically, whether Martian life is found to have had its own separate origin (and hence very likely different from terrestrial life) will probably be very important regarding the degree of value we place in that life. If, on the other hand, it is found to belong to the same phylogenetic tree as terrestrial life (the panspermia hypothesis) then we might be less conservative—although some will argue the scientific (and perhaps ethical) merit of allowing autonomous evolution to occur in quite different environment form that on earth.

As an example, peaceful co-existence is one long-term option to consider as a thought experiment. Ironically, Richard Taylor's slogan, "Move over microbe!" might apply.[33] That is, extraterrestrial microbes might be displaced, as often happens on earth, but they need not be harmed or destroyed. Can we co-exist with Martian life?[34] Would we combine into one ecosystem? Assuming we were careful, Martian life might not be destroyed. It could, however, change via the forces of its new ecosystem. Or perhaps we will decide to preserve that life in a kind of isolated conservatory with the indigenous Martian environment intact, so that, to some approximation, it will be allowed to evolve as it might have otherwise.[35] This could satisfy many people (although there will certainly be legitimate skepticism.) This may even satisfy those who believe that primitive extraterrestrial life should evolve autonomously. The caveat, of course, would be to exercise extreme caution in our interaction with that environment.

For those who would suggest that Martian life has "rights", this compromise might not be satisfactory. Only a non-interference policy would be acceptable.[36] However, we might consider Chris McKay's compelling view that the rights of Martian life "confer upon us the obligation to assist it in obtaining global diversity and stability."[37]

Clearly, as hinted at above, underlying many of these questions are issues of value, and policy will ultimately be driven by which values are made the priority and why.

3.3 Some Relevant Value Theory

Regarding the value dimension of this issue most generally, we want to ask: How much do we value the preservation of a primitive extraterrestrial life form and why?

There is much to be said in a rigorous treatment of such a question given the great body of work that exists on ethics and values. But there have been a few recent thinkers who have addressed some ethical issues associated with space exploration and their views will be represented in this section, along with brief discussion of the applicability of some general value theories. Much of the following comes from a previous work entitled, "Do We Need A Cosmocentric Ethic?".[38]

3.3.1 Scientific Value

Certainly there is instrumental value, or more specifically, scientific value associated with the preservation of extraterrestrial life. Clearly, masking the existence of such life and/or destroying it beyond recognition would be a scientific loss of immense proportion. Biology is desperate for a second data point. And as this paper indicates, there are many important questions that need consideration if we are to ensure the benefits associated with this scientific value. However, it isn't clear that scientific value will be enough to warrant the kind of conservative approach that may be needed to ensure the preservation of possible indigenous extraterrestrial life, thereby realizing that scientific value. As history has painfully demonstrated, the momentum of doing a thing, of accomplishing a goal to satisfy certain needs or desires, often overshadows contemplation of consequences and any potential policy action that might result thereof. The exploration and exploitation of the Americas, while certainly having some positive effects, is a poignant example of the harm we are capable of when we do not take pause to consider the consequences of our actions. Also, looking further ahead, we might also wish to consider how we will guide our actions when the scientific novelty wears off.

3.3.2 Anthropocentrism

Generally, anthropocentrists would not have much reservation about displacing or possibly destroying indigenous extraterrestrial life if it was required for human exploration and colonization of an extraterrestrial environment. Anthropocentric ethical views make humans needs and desires the priority, generally at the expense of all else.

As Robert Zubrin points out, an obvious problem for those who would answer no to whether human settlement of Mars should take priority over the continued existence of extraterrestrial microbes is to provide some explanation as to why such an answer wouldn't apply to terrestrial microbes which we wouldn't hesitate to kill with an antibiotic pill.[39] This is a reasonable challenge. However, at the same time, it also seems reasonable to suppose that extraterrestrial microbes should not be viewed the same as terrestrial microbes. Zubrin himself acknowledges their unique value.[40] An answer to Zubrin's challenge might be to point out that extraterrestrial microbes are not likely to be pro-actively destructive to our well-being, as are terrestrial microbes. Perhaps extraterrestrial microbes should be assumed innocent until proven otherwise. Also, perhaps more importantly, assuming Martian microbes are not of the same phylogenetic tree as life on earth, as a species, they would be unique in a way that terrestrial microbes are not. This significant uniqueness might imply some kind or degree of value, instrumental or otherwise, that might not necessarily be attributed to terrestrial microbes.[41]

Criticisms of anthropocentrism that it fails to consider ecological concerns and long-term effects are not so obvious since one can be concerned about the long-term ecological impacts on humans.[42] However, it has generally been the case that anthropocentrism has been more short-sighted than far-sighted. These complaints reflect a deeper instinct articulated by the philosopher Don MacNiven that theories biased towards humans are suspect.[43] This concern is supported by thousands of years of seeing our knowledge expand, constantly de-centralizing human beings—"The Great Demotions," as Ann Druyan has poignantly observed. It may ultimately be true, if we can even know such a thing, that anthropocentric value theories are valid, but we would be wise to heed the lessons of history and consider broader views.

3.3.3 Utilitarianism

A traditional utilitarian view has at its heart the concept of intrinsic value in the form of pleasure. Such a view, while used to justify respectful treatment of animals because they experience pleasure and pain, does not seem applicable to extraterrestrial microbes. We might consider, then, that the anthropocentric bias noted by MacNiven, although diluted by an expanded sphere of moral considerability in some utilitarian views, could still hold against a view that excludes primitive life forms that do not feel pain. Indeed, objective justification for the intrinsic value of pleasure requires much elucidation. In addition, appealing to happiness or pleasure as a variable for measuring value seems ultimately to involve much subjectivity, retaining a fundamental dilemma of assessing and/or measuring value.

3.3.4 A Geocentric Bias?

Robert Haynes, Chris McKay, and Don MacNiven have been prompted by the consideration of extraterrestrial activities to suggest the need for a "cosmocentric ethic". They conclude that existing ethical theories exclude the extraterrestrial environment because they are geocentric and cannot be applied to extraterrestrial environments, hence leaving a vacuum for a cosmocentric ethic.[44] Haynes says that anthropocentrism implies geocentrism because we know of no other sentient beings in the Universe.[45] Perhaps in some sense this is true for now because we only inhabit the earth, but can't we take our anthropocentrism with us anywhere we go? And can't we still be anthropocentrists if we were to discover extraterrestrial intelligence? Haynes' claim doesn't seem to apply in a general sense. McKay notes that ecological ethics has been "inextricably intertwined" with life on earth and so he comes to the same conclusion.[46] But this observation does not necessarily rule out the application of existing ethical theories to extraterrestrial considerations. If a theory excludes entities from moral consideration, it could very well be because the theory requires it, not necessarily because it's geocentric, or because it hasn't been applied to extraterrestrial considerations (although certainly these thinkers are right to question the applicability of existing views that do not address extraterrestrial considerations since such an omission might indeed be evidence of an incomplete theory.) MacNiven, while offering no additional reasons, agrees with Haynes and McKay, and further suggests that anthropocentrism, zoocentrism, and biocentrism would present no moral objection to activities such as terraforming.[47]

There may be, however, a deeper instinct being expressed by these thinkers that is more akin to realizing deficiencies in existing ethical views in general, not just as they apply to issues of space exploration—although it may be that the new context, or lens of space exploration, has rightly prompted the consideration of a new perspective—i.e. a cosmocentric perspective.[48] Nevertheless, some traditional ethical ideas have been applied to the issues at hand.

3.3.5 Rights, Intrinsic Value, and Bio/Ecocentrism

Carl Sagan's sentiment, noted in a previous footnote, is worth repeating: "If there is life on Mars, I believe we should do nothing with Mars. Mars then belongs to the Martians, even if they are only microbes."[49] Although the notion of rights is not directly invoked in Sagan's remark, his kind of view can be associated with such a rights based ideology. Similarly, Chris McKay's view is based on the intrinsic value of life principle and hence suggests that Martian microbes have a right to life—"to continue their existence even if their extinction would benefit the biota of Earth."[50] Haynes suggests that Tom Regan's arguments in his 1982, *All That Dwell Therein*, would ascribe "direct moral significance" to indigenous exoecosystems[51] (and hence presumably the resident microbes).

Such "rights" based views need to demonstrate why life should be considered intrinsically valuable and why microbes would have an absolute right to life. Rights are problematic because they are often seen as matters of degrees when difficult decisions have to be made. Degrees of rights, in the final analysis, ultimately seem no different than degrees of value. Indeed,

J. Baird Callicott writes: "The assertion of 'species rights' upon analysis appears to be the modern way to express what philosophers call 'intrinsic value' on behalf of nonhuman species. Thus the question, 'Do nonhumans species have a right to exist?' transposes to the question, 'Do nonhuman species have intrinsic value?'"[52] If one claims that other animals have rights and that there are no degrees of rights, how are we to assess those situations that involve conflict of rights and/or interests between humans and other life forms?[53]

Robin Attfield, and Paul Taylor each take similar approaches to justifying the intrinsic value of life. As beings with "natural fulfillments" or "goods of their own", organisms are "teleological centers of life" in that they are motivated by the goal of maintaining their existence. With such a good of their own, living organisms are thought to have intrinsic value, all in the same degree.[54]

Donald VanDeVeer explores, within the context of environmental ethics, possibilities involving *degrees* of value and rights. He writes that he is not aware of any plausible analysis of inherent or intrinsic value such that those very concepts preclude judgements of varying levels of value. His is a biocentric, stepped egalitarianism view roughly categorizing life into three broad groups: being alive, being sentient, and possessing rational autonomy. He cites some support from our intuitions regarding differential judgments and treatment and acknowledges the subtle complexities involved in assessing degrees of value.[55]

Freya Mathews, in *The Ecological Self* (1991), suggests degrees of intrinsic value might be associated with degrees of complexity when considering the worth of an individual. The greater the complexity, the greater the power and degree of self-maintenance, where self-realization is a fundamental element of intrinsic value. However, Mathews also goes on to articulate a view in which intrinsic value is thought to shift and flow in a systemic context, resulting in an "averaging" of intrinsic value throughout the whole.

3.3.6 A Hybrid View

Steve Gillett has suggested a hybrid view combining anthropocentrism as applied to terrestrial activity combined with biocentrism towards worlds with indigenous life.[56] Invoking such a patchwork of theories to help deal with different domains and circumstances could be considered acceptable, and perhaps even desirable, especially when dealing with something as varied and complex as ethics. Indeed, it has a practical common sense appeal. Andrew Brennan is critical of moral theory that attempts to encompass the complexity of life under a single principle and hence embraces a pluralistic approach to environmental ethics.[57] Alan Marshall writes of a "postmodern associationism" which, based on the deconstruction of metaphysical unity, "places an emphasis on respecting the other as arbiter of its own reality, without imposing metaphysical imperialism under the guise of a great organic unity," perhaps leading to an "appreciation of ontological meaninglessness, disunity, difference and respect for individuals as *others*—rather than as colonies of unity..."[58] We might also consider another view of this legitimate epistemological issue. Callicott writes: "But there is both a rational philosophical demand and a human psychological need for a self-consistent and all-embracing moral theory. We are neither good philosophers nor whole persons if for one purpose we adopt utilitarianism, another deontology, a third animal liberation, a fourth the land ethic, and so on. Such ethical eclecticism is not only rationally intolerable, it is morally suspect as it invites the suspicion of ad hoc rationalizations for merely expedient or self-serving actions."[59]

3.3.7 Anthropogenic Intrinsic Value

For Callicott, species possess a "truncated" version of the traditional definition of intrinsic value in that they have value "for" themselves, for their own sake, but not "in" themselves, independent of a valuing consciousness.[60] The basis for Callicott's perspective on intrinsic value is a Human/Darwinian emotive/bioempathic view which suggests that emotionally based value identification with other living things results from natural selection. Furthermore, relativism can be avoided by appealing to Hume's "consensus of feeling" which standardizes or fixes the human psychological profile and values that result thereof. Although value may not be focused solely on humans in this view, humans are indeed the source of value (i.e. value is *anthropogenic*) in that we recognize intrinsic value of other living things as their "standard" genetic make-up dictates. But is such recognition of the intrinsic value of nonhumans so standard or fixed? It appears not since there exists much intense, often violent, controversy over the value of nonhumans. Hence, there still appears to be an inherent subjectivity on an individual as well as a collective basis, since the feelings of humans are what generates the intrinsic value (making the invocation of the word 'intrinsic' somewhat suspect). This view, then, seems not to objectively justify intrinsic value or provide a way for measuring such value when difficult decisions have to be made.

3.3.8 Organic Unity

Robert Nozick draws from aesthetics the concept of organic unity in which to ground intrinsic value. He says: "The more diverse the material that gets unified (to a certain degree), the greater the value."[61] This unity in diversity is what Nozick suggests can be equated with intrinsic value. More precisely, he suggests that it might be the best *approximation* to value because our experience may be limited regarding what is valuable.[62] However, although Nozick gives a compelling account of how organic unity fits with our general perception of value, ultimately, what appears to be missing is a truly objective justification for why organic unity should be considered intrinsically valuable.[63]

3.3.9 Cosmocentrism

We have seen that various thinkers suggest the need for a cosmocentric ethic. Robert Haynes writes:

> These considerations suggest to me that we need from philosophers a new "cosmocentric" ethics, and perhaps a revised theory of intrinsic worth, if we are to evaluate the moral pros and cons of proposals for ecopoiesis (small-scale ecosystem construction) in an intelligent and sensitive way. As I see it, the first objective of such an ethic would be to resolve the dialectical contradiction that commonly arises between superficial views of "evolutionary progress" and "ecological harmony." If pushed to their obvious extremes these conflicting myths could lead the grossest kind of human environmental imperialism on the one hand, or to the destructive elimination of all technology, including modern medicine and agriculture, on the other. For me, a cosmocentric ethic would allow scope for human creativity in science and engineering throughout the solar system, but also recognize that at present we depend utterly on the vitality of the Earth's biosphere for our very existence. It would recognize also that the physical artifacts of humanity are as much a part of the Universe as are stars, planets, plants and animals.[64]

Although by no means well-defined, a cosmocentric ethic might be characterized as one which (1) places the Universe at the center, or establishes the Universe as the priority in a value system, (2) appeals to something characteristic of the Universe (physical and/or metaphysical) which might then (3) provide a justification of value (presumably intrinsic value), and

(4) allow for reasonably objective measurement of value. Related to this kind of ethic are views which appeal to "cosmologies" as the foundation for ethical views.[65]

At first glance, talk of a cosmocentric ethic might seem paradoxical. How can an ethical view be centered or focused on "all that is"? From egocentrism to eco/geocentrism, we are able to center, focus, and prioritize value because there is some other, generally larger frame of reference which is relatively de-valued. Nevertheless, as has been suggested by others noted above, such an ethic may be helpful in dealing with value based questions involving extraterrestrial issues such as interaction with indigenous primitive extraterrestrial life forms.

As with environmental ethics, the central issue for a cosmocentric value theory is justifying intrinsic value.[66] Indeed, the significance of appealing to the Universe as a basis for an ethical view is that an objective justification of intrinsic value might be realized to the greatest extent possible by basing it on the most compelling objective absolute we know—the Universe. In a pantheistic world-view, this is functionally equivalent to knowing the nature of, and perhaps doing the "will" of, God. In addition, we should like to have some way of objectively assessing, preferably measuring, value.

3.3.9.1 The Projective Universe's Formed Integrity

Holmes Rolston proffers a compelling view which appeals to the "formed integrity" of a "projective Universe." This view suggests that the Universe creates objects of formed integrity (e.g. objects worthy of a proper name) which have intrinsic value and which should be respected.[67] However, Haynes points out that Rolston's view appears to conflict with modifying the earth, even to the benefit of humans.[68] Rolston's view would certainly call for the preservation of primitive extraterrestrial life.

In Rolston's view, justification of intrinsic value might come from the creative processes of the Universe itself—that is, the creative process, and all that results from it, is intrinsic to the Universe.[69] However, in assigning value to the Universe's creative processes, we might be guilty of anthropomorphizing the Universe.[70] Indeed, we could ask why the Universe is a creative entity—which might shed light on the general requirement for more rigorous elucidation of how the Universe's creative process can give rise to a justification for intrinsic value.

Rolston's view also attempts to address the problem of assessing or measuring value by suggesting that if a thing has formed integrity, or is worthy of a proper name, it should be respected, which presumably means left alone. But how do we decide what has formed integrity so that it will be named? This is the value measurement problem in a different form. Conflict ultimately remains since personal subjective value judgments seem unavoidable in assessing what has formed integrity.

3.3.9.2 The Sanctity of Existence

MacNiven has suggested that a central tenet of a cosmocentric ethic would be the principle of the sanctity of existence, which, he notes, would make it difficult to justify the significant modification or destruction of indigenous life forms.[71] In a minimal sense, the principle of the sanctity of existence might satisfy criterion one and two for the definition of a cosmocentric ethic suggested previously because the Universe, and all therein, exists. However, we do not see a compelling articulation of why, specifically, all things have intrinsic value because they exist. We should prefer some justification of the principle itself as well as its invocation. MacNiven additionally suggests appealing to a "selective concept of uniqueness" as we sometimes do in considering terrestrial matters such as preserving the Grand Canyon.[72] Here, again, we might ask why uniqueness should have intrinsic value. Even in light of the notion of uniqueness, the issue of measuring value—or more specifically, of weighing the value of human activity against other forms of value such as the preservation of an extraterrestrial life form still appears to be without a firm theoretical foundation.

3.3.9.3 Connectedness

The systemic, interdependent connectedness of ecosystems is often cited as a foundation justifying the value of parts of the larger whole, since a subset contributes to the maintenance of the larger whole. Consider Leopold's egalitarian ecosystem ethic: "A thing is right when it tends to preserve the integrity, stability, and beauty of the biotic community. It is wrong if it tends to do otherwise."[73]

In *The Ecological Self*, Freya Mathews suggests that intrinsic value can be grounded in self-realization, which is a function of interconnectedness. The Universe qualifies for self-hood and hence self-realization (again, for which interconnectedness plays a critical role) and humans participate in this cosmic self-realization.[74]

Construed cosmically, then, connectedness may hold promise for a cosmocentric ethic. In particular, it may be that connectedness itself is a *necessary* property of the Universe, and that to actualize/instantiate a connection *necessarily* requires an interaction—hence connectedness gives rise to, or is instantiated via, interaction. Such a view might favor maximizing interaction and any other consequence of realizing robust actualizations of connectedness/interaction action (perhaps, for example, complexity, creativity, uniqueness, diversity, intensity, etc.) as the foundation of a cosmocentric ethic since it would contribute to the greatest realization of the nature of the Universe (i.e. its "self-realization"). Indeed, in making choices consistent with this view, humans might help propagate diversity here on earth and throughout the Universe, but not *necessarily* at the expense of other robust actualizations of connectedness (e.g. perhaps other "kinds" of life forms).[75] The trick would be to assess relative degrees of value corresponding to degrees of realizing connectedness/interaction.[76]

3.3.10 The Fact/Value Problem

It is important to acknowledge the importance of the fact/value (or "is/ought") dilemma which suggests, among other things, that knowing something about the way the Universe is cannot lead to a justification of value. Thankfully, this complex philosophical problem, although ultimately relevant, is beyond the scope of this paper. But, consider that this problem can also be understood as the idea that values do not necessarily follow from facts—not that values absolutely cannot follow from facts. That is, if we find a fact-based value theory compelling enough, we have the choice to associate and/or derive value (an "ought") from what "is".[77] Our value theories can be models just like physical theories. What's important, of course, is that they have broad explanatory and problem-solving power.

The ecologist Frank Golley has argued that activities in space such as the colonization and terraforming of Mars will be unavoidable since it is consistent with the dominant myths and metaphors of western civilization. Historically, these dominant myths and the exploration that results from them have not been concerned about the indigenous systems they effect, including the existence of human beings. Is this the kind of action that is unavoidable? Golley suggests that to turn away from these pursuits would require a fundamental reorientation of our culture.[78] If a lack of concern for indigenous systems is part of our dominant myths and exploratory pursuits, then perhaps a fundamental reorientation of our culture is exactly what's needed. Ironically perhaps, this would be consistent with Robert Zubrin's vision of Mars as an opportunity for the grand, noble experiment—a chance to realize new ways of life. Indeed, we could explore, create a new branch (or branches) of human civilization, terraform, etc.—all the while fostering and exercising a kind of respect and caution that has traditionally been absent. To some degree, it's already happening. This century's strong environmental and animal rights movements are powerful examples. We need only to continue to foster extend these concerns to extraterrestrial environments.

Finally, some may argue that the rational pursuit of ethics is futile—that rationality is slave to the passions, and/or that economics (competition for resources) is the primary motivation for human activity. Perhaps this is partly true. But, there is certainly a critical role for considered rational thought regarding what we value and why. Human beings are extremely diverse, and are motivated by many different forces. Ultimately, through a mix of reductive, creative, and ecological thinking, as favored by Frederick Turner,[79] a compromise among many diverse forces will likely strike a reasonable balance regarding how the status of extraterrestrial life will fit into our policies for exploring our solar system and beyond.

REFERENCE NOTES

1. Treaty on Principles Governing the Activities of States in the Exploration and Use of Outer Space, Jan 27, 1967, 18 U.S.T. 2410.
2. Treaty on Principles Governing the Activities of States on the Moon and Other Celestial Bodies, Dec. 9, 1979, 34 U.N. GAOR Supp. (No. 20), U.N. Doc. A/AC/105/L.113 Add. 4.
3. Darlene A. Cypser, International Law & Policy of Extraterrestrial Planetary Protection, *Jurimetrics* Vol. 33, p. 323 (1993).
4. N. Jasentuliyana, Environmental Impact of Space Activities: An International Law Perspective, *Proceedings of the 27th Colloquium on the Law of Outer Space, International Institute of Space Law of the International Astronautical Federation*, Oct. 7-13, p. 394 (1984).
5. Chris McKay and Wanda Davis write: "It is arguable that once humans land on Mars, attempts to maintain a strict policy of preventing the introduction of Earth life into the martian environment will become moot." Planetary Protection Issues in Advance of Human Exploration of Mars, *Advanced Space Research* Vol. 9, No. 6, p. 197 (1989).
6. Cypser argues convincingly that "harmful contamination to other states" (that which is to be avoided as called for by the Outer Space Treaty) can best be interpreted to mean interference with future life-detection experiments. p. 324-325, 338.
7. E. C. Levinthal, J. Lederberg, and C. Sagan, note that "the scientific value of detection and characterization of life was the overriding value to be considered." Relationship of Planetary Quarantine to Biological Search Strategy, *Life Sciences and Space Research VI*, eds. A. H. Brown and F. G. Favorite, Amsterdam, North-Holland, p. 136 (1967).
8. Space Studies Board (SSB), National Research Council, *Biological Contamination of Mars: Issues and Recommendations.* National Academy Press, Washington, D.C., p. 47, 49 (1992).
9. Space Studies Board (SSB), National Research Council, *Mars Sample Return Report.* National Academy Press, Washington, D.C., pp. 37-38 (1997).
10. Robert Zubrin and Richard Wagner, *The Case For Mars*, New York, Simon and Schuster, pp. 132-134, (1996). T. H. Jukes makes the case for pathogenic coevolutionary dependence in his paper, Evolution and Back Contamination, *Life Sciences and Space Research* XV, 9 (1977).
11. Prior to the Apollo missions, the Space Studies Board recommended a sterile drilling system. M. Werber, Objectives and Models of the Planetary Quarantine Program 13, NASA SP-344, Sup. Doc. No. NAS 1.21:344 (1974). Also, subsurface drilling on Earth has raised concern about whether or not organisms brought to the surface are indeed indigenous to the subsurface or whether they were transported there from surface contamination.
12. McKay and Davis note several sources of environmental impacts due to a human base that should be considered, including mechanical disturbances, life support system leakage, airborne pollution, and "seemingly innocuous perturbations" like water, heat, light, etc. P. 198.
13. For an analysis of social factors see: Margaret S. Race, Societal Issues as Mars Missions Impediments: Planetary Protection and Contamination Concerns. *Advanced Space Research* Vol. 15, p. 285 (1994).

14. H. P. Klein, *Planetary Protection Issues for the MESUR Mission: Probability of Growth (P_g)*. NASA Conference publication. NASA Ames Research Center, Moffett Field, California, (1991).
15. Space Studies Board, *Biological Contamination of Mars*, pp. 46-47.
16. For example, McKay and Davis suggest lubricants and refrigeration gases as possible sources industrial chemicals. P. 198.
17. McKay and Davis write: "It may be assumed, *a priori*, that all space suits and habitats will leak." P. 197. This is known to have been the case with Apollo since it is thought that there was "significant leakage of gases from the joints of the astronauts' suits". Victor Cohn, Lunar Contamination: Growing Worry, *Washington Post*, 28 May 1969, p. A12.
18. Space Studies Board, *Mars Sample Return Report*, p. 2.
19. Freeman Dyson brought this terrestrial analog to my attention in a personal conversation (Aug. 1998).
20. Indeed, many personal conversations with average laypersons, scientists, and senior NASA managers bear this out.
21. J. C. Sharp, Manned Mars Missions and Planetary Quarantine Considerations, *Manned Mars Missions*, NASA M002, NASA Washington, D.C., p. 553, (June 1986).
22. National Aeronautics and Space Administration, 12 May 1969, *Apollo Spacecraft Cleaning and Housekeeping Procedures Manual*, MSC-000 10, p. 3.
23. National Aeronautics and Space Administration, 12 May 1969, *Outbound Lunar Biological and Organic Contamination Control: Policy and Responsibility*, Washington, D.C., NASA Policy Directive 8020.8A.
24. James R. Murphy, Robert M Haberle, Owen B. Toon, James B. Pollack, Martian Global Dust Storms: Zonally Symmetric Numerical Simulations Including Size-Dependent Particle Transport, *Journal of Geophysical Research*, Vol. 98, No. E2, (1993).
25. McKay and Davis, p. 198.
26. McKay and Davis, p. 198.
27. Carl Sagan has written: "If there is life on Mars, I believe we should do nothing with Mars. Mars then belongs to the Martians, even if they are only microbes." Carl Sagan, *Cosmos*. Random House, New York, p. 130 (1980). It's not clear if this implies we stay off the surface completely, use sterilized robots only, or just prohibit colonization while allowing *in situ* experimentation via human explorers.
28. It should be mentioned that it will obviously be important to consider the scenario that contamination will be regional—that is, somewhere in between local and global. This will reduce the global biological status problem somewhat, but the fundamental challenges remain, including assessing the type, extent, and geographical range of contamination number and kind of missions that might be required.
29. Richard Randolph, Margaret Race, and Christopher McKay, in Reconsidering the Theological and Ethical Implications of Extraterrestrial Life, *The Center for Theology and the Natural Sciences* Vol. 17, No. 3, p. 6, (Summer 1997), write: "There is currently no NASA policy, or international protocol, for the proper handling of non-intelligent extraterrestrial life. We believe that such a policy should be developed now, before these discoveries are made. Such a policy would be informed by an ethical analysis concerning our obligations as space explorers."
30. Bruce Murray, "Chasing Mars—Great Expectations and Hard Choices." Presentation at the Mars Symposium: Life In the Universe. George Washington University, Space Policy Institute, 22 November 1996.
31. The Space Studies Board, *The Biological Contamination of Mars*, p. 50, notes: "Even if there is not organismal growth, local contamination is to be expected around an unsterilized spacecraft. Clearly a lander should not return to do life-detection experiments at a site where unsterilized spacecraft have landed previously." This could obviously have significant impacts on long-term mission planning and should be taken seriously when considering landing sites for all missions, near and long-term.
32. Don DeVincenzi notes in an abstract submitted at the Fourth Symposium on Chemical Evolution and the Origin and Evolution of Life (1990) that guidelines from a previous workshop on planetary protection suggest that "human landings are unlikely until it is demonstrated that there is no harmful effect of martian materials."
33. Martyn Fogg notes a radio interview with Richard Taylor. Martyn Fogg, *Terraforming: Engineering Planetary Environments*. SAE International, Warrendale, p. 494 (1995).

34. J. Baird Callicott notes that co-existence may be feasible since we will not have to consume indigenous life as we do on earth. This may true in the near-term, but longer term activities could cause the extinction of indigenous life via competition for resources. J. Baird Callicott, Moral Considerability and Extraterrestrial Life. E. C. Hargrove (Ed.), *Beyond Spaceship Earth: Environmental Ethics and the Solar System*. Sierra Club Books, San Francisco, pp. 250-251 (1990).

35. Robert Zubrin opens the door for such a compromise when he suggests that the polar regions will be available for indigenous life to predominate. Robert Zubrin, The Terraforming Debate, *Mars Underground News*, Vol. 3, pp. 3-4 (1993).

36. Alan Marshall, Ethics and the Extraterrestrial Environment. *Journal of Applied Philosophy* Vol. 10, No. 2, p. 233 (1993).

37. Chris McKay, Does Mars Have Rights? D. MacNiven (Ed.), *Moral Expertise*. Routledge, London, p. 194 (1990).

38. Mark Lupisella and John Logsdon, "Do We Need A Cosmocentric Ethic?" Paper IAA-97-IAA.9.2.09 presented at the 48th Congress of the International Astronautical Federation, Turin, Italy (October 1997).

39. Zubrin, The Terraforming Debate, pp. 3-4.

40. Zubrin and Wagner, *The Case For Mars*, p. 135.

41. Indeed, we will see that Don MacNiven and others cite the importance of uniqueness in determining value.

42. Indeed, many environmentalists are also anthropocentrists. Eugene Hargrove, in *Foundations of Environmental Ethics*, Prentice Hall, (1989), considers aesthetics as an anthropocentric foundation of environmental ethics. In *Why Preserve Natural Variety?* (1987), Bryan Norton advocates the "transformative" value (e.g. the effect of positively enhancing, ennobling, etc.) that other species have on humans. J. Baird Callicott, in *Moral Considerability and Extraterrestrial Life* (p. 252) applies Norton's "weak anthropocentrism" to the issue of preserving primitive extraterrestrial life by appealing to its transforming and ennobling effect on human nature. He says, "I can think of nothing so positively transforming of human consciousness as the discovery, study, and conservation of life somewhere off the earth." Witness also the poignant poster of the Earth under which reads, "Save The Humans."

43. Donald MacNiven, *Creative Morality*. Routledge, New York, pp. 202-203 (1993).

44. Robert Haynes, Ecopoiesis: Playing God On Mars. *Moral Expertise*, p. 177. See also Haynes and McKay, Should We Implant Life On Mars? *Scientific American*, p. 144 (December 1990) and MacNiven's, *Creative Morality*, p. 204.

45. Haynes, Playing God On Mars, p. 176.

46. McKay, Does Mars Have Rights? p. 196.

47. Donald MacNiven, Environmental Ethics and Planetary Engineering. *Journal of the British Interplanetary Society* Vol. 48, pp. 442-443 (1995). Also published as Paper AAS 97-389 in *From Imagination to Reality: Mars Exploration Studies of the Journal of the British Interplanetary Society* (Part II: Base Building, Colonization and Terraformation), R. M. Zubrin, ed., AAS Science and Technology Series, Vol. 92, pp. 303-307 (1997).

48. In addition, Martyn Fogg writes: "the concept of terraforming is inspiring enough to perhaps generate a formal effort toward extending environmental ethics to the cosmic stage." Martyn Fogg, *Terraforming: Engineering Planetary Environments*, p. 490.

49. Carl Sagan, *Cosmos*. Random House, New York, p. 130 (1980).

50. McKay, Does Mars Have Rights? p. 194.

51. Haynes, Playing God On Mars, p. 177. For this interpretation of Haynes, and for an excellent analysis of the ethical issues regarding terraforming, see Richard Miller's, independent study entitled, "The Greening of Mars: Ethics, Environment, and Society, Terraforming: An Ethical Perspective." University of Waterloo, Canada, pp. 14-15 (1996).

52. J. Baird Callicott, On The Intrinsic Value of Nonhuman Species. Bryan Norton (Ed.), *The Preservation of Species*. Princeton University Press, Princeton, p. 163.

53. Deep Ecology views tend to have as a central tenet, biological egalitarianism, according to which all organisms have an equal right to life. See Arne Naess, *Ecology, Community, and Lifestyle: Outline of an Ecosophy*, Cambridge, 1989.

54. Wayne Ouderkirk, Earthly Thoughts: An Essay on Environmental Philosophy, *Choice*, November 1997, p. 424. Robin Attfield, *The Ethics of Environmental Concern*, 2nd ed. Georgia Press (1991). Paul Taylor, *Respect for Nature: A Theory of Environmental Ethics*, Princeton (1989). Also, Bernard Rollin draws a line between protozoa which exhibit behavior that might indicate consciousness (albeit broadly defined) and bacteria and plants for which there is no such evidence. *Animal Rights and Human Morality*. Prometheus Books, Buffalo, pp. 39-42 (1981). Carl Sagan seems to articulate an even more general view when he says: "Consciousness has various meanings. If it means an awareness of the external world, and modifying your behavior to take account of the external world, then I think microbes are conscious." Carl Sagan, "The Age of Exploration", in *Carl Sagan's Universe*, ed. Y. Terzian and E. Bilson, p. 154 (1997).

55. Donald VanDeVeer, Interspecific Justice and Intrinsic Value, *The Electronic Journal of Analytic Philosophy*, 3, Spring 1995.

56. Steve Gillett, The Ethics of Terraforming, *Amazing*, pp. 72-74 (August 1992).

57. Andrew Brennan, *Thinking About Nature: An Investigation of Nature, Value, and Ecology*, (1988).

58. Alan Marshall, A Postmodern Natural History of the World: Eviscerating the GUTs from Ecology and Environmentalism, *Studies in History and Philosophy of Biological and Biomedical Sciences*, p. 158, 160 (1998).

59. Callicott, Moral Considerability and Extraterrestrial Life, p. 251.

60. Callicott, On the Intrinsic Value of Nonhuman Species, p. 143.

61. Robert Nozick, *Philosophical Explanations*, Cambridge: Harvard University Press, p. 416 (1981).

62. Nozick (1981), p. 442.

63. Diversity is found to be an important tenet in many world-views, especially environmentally sensitive world-views. Freeman Dyson writes: "Diversity is the great gift which life has brought to our planet and may one day bring to the rest of the Universe. The preservation and fostering of diversity is the great goal which I would like to see embodied in our ethical principles and in our political actions." *Infinite In All Directions*. New York: Harper & Row, (1988). Peter Miller offers the notion of "richness" (which can be roughly equated with diversity) as a "generalized normative concept" against which to assess value. Value as Richness, *Environmental Ethics*, vol. 4, no. 2, p. 106-107, (Summer 1982). Richard Hanley, in *The Metaphysics of Star Trek*, Harper Collins, 1997, p. 17, calls attention to the "well-known Vulcan IDIC credo ('Infinite Diversity in Infinite Combinations'). In *The Ecological Self* Barnes & Noble (1991), Freya Mathews argues along Leibnizian/Spinozian lines that "the actual world, must be the *fullest* possible world." In its barest form, this is Arthur Lovejoy's principle of plenitude (*The Great Chain of Being*, 1971), versions of which have been proposed since Plato's Demiurge (a "God" who "wanted the world to lack nothing").

64. Haynes, Ecopoiesis: Playing God On Mars, p. 177.

65. See, for example, the critical role of cosmology in Freya Mathews', *The Ecological Self* (1991), Arran Gare's *Postmodernism and the Environmental Crisis* (1995) for a criticism of postmodernism as an "inadequate guide for political action or how to live", suggesting the need for a kind of "postmodern cosmology", and Joseph Grange's, *Nature: An Environmental Cosmology* (1997). Carolyn Merchant, in *Radical Ecology: The Search for a Livable World* (1992) stresses that ecological ethics is a radically different form of ethics grounded in the cosmos.

66. Callicott writes: "In addition to human beings, does nature (or some of nature's parts) have intrinsic value? That is the central theoretical question in environmental ethics. Indeed, how to discover intrinsic value in nature is the defining problem for environmental ethics. For if no intrinsic value can be attributed to nature, then environmental ethics is nothing distinct. If nature, that is, lacks intrinsic value, then environmental ethics is but a particular application of human-to-human ethics. Or, putting the same point yet another way, if nature lacks intrinsic value, the nonanthropocentric environmental ethics is ruled out." Intrinsic Value in Nature, *The Electronic Journal of Analytic Philosophy*, 3, Spring 1995.

67. Holmes Rolston III, The Preservation of Natural Value in the Solar System, E. C. Hargrove (Ed.), *Beyond Spaceship Earth: Environmental Ethics and the Solar System*, Sierra Club Books, San Francisco, (1990).

68. Haynes, Playing God on Mars, p. 177.

69. Similar to MacNiven's view, this perspective has been referred to as "object-centered" by Richard Miller, in an independent study entitled, "The Greening of Mars: Ethics, Environment, and Society, Terraforming: An Ethical Perspective." University of Waterloo, Canada, (1996). Rolston also articulates a more specific view regarding the intrinsic value of primitive organisms whereby "A life is defended for what it is in itself, without further contributory reference...That is ipso facto value in both the biological and philosophical sense, intrinsic because it inheres in, has focus within, the organism itself." *Conserving Natural Value*, Columbia University Press, p. 173 (1994).
70. Generally, this anthropomorphizing tendency causes suspicion regarding a view's validity. However it could, ironically, be interpreted as evidence to support or justify intrinsic value—perhaps via some version or derivation of the Anthropic Principle, for example.
71. MacNiven, Environmental Ethics and Planetary Engineering, pp. 442-443 (1995).
72. MacNiven, pp. 442-443.
73. Aldo Leopold, *A Sand County Almanac*, New York, p. 262, (1966).
74. Mathews articulates self-hood and self-realization, generally, and in a cosmic sense, in Chapter 3, and the associated ethical implications in Chapter 4.
75. Indeed, the kind of life that might exist on Mars could play a critical role in what kind of value is assigned to it. The cosmocentric view suggested here might imply that a unique extraterrestrial life-form be assigned a higher value than primitive terrestrial organisms, since it would constitute a significantly different universal creation (where creation of new, robust forms of interaction are a central—perhaps intrinsic—value suggested by the cosmocentric ethic.)
76. In, From Biophysical Cosmology to Cosmocentrism, (*SETI In The 21st Century: Cultural and Scientific Aspects*, SETI Australia Centre, January 1998), I try to articulate the philosophical principles and implications of such a view.
77. Callicott claims that Hume's is/ought dichotomy can be bridged "in Hume's terms, meeting his own criteria for sound practical argument." Hume's Is/Ought Dichotomy and the Relation of Ecology to Leopold's Land Ethic, *Environmental Ethics*, Vol. 4, (Summer 1982).
78. F. B. Golley, "Environmental Ethics and Extraterrestrial Ecosystems," *Beyond Spaceship Earth: Environmental Ethics and the Solar System*, ed. E. C. Hargrove, San Francisco, Sierra Club Books, p. 225 (1986).
79. Frederick Turner, Life On Mars: Cultivating a Planet and Ourselves, Harper's Magazine (August 1989).

THE ETHICAL RAMIFICATIONS OF DISCOVERING LIFE ON MARS

Katherine Osborne

This convention is the start (most hopefully) of a concordance of human values. What we are doing here today (this week) sets the template for our most critical policies tomorrow. Probably most of us here in this room have seen at least one episode of Star Trek. We have been exposed to the idea that life, has great value and enormous potential, and that to interfere with its progression, evolutionary, culturally, socially, without its own explicitly expressed desire for such interference, is anathema. This is called the Prime Directive, and has been the cornerstone of the television series. Considering that there is no precedent for extra-terrestrial contact yet (or at least, any recorded contact), and considering that we may very well find life on Mars, it is perhaps time that we started thinking about drawing up our own prime directive. Don't get me wrong, I am not suggesting we will find any lifeforms being anything close to our limiting definition of intelligent—if we do find anything at all, it will most likely be rather microbial. But we will have to start somewhere and it might as well be on our journey to Mars. To start this process, we will need to ask ourselves a few questions:

- What should we do when we meet extraterrestrial life for the first time?
- Will we meet life on Mars?
- How will we react to the discovery?
- How do we build a combined Martian/Terran ecology? How do we integrate the two without destroying one in the process?
- What can we do now?

What should we do when we meet extraterrestrial life for the first time?
What we do depends on the following:

- Our definition of life
- How we value life (all life)
- Humanity's need for a stabilized earth
- Our need to expand
- Our need to discover and explore
- Our culture and past

How do we define "human life"? Does a "human life" begin at conception when we first have all of the chromosomes that we will have as adults, but are not able to survive independently of our mothers? Or does "human life" begin sometime during fetal development when there is a functioning nervous system (and the fetus can feel the pain of artificial termination)? Or is it at birth when we can breathe on our own and can communicate on a very rudimentary level, yet still cannot survive alone. Or perhaps it begins sometime later in childhood... To other forms of life on earth, we do not nearly give so much consideration and debate. To lifeforms not originating on earth, we risk not assigning any consideration at all with our current attitude. To be able to fully understand and respect non-terran life, we must have a good working notion of just what life is comprised of, because it may not be like anything we know.

This century has seen the development of western man's value of human life—the equality of man and woman, the equality of all varieties of people, and most recently, we have been strenuously debating the value of the unborn fetus. We have begun to pull at traditional definitions of value. Discovery of extraterrestrial life will almost certainly pull even harder at those values. Even within western society we are very polarized when it comes to the value of life. Pro-Choice vs. Pro-Life, Evolutionists vs. Creationists, Vegetarians vs. those of us who don't mind a thick slab of meat once in awhile, the battles rage on. How will we react to new forms of life when we can't even figure out how to react to those that we already know about?

In this century, we are experiencing a rapid loss of species on this planet. It is strongly speculated by scientists that this is due in large part to the actions of human civilization. The burning of fossil fuels, the practices of deforestation to make way for short-lived farms, human encroachment upon natural habitat, may all have contributed to this loss. There are those that argue that this isn't what is happening, at least not because of man. And they may be right. After all, science has never really been a closed book. There is evidence that massive losses similar to this have occurred in the past; the most dramatic of which being the extinction of the dinosaurs 140 million years ago due to an asteroid impact, another much earlier than that due to unknown processes. Perhaps our current one is merely part of a natural (albeit rare) cycle. However, I think that the evidence has piled up enough to indicate human influence. We never think about how we control our environment, the irrevocable effect we have on it. Other life on earth has an effect on the environment as well, but not as dramatically or swiftly as we have had. And so, it is necessary for us as a species to be able to learn how to control the impact of our civilization on our environment, if for no other reason but our own survival. We are beginning to choke on our own output. In devising processes to sustain human life on Mars, we will inevitably have the potential to apply those same processes back on Earth. And to the public, this is a very legitimate reason for going. Hopefully, in our experiments, to "get it right", we won't end up destroying more life than we already are.

Our species may need to expand away from Earth in the future. Perhaps not necessarily due to complete destruction of Earth's ecology, but perhaps so that we are not dependent on Earth to carry us. To some of us, there is a destiny to be fulfilled in leaving the 'cradle', as Carl Sagan would have put it. As any number of movies this summer have demonstrated, it would be an awful shame if we got this far only to be vaporized by intervening interplanetary debris. Perhaps in this way, we will be able to better understand and respect any form of life we may encounter.

In our western culture, we have exhibited a communal desire to explore unknown frontiers and new horizons. Granted, this desire has been fueled by political pressures After the Renaissance, when we began as a culture to push away from the traditional confines and control of the church and state, into realizing our potential as individuals, this desire began to snowball into a much-exhalted idealism, the "American Spirit". It is primarily because of this desire that we now have a global culture, rapidly advancing technology, and are even contemplating exploring Mars. This desire has also been fueled by incredible success. It seems that we have been able to accomplish whatever we have set out to do under this idealism. Discovering the "New World", seeing microbes, circumnavigating the globe, manned flight, splitting the atom, landing men on the moon. At some point, somewhere somebody said these things could not be done, but we took the initiative and we did them. Perhaps we have every right to feel that that the universe is open to us, brimming over with incredible unfathomable possibilities.

However, these incredible accomplishments have come at a terrible price. As just one example, when the New World was discovered, the discoverers almost immediately began to plunder its resources. In instances where the native populations resisted this, particularly because of the disparity in technological savvy in the area of weaponry, they were slaughtered. It was a genocidal bloodbath. We lost whole cultures because of this. And those that were not affected by war, were newly exposed to all manner of diseases. And again, sadly, many died. Hopefully, we will not approach new life in this manner, but in order to do so, we must keep in perspective how we value life.

Will we meet life on Mars?

I can just hear the reaction of the collective American public saying (after an initial shock of some sort) "big deal you can't even see them!" How can we foster respect for such small, seemingly inconsequential entities in the public, when the public for the most part, barely has respect for itself or its own environment? And that's even if there is anything there!

I am not a biologist, I couldn't tell you the probability of discovering life on Mars. But I feel it would be incredibly ignorant to simply ignore the possibility that life on Mars may very well exist. Even if the life we discover is incredibly tiny, or perceived to be inconsequential, it is not, for it means that life could very easily exist in abundance throughout the universe or that the universe is ready to receive and nurture it. It would be very irresponsible to push ahead human expansion without first giving thought to what we may be expanding into, especially so to ourselves. If not on Mars, then certainly we will meet other life eventually through our journeys. We owe it to ourselves enter such a situation at least a little prepared for what may lie ahead. In the end, it is a big deal. It is our future at stake.

How do we build a combined Martian/Terran ecology?
How do we integrate the two without destroying one in the process?

We have the following scenarios to consider regarding native life if we become able to settle Mars:

1. Disregard it, especially if it doesn't have a chance at evolving into anything resembling us. Colonize, terraform, our human needs are just more important.

2. Survey the lifeforms. Catalog, and preserve information on them for future reference. Colonize and terraform. It is important to know as much as possible about this life, but human needs must come first to ensure our own survival.

3. Research the lifeforms. Understand the existing ecology. Colonize and adapt to the environment, with as little disruption as possible to the existing ecology. Perhaps in time, we can begin to terraform gradually, and in balance.

4. Make Mars completely off-limits to human intervention. We have no right to destroy the Martian ecology as well.

In my opinion, option number 3 is the best. The concept of balance has been difficult for humans to implement in the influence we have over our earth's ecology, especially by the west and especially in this last century or so. But balance is the key to stabilizing Earth's ecology, and the key to forging a combined Martian/Terran ecology on Mars. Just as on Earth (as a planet is a fairly closed system), maintaining a diverse and systemically balanced biosphere promotes stability in the environment to a certain degree.

Option number 1 would be foolish and selfish. First of all, we would miss out on a wonderful opportunity to study life arising in an environment that is currently very alien to the Earth's, but which at one time, may have been very much like our planet's own beginnings. We may be able to learn how life started on earth, whether the two planets have shared life over time through a possible panspermia scenario, or learn the answers to many other myriad questions. Secondly, if we truly want to forge our own version of the prime directive, this would a very bad way to start.

Option number 4 may just not be possible. If we start a headlong rush towards manned-exploration of Mars, we may find it difficult to backtrack to complete non-interference. And again, we would lose out on a wonderful opportunity to study the origins of life in our solar system.

All things considered, option number 2 is more likely to occur, if we are lucky. Due to probable lack of resources (mainly people), it may not be feasible to attempt a comprehensive survey of life on Mars. We may have to make due with simple specimen collections at first, and then perhaps if interest peaks among the public back on Earth further funding and resources may be committed to enlarge the research. If we do encounter an existing ecology, it will probably be rather frail considering the harsh conditions of the environment, and we are more likely be able to adapt to it than it would be able to adapt to us.

What can we do now?

Many people, when they purchase a lottery ticket, believe in at least some small way, that they have already won the money, and begin to imagine ways that they could spend it. Mars may be out there beckoning us, it may be incredibly convenient, so close as it is, but its not "ours", it is not anybody's. It would be very foolish to spend something that we have not. On the other hand, if you never play a lottery, you will never win one.

What we can do now, is think about what Martian life would mean to us personally. We must actively seek out discussion, discover what others think about the concept, especially those who are not predisposed to the idea, or who originate from a non-western culture and have different ways to view the value of life. What we learn from others who have been removed from thinking about constant Western progression as a way of life may surprise us. We must begin to develop methods of coping with the idea, should it become a reality. This will help us become more responsible in the future, hopefully never again able to commit the atrocities we as a civilization have committed in the past.

We are bound to make mistakes in our quest for Mars, but thinking ahead, starting now, will help minimize the effects of those mistakes. Let us proceed in a manner that will make future generations proud, and past generations forgiven. I would like to invite all of you to participate in a continuing discussion. Please feel free to email me with your ideas and comments and I will post them on my webpage. If demand warrants it, I will set up an online forum.

Thank you. Are there any questions or comments?

INTERPLANETARY BIOLOGICAL TRANSFER OF BACTERIA ENTRAPPED IN SMALL METEORITES: ANALYSIS OF BACTERIAL RESISTANCE TO IMPACT IN BALLISTIC EXPERIMENTS

C.-A. H. Roten,* A. Gallusser,† G. D. Borruat,‡ S. D. Udry,**
G. Niederhäuser,‡ A. Croxatto,‡ O. Blanc,‡ S. De Carlo,‡
C. K. Mubenga-Kabambi‡ and D. Karamata‡

The Martian origin of at least twelve meteorites, ejected into a solar orbit after a primary hypervelocity meteorite impact and subsequently captured by the Earth, clearly demonstrates that regular exchanges of crust fragments between planets take place in the solar system. Recently described, putative biological traces in one of these ejecta led to the debated proposal that life was present in Martian surface rocks. As interplanetary transfers of biological know-how may provide an explanation for the presence of life on at least two solar system bodies, survival in conditions mimicking final steps of interplanetary transfer of life forms entrapped in crust fragments was investigated with respect to small meteoroids.

From observations on the free fall of small impactors, analogous ballistic experiments can be designed and help to investigate if living cells can withstand the terminal low velocity impact. We have established that several different microorganisms survived an initial acceleration of 100'000 g and an impact in sand with a velocity of 300 to 800 m/s. Based on these experiments and on the observation that the interior of small meteoroids remains cold during the fall, we demonstrate, for. the first time, that various kinds of organisms entrapped in small ejecta can withstand (i) the heat produced by the Earth's aerobraking, reducing the preatmospheric velocity (usually between 11 and 70 km/s) to that of a free fall (125 to 250 m/s), as well as (ii) the subsequent non-explosive impact. Moreover, survival to bullet acceleration also demonstrates that bacteria entrapped in rocks can withstand the launch of a planet ejecta. The significance of our observations for the origin and the early development of life in the solar system will be discussed.

Key words: interplanetary transfer, stone-borne transfer, Mars, Earth, meteorite, impact, ballistics, bacteria, bacteriophage, lithopanspermia, panspermia.

---

* Institut de Génétique et de Biologie Microbiennes, Université de Lausanne, rue César-Roux 19, CH-1005 Lausanne, Switzerland. Http://www.unil.ch/igbm/. E-mail: claude-alain.roten@igbm.unil.ch.
† Institut de Police Scientifique et de Criminologie, Université de Lausanne, Bâdtiment de Chimie, CH-1015 Lausanne, Switzerland.
‡ Institut de Génétique et de Biologie Microbiennes, Université de Lausanne, rue César-Roux 19, CH-1005 Lausanne, Switzerland.
** Observatoire de Genève, chemin des Maillettes 51, CH-1290 Sauverny, Switzerland. http://obswww.unige.ch/

1. INTRODUCTION

1.1. Early life on the Earth.

During the first half of the period characterized by the continuous presence of life on the Earth, prokaryotes, such as bacteria and archaebacteria, were probably the only inhabitants of the planet. They were, and still are, spread all over the Earth's surface and throughout its crust [1,2]. Two billion years ago these true terraformers gradually transformed the terrestrial carbon dioxide atmosphere, similar to the current atmospheres of Venus or Mars, into the oxygen containing atmosphere of the present day Earth. Today, about 20 per cent of the greenhouse effect of the Earth's atmosphere is attributed to methane produced by archaebacteria [3].

Considering the importance of microbes for life on the Earth, microbiology and more particularly bacteriology, could provide some clues to the early stages of life on the Earth as well as on other planets.

1.2. Early life on Mars?

It was claimed that possible biological relics were present on a Martian meteorite [4], leading to the still debated proposal that bacteria were also present on early Mars [5,6,7]. The most direct way to conclusively answer this question would be to visit the red planet and analyze its sediments for the presence of bacterial tracers such as fossils or chemical compounds. Had life indeed been present on two planets of the solar system, its occurrence could be accounted for in two different ways.

Like the nineteenth century theory of the spontaneous generation of microorganisms, the first proposal argues that life, forcibly obeying the physico-chemical laws of the universe, appeared spontaneously wherever favorable conditions were present. Panspermia, the second hypothesis, considers that spontaneous generation could have been a rare, even a unique event. Presence of life on different celestial bodies would be accounted for by a continuous natural spreading of life. Presently, considering the origin of life in the universe, we are no more advanced than Louis Pasteur who at the end of the nineteenth century challenged the hypothesis of widespread spontaneous generation. Investigation of extraterrestrial, possibly biological fossil material, should shed some light on this problem. For instance, presence of terrestrial cell-building blocks, such amino acids or nucleic acid bases, in extraterrestrial samples would favor the uniqueness of spontaneous generation but also imply that interplanetary transfer of biological material must have occurred.

1.3. Is there an extraterrestrial clue to terrestrial life?

All living forms on the Earth are supposed to derive from a unique prokaryotic ancestor (Figure 1): the cenancestor [8,9]. Three major subdivisions are now recognized, i. e. i) the Eukaria or eukaryotes, comprising animals, plants, algae, fungi, and protists, ii) the Archaea or archaebacteria and iii) the Bacteria or eubacteria. The latter two taxons form the prokaryotes.

The oldest bacterial remnants are 3.85 billion years old [10]. Since the solidification of the planet's surface has probably occurred 3.9 to 4 billion years ago it is easy to understand why older fossils were never found. The only organisms responsible for the presence of oxygen in the Earth's atmosphere are cyanobacteria, previously designated as blue green bacteria. Presently, oxygen is still produced from these particular bacteria, as well as from related [11,12] endosymbiotic photosynthetic organelles called chloroplasts, or more generally plastids. Analysis of the mutation rate of RNA genes of these photosynthetic bacteria, comprised within the range of that of the entire bacterial taxon [13], sheds some light on early evolution events.

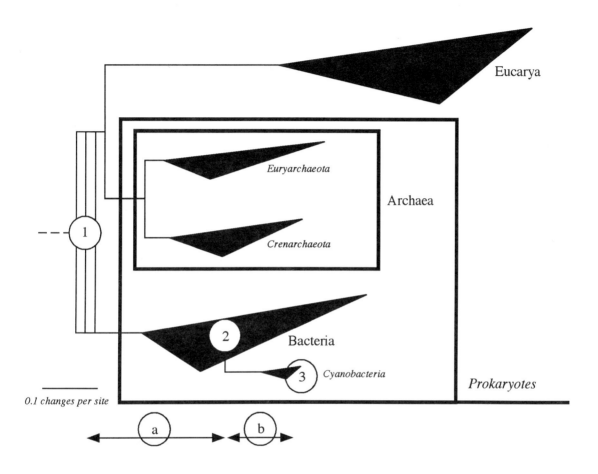

Figure 1 Universal phylogenetic tree.

This figure schematically summarizes [16] the recently published universal phylogenetic tree [13], confirmed by comparative analyses [70], and derived from the gene sequence of the small ribosomal subunit present in any living cell. Three major subdivisions are now recognized, i.e. i) the Eucaria or eukaryotes, comprising animals, plants, algae, fungi, and protists, ii) the Archaea or archaebacteria and iii) the Bacteria or eubacteria.

Only horizontal distances are meaningful and the left scale bar corresponds to a 0.1 change per nucleotide. A triangle corresponds to the number of changes for a specific taxon. Horizontal projections of the lower and upper left sides of each triangle indicate the minimum and the maximum of change rates, respectively, from the last common ancestor of the phylum to present day species.

It appears that the evolutionary distance from the cenancestor to present day cyanobacteria could be divided into two unequal parts: i) the larger one, unspecific of the blue green bacteria, represent the early evolutionary distance (a on the graph) covering the gradual evolution from the cenancestor at the origin of all living beings (number 1 on the graph) to the last ancestor non-specific of cyanobacteria (number 2) and ii) the final distance (b), specific to the blue green bacteria, which measures all the steps from the branching point (number 2) to present day cyanobacteria (number 3).

Figure 1 shows that the evolutionary distance between the cenancestor and present day cyanobacteria can be divided into two unequal parts: i) the earlier part, unspecific to blue green bacteria, is twice as long as ii) the later part, specific to cyanobacteria.

How can we measure the length of these two, clearly, different periods? The age of the oldest cyanobacteria fossils provides an estimate of the duration of the second period, *i.e.* that specific to blue green bacteria: 3.5 billion years [14].

If we suppose that life arose on the Earth it would follow that the first period, leading from inorganic matter to a living cell, lasted only 400 million years [15]. Such a conclusion would certainly imply that the mutation rate during the earlier stages of evolution would have been at least 20 times faster than the present day one.

However, if we consider that the mutation rate remained roughly constant throughout the evolution, it would appear that the cenancestor from which all Earth's life has derived, would be as old as, if not older, than our galaxy, *i.e.* over 10 billion years, implying that: i) the early evolution could not possibly have taken place on our 4.6 billion year old planet [16], ii) bacteria could be widespread at a galactic level, explaining the observed galactic distribution of biological compounds [17,18].

1.4. How could life be spread naturally throughout the universe?

Two possible kinds of biological transfers were recently summarized [19,20]. Radiopanspermia developed by Arrhenius [21,22,23], argues that unicellular organisms from the uppermost level of the atmosphere [24,25] can escape planetary gravitation through the action of electromagnetic forces and, moved by radiation pressures, travel to other stars. Lithopanspermia, proposed by H. Richter, H. von Helmholtz [26], and Lord Kelvin [27], considers that life was first brought to the Earth by contaminated meteorites.

Due to the difficulty in explaining how bacteria can escape the Earth's atmosphere and how a very low number of cells can reach a planet belonging to another stellar system, we will focus our present investigations on constrains inherent to a stone-borne transfer of biological material within our solar system.

1.4.1. Launch of stony interplanetary vehicles.

To date 12 Martian meteorites have been found on the Earth. Five hundred kilograms of rocks of Martian origin fall on the Earth each year. Recently, it has been envisaged by Zubrin and Wagner that bacterial spores could withstand a stone-borne transfer from Mars to Earth [28]:

...Despite the fact that in general each SNC meteorite must wander through space for millions of years before arrival at Earth, it is the opinion of experts in the area that neither this extended period traveling through hard vacuum, nor the trauma associated with either the initial ejection from Mars or re-entry at Earth would have been sufficient to sterilize these objects, if they had originally contained bacterial spores...

Analyses of Martian meteorites revealed that they could have been accelerated to the red planet escape velocity (5 km/s) by explosive hypervelocity impacts [29,30,31]. Using simulations [32], this acceleration is estimated to be about 15-20'000 g during 30 ms (Figure 2a, Table 1).

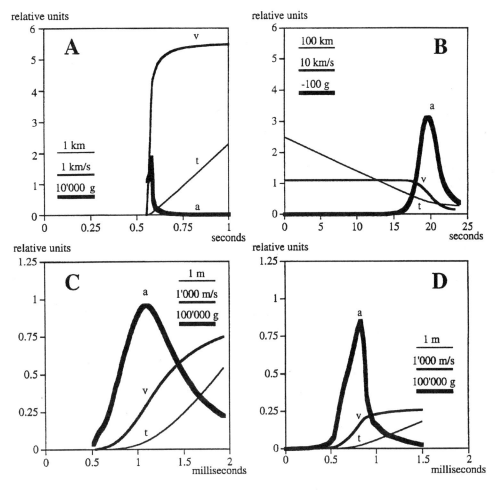

Figure 2   Travel, velocity and acceleration curves of planet ejecta and bullets.
In each graph, light, plain, and bold lines represent the travel (t), velocity (v), and acceleration (a) curves, respectively. The values of the relative unit are indicated in each figure.

- **2a: Launch of Martian ejecta.** The travel and acceleration curves were deduced from the calculated velocity curve [32] of Martian ejecta of 0.5 m of diameter, located primarily at 3 km of the impact point and launched after a 25 km/s impact of ice on andesite.
- **2b: Re-entry of a small object.** From published data [38], we have drawn the travel, velocity and acceleration curves of a small satellite (10 kg, 0.25 cm of radius) vertically penetrating the Earth's atmosphere. The velocity curve is similar to that calculated for meteoroids [35].
- **2c: Travel, velocity and acceleration of a bullet in a military rifle.** Kindly provided by Kneubuehl, a pressure curve of a GP11 cartridge was used to estimate [16] the bullet acceleration in the barrel (0.76 m) of the infantry rifle model 11 [71] routinely used in our experiments (Table 1). The shape of the pressure curve, usually similar to the acceleration curve [72] allowed us to derive a relative velocity curve. From the muzzle velocity of 805 m/s [43,71], an absolute velocity curve can be deduced. The latter, in turn, allows to estimate the absolute travel and acceleration curves of the bullet in a barrel.
- **2d: Travel, velocity and acceleration of a bullet in an air rifle.** As in 2c, an observed muzzle velocity of 260 m/s and a pressure curve, kindly provided by A. Wirth [73] and similar to a published one [74] has allowed us to estimate the absolute travel, velocity, and acceleration of a diabolo 4.5 mm pellet in a Diana 45 air rifle, a little less powerful than the Diana 48 used in our experiment (Table 1).

Table 1
COMPARISON BETWEEN ACCELERATIONS AND VELOCITIES INVOLVED IN SMALL STONY INTERPLANETARY TRANSFERS AND BALLISTICS EXPERIMENTS

	Maximum acceleration:	Initial velocity: (m/s)	Terminal velocity: (m/s)
Meteorites:			
Launch of small Martian ejecta [32].	20'000 g (30 ms)[1]	0	5'500
Terrestrial aerobraking during the re-entry of a small satellite, similar to a meteoroid, striking perpendicularly the crust[2] [38].	-300 g (3 s)[1]	11'000	125-250
Supposing a constant[3] deceleration in sand during a free fall impact of a rock having a 3.3 density and a 9 cm radius.	-5'300 g (4.8 ms)[4]	250	0
Ballistic projectiles:			
Acceleration of a 7.5 mm bullet in a rifle: Swiss military rifle 1889, GP90 bullet[5].	not measured	0	590 [43,71]
Infantry rifle model 11, GP11 bullet[5].	100'000 g (0.8 ms)[1]	0	805 [43,71]
Acceleration of a pellet in an airgun using 4.5 mm diabolo pellets: Diana 45[6], (495 mm barrel)	85'000 g (0.25 ms)[1]	0	260[7] [73]
Diana 48[6][8], (440 mm barrel)	not measured	0	298[9] ± 3.6[10]
Supposing a constant[3] deceleration in sand of a pellet fired by a Diana 48.	-79'000 g (0.39 ms)[4]	300	0

This table summarises the kinetic parameters from the literature and Figure 2.

1) The times during which 50 % of the maximal acceleration has lasted was estimated from Figure 2.
2) Similar for the same object, the Earth's and Venus aerobraking decelerations are higher than that of Mars [38].
3) The stop distance for the meteorite and the pellet amounting to 0.6 and 0.058 m, respectively, was determinated according to the formula of Didion [52].
4) Calculated period of constant deceleration.
5) GP stand for Gewehrpatrone: GP 90 or GP11 cartridges are loaded with 2 or 3.5 g of smokeless powder and their oblong missiles are weighing 13.8 and 13.2 g, respectively.
6) Airguns are made by Mayer & Grammelspacher, Dianawerk GmbH & Co, Rastatt, Germany [76].
7) The measured muzzle velocity differs from the maximum theoretical speed of 280 m/s [73].
8) In all our experiments we have used pellets produced by Haendler & Natermann GmbH, Hannover Münden, Germany [76].
9) The theoretical maximum muzzle velocity of 320 m/s is lower than that measured experimentally with a specialised device using infrared beams (LS energy, Lahrmann & Schümann Sondergerätebau GbR, Hemer, Germany).
10) The standard error was determined after 15 shots.

1.4.2. Space mechanics of meteoroids.

Computer simulations suggest that Martian ejecta could remain in space for several million years before falling on the Earth [33]. All objects crossing the Earth's orbit generally have a velocity of 40 km/s. As the Earth is orbiting the Sun at about 30 km/s, an impactor could hit the terrestrial surface at a speed ranging from 11 to 70 km/s, depending on the relative direction of the Earth and the impactor movements, the minimum speed being equivalent to the escape velocity of the targeted planet (2.4, 5 and 11.2 km/s, for the Moon, Mars, and the Earth, respectively).

1.4.3. Kinetics of the fall.

In the case of an atmosphereless body of the solar system, such as the Moon, future impactors with diameters ranging from microns to kilometers will impact the surface at a speed of a few km/s. A meteoroid moving at a speed of 3 km/s possesses a kinetic energy equal to that released by the explosion of an equivalent weight of TNT. These explosive impacts create round craters independent of the fall angle [34].

The presence of a terrestrial atmosphere does not affect the initial velocity of very massive falling objects [35]. Hypervelocity impacts of that type occur on Mars and are responsible for the ejection of Martian meteorites recently found on the Earth. Smaller objects weighing more than several tons have their fall velocity affected by the transit through the Earth's atmosphere but nevertheless strike the ground at a hypersonic speed of a few km/s [36,37].

On the Earth, falling objects weighing less than a few tons have their initial preatmospheric velocity reduced to that of a free fall. This important deceleration represents a few hundred g's [38]. During the free fall, the aerobraking of these smaller masses is able to counterbalance gravity by drag deceleration [37]. Several observations confirmed that these boulders hit the surface at a subsonic velocity of a free fall [35,36,39], ranging from 125 to 250 m/s (Figure 2b, Table 1). Explosions are not produced by these low-velocity impacts. The final deceleration needed to stop such a small impactor is estimated to represent a few thousand g's (Table 1).

Due to severe stress during the deceleration, small friable meteoroids disintegrate while tougher boulders can be dismantled along pre-existing cracks and spread as a meteorite shower. Finally, no terrestrial impacts are produced by objects less than a few centimeters wide [40].

2. STONE-BORNE BIOLOGICAL TRANSFER?

Could planet ejecta be a possible vehicle for life? Since spores of bacilli, known to be the living forms most resistant to mechanical stress, were unable to withstand the explosion of an impact of several km/s produced by a powder-gas gun [41], only the free fall impacts of small impactors ranging from 125 to 250 m/s will be considered in this communication. A survey of the literature reveals that ballistic experiments are appropriate to address this question [16,42,43]: projectiles undergo in a shorter time, i) accelerations (75'000 to 100'000 g) higher than those calculated for the launch of small Martian ejecta or during their final terrestrial aerobraking and impact, and ii) velocities (300 to 800 m/s) greater than those involved in a free fall impact (Figure 2c, 2d, Table 1).

In summary, if bacteria entrapped in rocks are able to withstand the mechanical stress of the ballistic experiments, they ought to be able to survive the launch and a low-velocity impact of a small planetary meteoroid.

2.1. Survival of organisms subjected to the mechanical stress experienced by small planet ejecta falling on the Earth

2.1.1. Bullet experiments.

Already towards the end of the nineteenth century, a medical report pointed out that eukaryotic cells could survive a spreading by a bullet [44,45]. A few years later, several recently reviewed [16] medical studies, have systematically addressed this question. In 1892 in the USA, Lagarde was the first to observe that bacteria were able to withstand the impact of a bullet [46]. However, the possible presence of spores could have affected the outcome of the experiment. At the same time in Germany, Messner working with purified non-sporogenic bacteria, was apparently the first scientist able to unambiguously demonstrate that vegetative bacterial cells do resist such an impact [47]. In 1895 in Switzerland, Pustoschkin confirmed

Messner's results [48,49]. Due to a precise description of the ballistic experiment, an up to date analysis of the physical conditions endured by bacteria was possible [16]. Actually Pustoschkin showed: i) that bacteria spread on a bullet withstand an impact performed at 590 m/s, ii) when bacteria are spread in the barrel, a sterile bullet is capable of contaminating the target, iii) bacteria survived the impact with or without the preheating of the barrel, and finally iv) Gram-positives are more resistant than Gram-negative Bacteria which correlates well with the relatively stronger cell wall of the former [16]. These experiments clearly revealed that bacteria can easily survive bullet impacts. Results of the earlier experiments were recently confirmed in several studies reviewed in Table 2 [50,51].

Table 2
COMPARISON OF DIFFERENT BALLISTIC EXPERIMENTS MEASURING THE IMPACT RESISTANCE OF BACTERIA

Bullet velocity (m/s)			430, 670	590	490		805
Calibre (mm)			7, 11	7.5			7.5
Firing distance (m)		3	125, 250	7	7.6	1	7
Target		iron		gm	gm	sand	gm
References		[46]	[47]	[48,49]	[50]	[51]	this work
Organisms:							
Gram-positive bacteria:							
Bacillus subtilis		nd	nd	+	nd	nd	+ (168)
Bacillus anthracis		+	nd	nd	nd	nd	nd
Streptococcus pyogenes		nd	nd	+	nd	nd	nd
Staphylococcus aureus (*Streptococcus pyogenes*) [47,77]		nd	+	nd	nd	+	nd
Gram-negative bacteria:							
Escherichia coli (*Bacillus coli*) [48,78]		nd	nd	+	nd	nd	+ (DH5α)
Serratia rubidaea (*Bacillus prodigiosus*) [47,77]		nd	+	nd	+	nd	nd
Serratia plymuthica (*Bacillus ruber*) [48,77]		nd	nd	+	nd	nd	nd
Putative *Pseudomonas* isolated from green pus [47]		nd	+	nd	nd	nd	nd
Eukaryotes:							
The yeast *Saccharomyces cerevisiae*		nd	nd	nd	nd	nd	+ (FY1679)

The table summarises results already published or obtained in this contribution. Velocity, calibre, firing distance, and type of target used in different ballistic experiments are indicated in the upper part of the table. Organisms are shown in the lower part. When relevant previous names of bacteria and relevant references.

In our conditions, the bullet and the barrel of the infantry rifle model 11 (Table 1) were systematically cleaned by a solution of 50 % ethanol. Living cells were spread on the very tip of the GP11 bullet and fired into a 7-metres-distant sterilised target containing, trapped in 2 aluminium foils, a 3-cm-thick deposit of sand and the same width of a solid culture medium (7.5% agar). Decelerated by the passage through sand layer, the bullet flew through the nutrient medium: LB or YPD [79] for bacteria or yeast respectively. Targets were collected sterilely and incubated in an oven at 30°C.

Strains used for our experiments are indicated in brackets. Three shots for each organism were routinely performed. After each shot, except when a sterile bullet was used, colonies had appeared on the target medium.

Abbreviations: gm, growth medium; nd, not determined; +, growth after the firing.

We have repeated Pustoschkin's experiments under even more stringent conditions. Similar in shape and in caliber, the GP11 cartridges are more powerful than the GP90 used in her shots (Table 1) giving, in our conditions, a higher muzzle velocity, i. e. 805 instead of 590 m/s [43]. As shown on Figure 2c, the GP11 bullet undergoes a 100'000 g acceleration half a millisecond after the firing.

All microorganisms used in our experiments, *i.e.* the Gram-positive *Bacillus subtilis*, the Gram-negative *Escherichia coli*, and yeast, an eukaryote, were able to withstand this harsher treatment (Table 2, last column), confirming earlier observations. They survive i) the acceleration in a military rifle which is 5 times greater than that withstood by a boulder escaping Mars, as well as ii) an impact performed at a velocity 3 to 5 times that of a small meteoroid impacting the Earth.

However, these observations do not allow a quantitation of the survival, specific to different organisms.

2.1.2. Pellet experiments.

Comparing the survival of different species requires the bullet to be completely stopped in a sterile target which can be achieved only by the use of weaker energies. We designed another experiment employing a powerful airgun, firing lead pellets of 4.5 mm caliber and weighing 0.56 g at velocities higher than that of a free fall. Its muzzle velocity of 300 m/s mimics the impact velocity of a small meteoroid. Using a pressure curve measured with a little less powerful airgun (Figure 2d), we estimate that in our air rifle, the projectile is first accelerated to more than 85'000 g in less than 0.5 ms (Figure 2d). In our conditions, the pellet is stopped by a negative deceleration of at least 75'000 g in less than 0.4 ms (Table 1) in an approximately 5-centimeter-thick sand layer, corresponding to the calculated stopping distance [52].

As summarized in Table 3, *B. subtilis* and yeast exhibit a significant level of resistance i) to acceleration 4 times stronger than those involved in the satellization of Martian ejecta and ii) to impacts with velocities above those of a small impactor striking the Earth's surface.

Table 3

COMPARISON OF THE SURVIVAL OF DIFFERENT ORGANISMS TO BALLISTIC IMPACTS

	Percent survival:
Eukaryotes:	
Saccharomyces Cerevisiae FY1679	8.9 ± 3.4
Gram-positive bacteria:	
Bacillus subtilis 168	
Spores	100 ± 12.6
Vegetative cells	52 ± 18.5
Protoplasts	0.21 ± 0.03
Gram-negative bacteria:	
Escherichia coli DH5a	7.7 ± 5.5
Bacterial phages:	
Bacteriopages T4D infecting *E. coli*	0.25 ± 0.17

Ribbed-skirt pointed-head diabolo pellets of 4.5 mm (0.177 in) calibre were used. A hole was drilled at the very tip of pellets, allowing to load 1 million of microorganisms in 1 microlitre of a cell culture. Organisms were sealed with a 40 % gelatine liquid solution. To solidify the gelatine, pellets were kept during 10 s in liquid nitrogen before being fired onto a sterile target. Pellets were loaded in the Diana 48 airgun. For each organism 3 shots were routinely performed. Pellets were shot into autoclaved bottles containing small gravels. Cells were recovered by adding a growth medium and subsequently collecting the samples. An absolute viable count was obtained by counting colony forming units. Viability controls were performed by manually introducing pellets directly in a gravel containing bottle. Survival was estimated by comparing the number of cells transferred by shooting to that obtained without shooting. Routinely checked, sterility did not reveal any contamination. Protoplast preparations of *B. subtilis* and their titration on DM3 plates were performed according to standard procedures [80]. Concerning the preparation of T4D phages see Figure 3.

Loaded at tip of the pellet, cells were ejected. Knowing that bacterial spores were able to withstand impacts at a few km/s [Wh43], spore preparations were used to estimate the transfer rate which was found to be 12 %. Therefore absolute counts were normalised with respect to the latter figure.

B. subtilis spores are the toughest living forms used in these experiments. Their vegetative cells are 5 times more resistant than those of other organisms used in our experiment, confirming Pustoschkin's results [16,48]. This difference is most likely due to the robust cell wall of Gram-positives. No survival was revealed in control experiments performed i) with *B. subtilis* protoplasts, *i.e.* viable wall-less forms of this bacterium, able to form colonies under cell wall regeneration conditions or ii) with bacteriophage T4, an *E. coli* bacterial virus. The latter observations are clearly confirmed by electron micrographs which reveal that this mechanical treatment can disrupt phage heads (Figure 3).

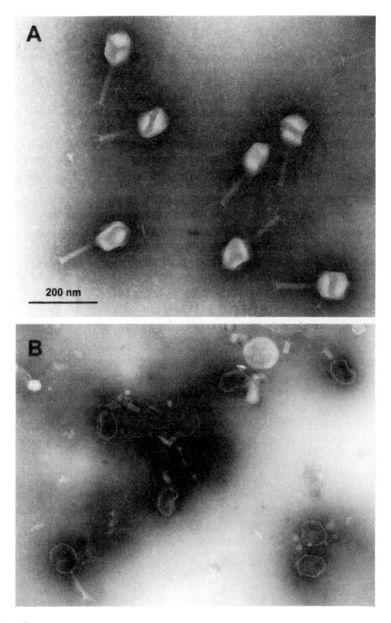

Figure 3

*2.1.3. Survival of organisms entrapped in small rocks,
ejected from Mars and falling on the Earth.*

Ballistic experiments clearly demonstrate that, when entrapped in rocks, living cells endowed with cell wall, such as bacteria or yeast, have the capacity to survive i) accelerations stronger than those experienced by small ejecta generated from a terrestrial planet by explosive impacts, ii) the aerobraking in the atmosphere of the possible impacted planet, iii) the low-velocity impact of a small meteoroid, and iv) impact velocities greater than those of small Martian ejecta striking the Earth's surface at a free fall speed.

These observations, taking into account the different severe conditions withstood by extraterrestrial planetary meteorites, provide, for the first time, the demonstration that cells possessing a cell wall, entrapped in small rocks, are able to survive the mechanical stress of a stone-borne interplanetary transfer.

Moreover, our proposal that living forms entrapped in rocks are resistant to accelerations comparable to those involved in the ejection of Martian meteorites has been clearly confirmed by independent experiments [53]. The latter have shown that bacteria such as *B. subtilis* and *Deinococcus radiodurans*, sterilely loaded into cannon projectiles, fired in closed tubes, and gently stopped by an air cushion, are able to withstand an acceleration ranging from 11'500 to 33'800 g.

2.2. Survival in space of organisms entrapped in planet ejecta

During a several million years' stay in space [33] and in vacuum, prokaryotes could well remain viable [16]. Indeed, i) staphylococci and spores of bacilli remained viable for 25 to 40 million years in amber [54,55], whereas ii) freeze-drying is very efficient for the storage of prokaryotes [56,57], and iii) their presence inside a rock should protect them from external radiation [58,59,60].

2.3. Survival of organisms to the heat generated by aerobraking of planet ejecta

During the entry into the atmosphere, only a small part of the friction energy heats the small meteoroid, while most of it warms the air. The surface melts at initial high velocities and fusion drops are lost in the air. After the cooling down of the surface during the free fall, a fusion crust appears. Easily recognizable, thermal effects are usually several millimeters deep in a poorly heat-conducting stone. Due to the insulation of the surrounding material it was observed that the meteorite interior remained cool even if its surface was melting [35,40]. Therefore, cells entrapped in a boulder could easily withstand the heat generated by aerobraking [16].

3. DISCUSSION AND CONCLUSION

3.1. Importance of interplanetary transfer resistance for the early life in the solar system

In conclusion, these experiments and observations confirm our earlier theoretical proposal [16] that bacteria are able to withstand the entire journey of a stone-borne interplanetary transfer. The same conclusion, drawn independently, is presented in an accompanying paper [60]. Moreover, our data show that this property seems to be widespread in the living world. Simple eukaryotes, like yeast, behave like vegetative cells of various bacteria in being able to withstand the various kinds of mechanical stress occurring during stony interplanetary transfer.

If the bacterial or yeast cell wall, primarily composed of polymerized peptidoglycan or chitin fibers respectively, can protect these organisms during a stone-borne space travel, the occurrence on the Earth of bacteria devoid of peptidoglycan such as mycoplasma is intriguing. The simplest explanation would be that the loss of the cell wall occurred on the Earth.

Failing to resist the impact experiments when transferred as free particles, bacteriophages could have well traveled in space in lysogenic form, their genomes being either integrated in a bacterial chromosome or present as plasmids, thus mechanically protected by the host cell wall.

3.2. Importance of biological planetary transfer for the early life in the solar system

The ability of single cells to survive planetary transfer strongly pleads in favor of similar life forms on the Earth and on Mars. Life could have spread through the solar system by transfer from planet to planet.

For the early life on Earth, resistance to impact was certainly an important property. It was estimated that more than ten impact events could have vaporized the early terrestrial oceans [61,62]. However, bacteria present on Earth's ejecta could have survived in space and later, upon falling back onto the terrestrial surface, recolonized the planet under more appropriate conditions [16].

As discussed in the introduction and elsewhere [16], resistance of bacteria to acceleration, space travel, and impact could well explain why life is likely to be older than the solar system, a property relevant to the origin of life in our solar system. If life appeared before the latter, we would also predict that bacterial forms as we know them could be widespread in our galaxy. As for the Earth, such bacteria could terraform extrasolar telluric planets or satellites [63] orbiting extrasolar giant gaseous planets [64,65]. Discovery of an oxygen rich atmosphere on an extrasolar planet in a near future [66] may provide one of the first clues of a galactic distribution of bacteria similar to their terrestrial counterparts.

3.3. Relevance of biological planetary transfer for future explorations of Mars

If putative bacterial life on Mars has had a common origin to that of the Earth while being constantly transferred to our planet without apparently affecting in a major way the Earth's inhabitants, we would predict, in opposition to an alarmist communication [67], that putative Martian life forms would be unlikely to threaten the health of human explorers of the red planet. Moreover, terraforming of Mars using terrestrial bacteria would have to be simply considered as an acceleration of a process occurring naturally in the solar system and in the universe.

ACKNOWLEDGMENTS

During the 1996-1997 academic year GN, AC, OB, and SDC were undergraduate students preparing a B. Sc. degree in biology at Lausanne University [68,69]. We are particularly grateful to C. Mileikowsky [53,60] for sharing unpublished results in spring 1998, M. Morel for technical assistance, and H. M. P. Pooley for coinage of "stone-borne", qualifying our transfer concept.

REFERENCES

1. T. Gold, "The Deep, Hot Biosphere," *Proc. Natl. Acad. Sci. USA*, 89, 6045-6049 (1992).
2. K. Pedersen, "Microbial Life in Deep Granitic Rock," *FEMS Microbiology Reviews*, 20, 399-414 (1997).
3. L. E. Joseph, *Gaia, the Growth of an Idea*, St Martin's Press, New York, 1990, 276 p.
4. D. S. McKay, E. K. Jr. Gibson, K. L. Thomas-Keprta, H. Vali, C. S. Romanek, S. J. Clemett, X. D. F. Chillier, C. R. Maechling, and R. N. Zare, "Search for Past Life on Mars: Possible Relic Biogenic Activity in Martian Meteorite ALH84001," *Science*, 273, 924-930 (1996).
5. J. P. Bradley, R. P. Harvey, and H. Y. Jr. McSween, "No 'Nanofossils' in Martian Meteorite," *Nature*, 390, 454 (1997).
6. D. S. McKay, E. Jr. Gibson, K. Thomas-Keprta, and H. Vali, "Reply to: No 'Nanofossils' in Martian Meteorite," *Nature*, 390, 455 (1997).
7. J. L. Bada, D. P. Glavin, G. D. McDonald, and L. Becker, "A Search for Endogenous Amino Acids in Martian Meteorite ALH84001," *Science*, 279, 362-365 (1998).
8. W. M. Fitch and K. Upper, "The Phylogeny of tRNA Sequences Provides Evidence for Ambiguity Reduction in the Origin of the Genetic Code," *Cold Spring Harbor Symposia on Quantitative Biology*, 52, 759-767 (1987).
9. S. Freeman and J. C. Herron, *Evolutionary Analysis*, Prentice Hall, Upper Saddle River, NJ, 1998, 786 p.
10. S. J. Mojzsis, G. Arrhenius, K. D. McKeegan, T. M. Harrison, A. P. Nutman, and C. R. L. Friend, "Evidence for Life on Earth before 3'800 Million Years ago," *Nature*, 384, 55-59 (1996).
11. L. Margulis, *Symbiosis in Cell Evolution: Life and its Environment on the Early Earth*, W. H. Freeman and Co., San Francisco, 1981, 419 p.
12. L. Margulis, *Symbiosis in Cell Evolution; Microbial Communities in the Archean and Proterozoic Eons*, W. H. Freeman and Co, New York, 1993, 452 p.
13. N. R. Pace, "A Molecular View of Microbial Diversity and the Biosphere," *Science*, 276, 734-740 (1997).
14. J. W. Schopf, "Microfossils of the Early Archean Apex Chert: New Evidence of the Antiquity of Life," *Science*, 260, 640-646 (1993).
15. A. Lazcano, "The Tempo and Mode(s) of Prebiotic Evolution," in *Astronomical and Biochemical Origins and the Search for Life in the Universe, Proceedings of the 5th International Conference on Bioastronomy*; IAU Colloquium No. 161, Capri, July 1-5, 1996, ed. by C. B. Cosmovici, S. Bowyer, and D. Werthimer, Editrice Compositori, Bologna, 419-429 (1997), 814 p.
16. C.-A. H. Roten, A. Gallusser, G. D. Borruat, S. D. Udry, and D. Karamata, "Impact Resistance of Bacteria Entrapped in Small Meteorites," *Bull. Soc. Vaud. Sc. Nat.*, 86.1, 1-17 (1998). See also: http://www.unil.ch/igbm/bsvsn_bact&impact.html.
17. F. Hoyle and C. Wickramasinghe, "Where Microbes Boldly Went," *New Scientist*, 91, 412-415 (1981).
18. F. Hoyle and C. Wickramasinghe, *Space Travellers: The Bringers of Life*, University College Cardiff Press, 1981, 197 p.
19. P. Parsons, "Dusting off Panspermia," *Nature*, 383, 221-222 (1996).
20. S. Glasstone, *The Book of Mars*, NASA SP-179, Washington D. C. 1968, p. 315.
21. S. A. Arrhenius, "Die Verbreitung des Lebens im Weltenraum," Die Umschau, 7, 481-485 (1903), or see also "The Propagation of Life in Space," translated by D. Goldsmith, in *The Quest for Extraterrestrial Life*, ed. by D. Goldsmith, University Science Books, Mill Valley, Ca, 32-33 (1980), 308 p.
22. S. A. Arrhenius, "Panspermy: the Transmission of Life from Star to Star," *Scientific American*, 96, 196 (March 2, 1907).
23. S. A. Arrhenius, *Worlds in the Making*, Harper & Bros., London & New York (1908), 230 p., translated by Dr. H. Borns from *Världarnas Utveckling*, Gebers, Stockholm 1906, 181 p.
24. P. H. Gregory, *The Microbiology of the Atmosphere; Plant Science Monographs*, Leonard Hill [Books] Limited, London, Interscience Publishers, Inc, New York, 1961, 251 p.
25. A. A. Imshenetsky, S. V. Lysenko, and G. A. Kazakov, "Upper Boundary of the Biosphere," *Applied and Environmental Microbiology*, 35, 1-5 (1978).
26. H. von Helmholtz, "Helmholtz on the Use and Abuse of the Deductive Method in Physical Science," preface to the second part of the German edition of Thomson and Tait's *Natural Philosophy*, vol. I, translated by Crum Brown, *Nature*, 11, 149-151 (December 24, 1874), 11, 211-212 (January 14, 1875).

27. W. Thomson (later Lord Kelvin), "At the British Association for the Advancement of Science. Inaugural Address of Sir William Thomson, LL. D., F. R. S., President," *Nature*, 4, 262-270 (1871).
28. R. M. Zubrin, and R. S. Wagner, *The Case for Mars: The Plan to Settle the Red Planet and Why We Must*, The Free Press, New York, 1996, p. 328.
29. H. J. Melosh, "Impact Ejection, Spallation, and the Origin of Meteorites," *Icarus*, 59, 234-260 (1984).
30. J. D. O'Keefe, and T. J. Ahrens, "Oblique Impact: A Process for Obtaining Meteorite Samples from Other Planets," *Science*, 234, 346-349 (1986).
31. A. M. Vickery and H. J. Melosh, "The Large Crater Origin of SNC Meteorites," *Science*, 237, 738-743 (1987).
32. A. Vickery, "Effect of an Impact-Generated Gas Cloud on the Acceleration of Solid Ejecta," *J. of Geophys. Res.*, 91, B14, 14'139-14'160 (1986).
33. B. J. Gladman, J. A. Burns, M. Duncan, P. Lee, and H. F. Levison, "The Exchange of Impact Ejecta between Terrestrial Planets," *Science*, 271, 1387-1392 (1996).
34. K. Mark, *Meteorite Craters*, The University of Arizona Press, Tucson, 1987, 288 p.
35. J. T. Wasson, *Meteorites: Their Record of Early-Solar System History*, W. H. Freeman and Co, New York, 1985, 267 p.
36. V. F. Buchwald, "Physics of the Fall," in *Iron Meteorites*, vol. 1, University of California Press, Berkeley, Los Angeles, London, 15-23 (1975), 243 p.
37. F. Heide and F. Wlotzka, *Meteorites, Messengers from Space*, Springer-Verlag, Berlin, New York, 1995, 231 p.
38. W.H.T. Loh, *Dynamics and Thermodynamics of Planetary Entry*, Prentice-Hall, Englewood Cliffs, NJ, 1963, 268 p.
39. H. Y. Jr McSween, *Meteorites and their Parent Planets*, Cambridge University Press, Cambridge, UK, 1987, 237 p.
40. J. K. Wagner, *Introduction to the Solar System*, Saunders College Publishing, Philadelphia, 1991, 453 p.
41. O. Whitfield, E. L. Merek, and V. I. Oyama, "Effect of Simulated Lunar Impact on the Survival of Bacterial Spores," *Space Life Science*, 4, 291-294 (1973).
42. H. Thadepalli and A. K. Mandal, *Antimicrobial Therapy in Abdominal Surgery; Precepts and Practices*, CRC Press Boca Raton, Ann Arbor, 1991, 105 p.
43. K. G. Sellier and B. P. Kneubuehl, *Wound ballistics and the Scientific Background*, Elsevier, Amsterdam, 1994, 479 p.
44. "A Veracious Chronicle," *Lancet*, 1, 35, (1875).
45. A. C. Gordon and R. D. Spicer, "Impregnated via a Bullet?" *Lancet*, 1, 737 (1989).
46. L. A. Lagarde, "Can a Septic Bullet Infect a Gunshot Wound ?" *The New York Med. J.*, 56, 458-464 (1892).
47. Messner (München), Comments to Bruns (Tübingen). "Über die kriegschirurgische Bedeutung der neuen Feuerwaffen", (Referat.), Bericht über die Verhandlungen der Deutschen Gesellschaft für Chirurgie, XXI. Kongress, 8-11 June 1892, Langenbeck-Hause. *Beilage zum Centralblatt für Chirurgie*, 32, 1-25 (1892).
48. N. Pustoschkin, *Versuche über Infektion durch Geschosse*, Obrecht & Käser, Bern, Switzerland, 1895, 24 p.
49. L. A. Lagarde, "Poisoned Wounds by the Implements of Warfare," *J. of the Amer. Med. Assoc.*, 40, 984-990 (April 11, 1903) and 40, 1062-1069 (April 18, 1903).
50. F. P. Thoresby and H. M. Darlow, "The Mechanism of Primary Infection of Bullet Wounds," *British J. of Surgery*, 54, 359-361 (1967).
51. A. W. Wolf, D. R. Benson, H. Shoji, P. Hoeprich, and A. Gilmore, "Autosterilization in Low Velocity Bullets", *Journal of Trauma*, 18, 63 (1978).
52. R. B. Baldwin, *The Face of the Moon*, The University of Chicago Press, 1949, 239 p.
53. C. Mileikowsky, E. Larsson, and B. Eiderfors, "Experimental Investigation of the Survival of *Bacillus subtilis* Spores and Vegetative Cells and of *Deinococcus radiodurans*, Accelerated with Short Rise Times to Peak Accelerations of 11'500 g, 17'700 g and 33'800 g,", (paper presented at the August 13-16, 1998 Mars Society Founding Convention, Boulder, CO).

54. R. J. Cano and M. K. Borucki, "Revival and Identification of Bacterial Spores in 25- to 40-Million-Year-Old Dominican Amber," *Science*, 268, 1060-1064 (1995).
55. L. H. Lambert, T. Cox, K. Mitchell, R. A. Rossello-Mora, C. D. Cueto, D. E. Dodge, P. Orkand, and R. J. Cano, "Staphylococcus succinus sp. nov., Isolated from Dominican Amber," *Int. J. Syst. Bacteriol.*, 48, 511-518 (1998).
56. D. M. Portner, D. R. Spiner, R. K. Hoffman, and C. R. Phillips, "Effect of Ultrahigh Vacuum on Viability of Microorganisms," *Science*, 134, 2047 (1961).
57. F. J. Mitchell and W. L. Ellis, "Surveyor 3: Bacterium Isolated from Lunar-retrieved Television Camera," in *Analysis of Surveyor 3 Material and Photographs Returned by Apollo 12*, National Aeronautics and Space Administration, Washington D. C., 239-248 (1972), 295 p.
58. P. Weber and J. M. Greenberg, "Can Spores Survive in Interstellar Space ?" *Nature*, 316, 403-407 (1985).
59. C. Mileikowsky, "Can Spores Survive a Million Years in the Radiation in Outer Space? Part I: Protection against Photons above 0.5 keV," in *Astronomical and Biochemical Origins and the Search for Life in the Universe, Proceedings of the 5th International Conference on Bioastronomy*, IAU Colloquium No. 161, Capri, July 1-5, 1996, ed. by C. B. Cosmovici, S. Bowyer, and D. Werthimer, Editrice Compositori, Bologna, 545-552 (1997), 814 p.
60. C. Mileikowsky, F. A. Cucinotta, J. W. Wilson, B. Gladman, G. Horneck, L. Lindegren, J. Melosh, H. Rickman, and M. Valtonen, "Natural Transfer of Viable Microbes in Space; Part 1: From Mars to Earth and Earth to Mars," (paper presented at the August 13-16, 1998 Mars Society Founding Convention, Boulder, CO).
61. K. A. Maher and D. J. Stevenson, "Impact Frustration of the Origin of Life," *Nature*, 331, 612-614 (1988).
62. N. H. Sleep, K. J. Zahnle, J. F. Kasting, and H. J. Morowitz, "Annihilation of Ecosystems by Large Asteroid Impacts on the Early Earth," *Nature*, 342, 139-142 (1989).
63. D. M. Williams, J. F. Kasting, and R. A. Wade, "Habitable Moons around Extrasolar Giant Planets," *Nature*, 385, 234-236 (1997).
64. M. Mayor and D. Queloz, "A Jupiter-Mass Companion to a Solar-type Star," *Nature*, 378, 355-359 (1995).
65. M. Mayor, D. Queloz, S. Udry, and J.-L. Halbwachs, "From Brown Dwarfs to Planets," in *Astronomical and Biochemical Origins and the Search for Life in the Universe, Proceedings of the 5th International Conference on Bioastronomy*, IAU Colloquium No. 161, Capri, July 1-5, 1996, ed. by C. B. Cosmovici, S. Bowyer, and D. Werthimer, Editrice Compositori, Bologna, 313-330 (1997), 814 p.
66. J. R. P. Angel and N. J. Woolf, "Searching for Life on Other Planets," *Scientific American*, 274, 46-52 (April 1996).
67. J. Kaiser, "Preventing a Mars Attack," *Science*, 279, 1309 (1998).
68. O. Blanc, A. Croxatto, and G. Niederäuser, Etude de la Résistance Mécanique des Bactéries à un Impact de 300 m/s, supervised by Claude-Alain Roten and Dimitri Karamata, scientific research report of the Institut de Génétique et de Biologie Microbiennes, Lausanne University, Switzerland, 1997, 24 p.
69. S. De Carlo, Le rôle des Fibres du Bacteriophage T4 lors de la Sédimentation. Étudié par Ultracentrifugation Analytique et Microscopie Électronique, coworker J.-L. Barblan, supervised by E. Kellenberger, and D. Karamata, scientific research report of the Institut de Génétique et de Biologie Microbiennes, Lausanne University, Switzerland, 1997, 50 p.
70. H. P. Klenk, L. Zhou, and J. C. Venter, "Understanding Life on this Planet in the Age of Genomics," in *Instruments, Methods, and Missions for the investigation of Extraterrestrial Microorganism*, conference held in San Diego, CA, 29 July-1 August 1997, ed. by R. B. Hoover, Proceedings of SPIE, 3111, 306-317 (1997), 530 p.
71. F. Pellaton, *Aide-Mémoire sur l'Armement Fédéral 1817-1975*, 3rd edition (1985), 94 p.
72. L. Stiefel, "Pressure-Time-Velocity-Travel Relationship in Typical Gun Systems," in *Gun Propulsion Technology*, ed. by L. Stiefel, *Progress in Astronautics and Aeronautics*, 109, American Institute of Aeronautics and Astronautics Inc., Washington, D. C., 61-74 (1988), 563 p.
73. A. Wirth, Mayer & Grammelspacher, Dianawerk GmbH & Co, Rastatt, Germany, Personal communication, March 1998.
74. G. V. Cardew and G. M. Cardew, *The Airgun, from Trigger to Target*, published by G. V. & G. M. Gardew, 1995, 235 p.
75. S. Brenner and R. W. Horne, "A Negative Staining Method for High Resolution Electron Microscopy of Viruses," *Biochim. Biophys. Acta* 34, 103-110 (1959).
76. J. Walter, *The Airgun Book*, Arms and Armour Press, London, 1981, 146 p.

77. ATCC, web address: http://www.atcc.org/ (1998).
78. F. Ørskov, "*Escherichia*, Castellani and Chalmers 1919," in *Bergey's Manual of Determinative Bacteriology*, 8th edition, ed. by R. E. Buchanan, and N. E. Gibbons, S. T. Cowan, J. G. Holt, J. Liston, R. G. E. Murray, C. F. Niven, A. W. Ravin, and R. Y. Stanier, William & Wilkins Company, Baltimore, 293-296 (1974), 1268 p.
79. *Current Protocols in Molecular Biology*, ed. by F. M. Ausubel, R. Brent, R. E. Kingston, D. S. Moore, J. G. Seidman, J. A. Smith, and K. Struhl, John Wiley & Sons, USA, (1997).
80. S. Bron, A. D. Gruss, P. Haima, C. R. Harwood, S. Holsappel, L. Jannière, J. Kok, L. Oskam, A. Palva, W. Quax, J. Vehmaanperä, and M. Young, "Plamids", in *Molecular Biology Methods for Bacillus*, ed. by C. R. Harwood and S. M. Cutting, John Wiley &Sons Ltd, Baffins Lane, UK, 75-174 (1990), 581 p.

Chapter 11
TECHNOLOGIES FOR HUMAN EXPLORATION

Scott Lowther lays out the possibilities for Mars mission heavy lift boosters.

Pathfinder Chief Scientist Matt Golombek explains his mission's results.

MAR 98-051

MARSSAT: ASSURED COMMUNICATION WITH MARS*

Thomas Gangale

In the past, robotic missions to Mars have accepted the inevitable communications blackout that occurs when Mars is in solar conjunction. This interruption, which lasts several weeks, would seem to be unacceptable during a human Mars mission. This paper proposes a relay satellite as a means of maintaining vital communications links during conjunction, and explores candidate orbits for such a spacecraft.

The basic approach to system design is to minimize size, weight, and power of spaceborne elements of the communications system, since it is more economical to compensate with large, heavy, and power-consuming elements on Earth. Ideally, it is the Earth-to-Mars link which should drive the overall system design, with the Earth-to-relay and Mars-to-relay links impacting system design as little as possible. This ideal is approached by minimizing the length of the link between the relay spacecraft and Mars. An orbit whose period is one Martian year, but whose eccentricity and inclination both differ from that of Mars, assures communications between Earth and Mars during conjunction while minimizing the length of the link between the communications satellite and the Mars mission.

1.0 STATEMENT OF NEED

At the end of April of this year, the Deep Space Network (DSN) lost contact with the *Mars Global Surveyor* spacecraft in orbit around Mars. An entire month passed before communications were reestablished with the vehicle. This hiatus was not caused by any hardware or software failure; rather, it was an inevitable consequence of planetary orbital mechanics. During this period in May 1998, Mars passed behind the Sun as seen from Earth, a planetary configuration known as solar conjunction. Fortunately, links were re-established at the end of May, and *Mars Global Surveyor* is continuing its mission. During the nearly three Mars years over which the *Viking 1* Lander operated, there were three such lengthy communication blackouts due to solar conjunctions. This Mars-Sun-Earth alignment occurs at 780-day intervals on the average, varying from 766 to 803 days due principally to the eccentricity of Mars's orbit.

Various mission profiles have been proposed for human expeditions to Mars. Among these is the conjunction class mission, which utilizes minimum-energy Hohmann trajectories both to and from Mars. However, the use of Hohmann transfers on both the outbound and inbound legs of the mission requires roughly a 500-sol layover on Mars to await the proper planetary configuration for the return flight. As can be seen in Figure 1, solar conjunction occurs near the midpoint of this 500-sol stay on Mars. Although the use of the conjunction class scenario on initial human Mars missions is an issue yet to be decided, it is likely that this type of mission profile will be flown at some point in a human Mars exploration program, not only because it is the most propellant-efficient profile, but also because it maximizes stay time on Mars while minimizing travel time to and from Mars with respect to other mission profiles using the same class of propulsion systems.

---

* Copyright © 1992, 1997, 1998 by Thomas Gangale, 430 Pinewood Drive, San Rafael, California 94903. E-mail: gangale@jps.net. The Martian Time Web Site: www.jps.net/gangale/mars/calendar.htm.

Conjunction Class Mission Profile

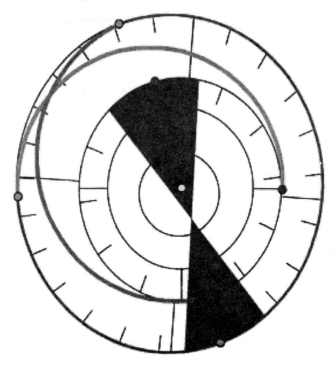

Figure 1

Until the Tracking and Data Relay Satellite System was completed in the 1980s, human missions historically endured short-duration interruptions in communications. These blackouts would last for a few minutes either while passing between ground stations or during reentry. There were communications losses of as much as an hour during the Apollo program when vehicles passed behind the Moon. But it is hard to imagine that a communications outage on the order of one month will be tolerated on a human Mars mission. Now, the specific duration of the interruption during a solar conjunction depends on several factors, such as the amount of link margin designed into the communications system, as well as the minimum data rate that is acceptable from a mission standpoint. Still, regardless of how much robustness is designed into the communications links, the minimum blackout period will always be on the order of weeks, not the few minutes or hours that have been experienced on past human space missions.

Communications interruption by the Sun will become even more of a problem as human Mars operations build up to permanent bases. The conjunction blackout will of course hold true for a Mars base as well as for a conjunction-class mission, but furthermore, such a base will also have to contend with oppositions, which are, from the Martian point of view, inferior conjunctions of Earth, i.e., when Earth passes in front of the Sun as seen from Mars. During oppositions, Earth will be able to receive signals from Mars, but Earth's transmissions to Mars will be drowned out by the Sun's radio noise.

Some means of assuring uninterrupted communications between Earth and Mars will have to be included in human Mars program planning. To satisfy this need, I propose a relay satellite concept which I call MARSSAT.

2.0 PRELIMINARY SYSTEM TRADE-OFFS

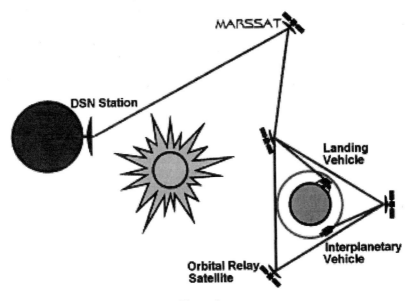

Figure 2

Figure 2 depicts the connectivity for a Mars communications system designed to circumvent the solar conjunction blackout, including Mars surface stations, low Mars orbit vehicles, a constellation of communications relays in Mars orbit, MARSSAT, and the DSN.

The basic considerations in sizing a satellite communications system are embodied in the link budget, which includes the following parameters:

P_t Transmitter power
G_t Transmitter antenna gain
SNR   Signal-to-Noise Ratio
P_L Path loss
P_s Received signal level
G_r Receiver antenna gain
N_f Receiver Noise Floor
P_n Noise in receiver bandwidth

On a conceptual level, spacecraft weight (and therefore cost) is driven by the size of the antenna and the power requirements of the transmitter and receiver. These in turn are driven by the range over which the link is required to operate. Exact numbers for the size, weight, and power of specific mission elements are a subject for detailed system engineering, and thus beyond the scope of this presentation. However, the basic approach is to minimize size, weight, and power of spaceborne elements of the communications system, since it is more economical to compensate with large, heavy, and power-consuming elements on Earth.

Ideally, it is the Earth-to-Mars link which should drive the overall system design, with the Earth-to-MARSSAT and Mars-to-MARSSAT links impacting system design as little as possible, since these alternative links represent additional costs. Assuming that the range of the Earth-to-MARSSAT link will be on the same order as that of the Earth-to-Mars link, the MARSSAT communications equipment need only be comparable to the equipment on the near-Mars elements. The stressing case, however, will be the Mars-to-MARSSAT link, for size, weight, and power will be at a premium on all spaceborne mission elements. It is for this link that the system must be optimized, since in this case, there are no large ground stations to figure into the link budget. To achieve minimum system impact, the maximum range over which the Mars-to-MARSSAT link must operate should therefore be as short as possible. At the same time, however, a minimum angular separation between the MARSSAT spacecraft and Mars, as seen from Earth during solar conjunction, must be maintained in order to reduce the impact of solar noise on the links. In general terms, this identifies the trade space to be investigated in the system engineering process.

For *Mars Global Surveyor*, mission planners and telemetry engineers defined the solar communications outage as occurring when the Sun-Earth-Mars angle was within seven degrees, and they planned for a loss of signal from 30 April to 26 May during the 1998 conjunction. It was also noted that a quiescent sun could have reduced this angle to five degrees. Several factors could reduce this minimum solar separation angle for human missions. The link throughput requirements might be limited to voice communication and only that telemetry necessary to the safety of the crew, while science data taken during the conjunction period could be recorded in situ and transmitted to Earth after the conjunction. Also, the use of a laser communications system might offer advantages over a conventional radio system. In my preliminary analysis, required minimum solar separation angles between two and three degrees were assumed. These may be unrealistically small angles from the mission operations perspective, but they provided me with some stressing cases for evaluating orbit stability.

This leads to my next point. Another consideration in the location of the relay spacecraft is the stability of its position relative to Mars over the design life of the satellite, since the more stable the orbit, the less fuel must be expended to maintain the spacecraft's position — yet another factor affecting design weight. In this paper, the eight-Mars-year (fifteen-Earth-year) cycle over which the relative positions of Earth and Mars more or less repeat is defined as the mission life of the MARSSAT vehicle. Thus the spacecraft would be required to provide communications through seven solar conjunctions.

3.0 ORBIT SIMULATION AND EVALUATION CRITERIA

3.1 MARSSAT Simulation

To investigate candidate orbits, I developed a simulation which provides two split-screen graphic displays. Display 1 (Figure 6) consists of the familiar solar system "overhead" view from above (north of) the ecliptic, and the in-the-ecliptic view along the vector of the vernal equinox. Display 2 (Figure 7) provides two views edge-on to the ecliptic plane. Both views are boresighted on Mars. The left side of the screen is a view from the perspective of the sun and sighted along the acceleration vector of Mars (the Sun-Mars line), while the right side of the screen is a view sighted along the velocity vector of Mars (perpendicular to the Sun-Mars line). The large divisions along the XY axes of the two in-the-ecliptic views represent one degree of arc, measured from 400 million kilometers, which is roughly the distance between Earth and Mars during conjunction. It is these two Mars-centered views that generate information that is useful in evaluating potential MARSSAT orbits.

3.2 Evaluation Criteria

Some criteria for evaluating candidate MARSSAT orbits are:

- MARSMAX — Maximum range from Mars in kilometers.

- SUNMIN — Minimum angular separation in degrees from the Sun-Mars line for an observer at a distance of 400 million kilometers.

- A figure of merit, defined as SUNMIN * 10^7 / MARSMAX, expresses the optimization of minimum solar angular separation and maximum range from Mars.
  Orbital drift (in kilometers).

- A second figure of merit, defined as FMERIT * 10^6/DRIFT, expresses the optimization of minimum solar angular separation, maximum range from Mars, and minimum orbital drift.

Not investigated were ΔV requirements to insert satellites into candidate orbits, although it should be pointed out that this would be an important consideration in the selection of a MARSSAT orbit.

4.0 ORBIT CONCEPTS

4.1 Co-Orbital Leader/Trailer Satellites

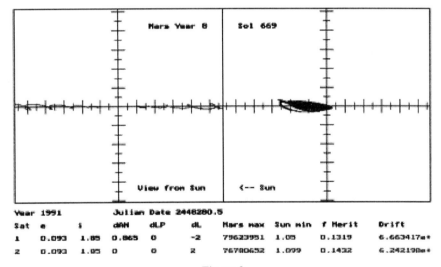

Figure 3

The simplest orbit for a MARSSAT vehicle would be one having the same parameters as that of Mars itself but slightly out of phase, i.e., either leading or trailing Mars by a few degrees in its orbit around the Sun. Ideally, a spacecraft in this orbit would remain stationary with respect to the Sun-Mars system. Unfortunately, such a relationship can only be stable if the phase angle of the orbit is either -60° or +60° with respect to Mars, corresponding to the equilateral Lagrange points L_4 and L_5. As seen in Figure 3, satellites with phase angles of -2° and +2° exhibit very poor stability, quickly departing from their assigned stations in the vicinity of Mars toward the L_4 and L_5 points.

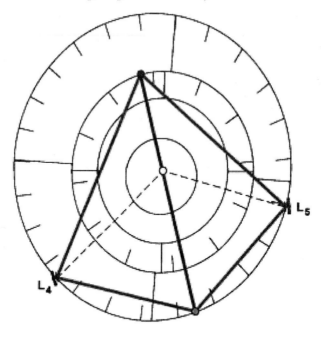

Figure 4

Of course, the equilateral Lagrange points themselves are too far from Mars to be suitable stations for MARSSAT, since links over this 230 million kilometer range would require spaceborne communications elements whose size, weight, and power would be on the order of a DSN ground station (Figure 4).

4.2 Co-Period In-Plane Oscillating Satellites

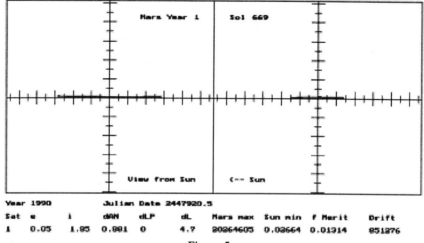

Figure 5

Another simple idea, one which would avoid the instability problem of the co-orbital leader/trailer concept, is an orbit in which a satellite would alternately lead and then trail Mars, thus tending to balance out the gravitational influence of Mars. Such a scheme can be achieved by having the satellite in an orbit whose period is the same as that of Mars, but whose eccentricity is either greater or less than that of Mars. Figure 5 depicts the behavior of a spacecraft whose orbital elements are the same as Mars, except for an eccentricity of 0.05. It can be noted that from the Martian point of view, a spacecraft in this orbit appears to orbit around Mars, although in reality it is gravitationally bound to the Sun. The MARSSAT simulation indicates excellent stability for this orbit over an eight-Mars-year period. Since the satellite circles Mars once every Martian year, the perturbation effects of Mars's gravity tend to cancel out.

An obvious disadvantage of this type of orbit is that since it is in the same plane as Mars's orbit, as seen from the Sun the satellite transits Mars twice each Martian year, and thus an unobstructed line-of sight with Earth during solar conjunction is not assured by a single satellite. A second satellite, one whose longitude of perihelion were 90° out of phase, would be necessary.

4.3 Co-Period Out-of-Plane Oscillating Satellite

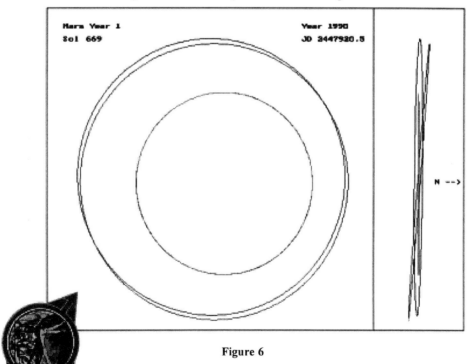

Figure 6

Co-period Out-of-plane Oscillating Satellite

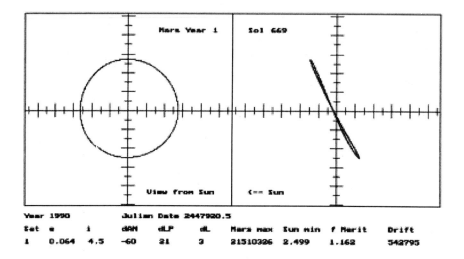

Figure 7

Assured Line-of-sight During any Conjunction

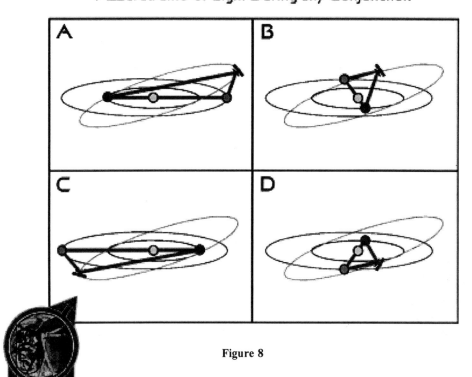

Figure 8

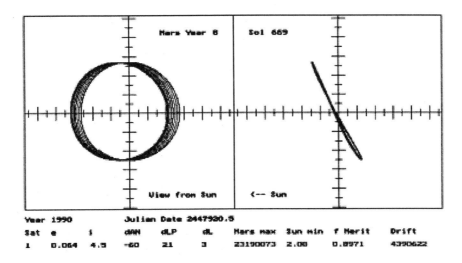

Figure 9

Figure 6 and Figure 7 illustrate an orbit whose period is one Martian year, but whose eccentricity and inclination both differ from that of Mars. As with the in-plane oscillating satellite, this out-of-plane orbit mimics the motion of a satellite in orbit around Mars, although the satellite is not bound by Mars's gravity, but is actually in solar orbit. In contrast to the in-plane oscillator, however, a minimum solar angular separation from Mars is maintained as, referring to Figure 8, MARSSAT A) rises north of, B) trails, C) drops south of, and D) leads Mars in their journey together around the Sun. In such an orbit, only one satellite is necessary to assure line-of-sight with Earth during any solar conjunction. As with the in-plane oscillator orbit, excellent orbit stability is demonstrated over a fifteen-year period (Figure 9). This type of orbit therefore seems well-suited for the MARSSAT concept. Now, if we need to increase the solar separation angle over what I assumed in this study, orbit stability improves, since we increase our distance from the gravitational influence of Mars.

Figure 7 also shows that a minimum angular separation, as seen from Earth, between MARSSAT and Mars, of 2.5 degrees is achieved. The line-of-sight distance from MARSSAT and Mars is on the order of 22 million kilometers. Note that this is only one-tenth the distance that a relay satellite stationed at a Lagrange point would be from Mars. All other things being equal, signal strength is inversely proportional to the square of the distance, thus to gain the same signal strength across ten times the distance, a Lagrange point satellite would need to be 100 times more powerful, affecting the weight of the spacecraft accordingly (Figure 10). So, if we have to double the separation angle to five degrees, the MARSSAT link is still only one-fifth the range of a Lagrange point link, which represents a link budget savings of 25 to 1.

MARSSAT vs. Lagrange Point Relays

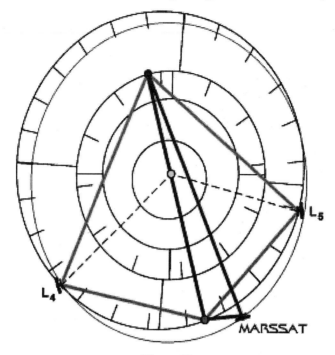

Figure 10

5.0 ANALYSIS

The following parameters are used to characterize MARSSAT orbits:

e	Eccentricity
i	Inclination
$\Delta\Omega$	Longitude of ascending node, referenced to that of Mars
$\Delta\varpi$	Longitude of perihelion, referenced to that of Mars
ΔL	Longitude at epoch, referenced to that of Mars
a	Semimajor axis of Mars

The sixth orbital parameter — a — is understood to be equal to the semimajor axis of Mars's orbit, since in all cases we want the period of the MARSSAT orbit to be one Martian year.

Several general observations can be made concerning MARSSAT orbits:

- Increasing the delta eccentricity with respect to Mars increases the horizontal (in-plane) travel of the spacecraft as seen along the Sun-Mars line. Increasing the delta inclination with respect to Mars increases the vertical (normal to plane) travel of the spacecraft. To increase the minimum solar separation angle (SUNMIN) requires simultaneous increases in delta eccentricity and inclination.

- For $e_{MARSSAT} < e_{Mars}$ and $\Delta\Omega < 0$

  the spacecraft rotates around Mars in a counterclockwise direction as viewed from the Sun.

- For $e_{MARSSAT} < e_{Mars}$ and $\Delta\Omega > 0$

  the spacecraft rotates around Mars in a clockwise direction as viewed from the Sun.

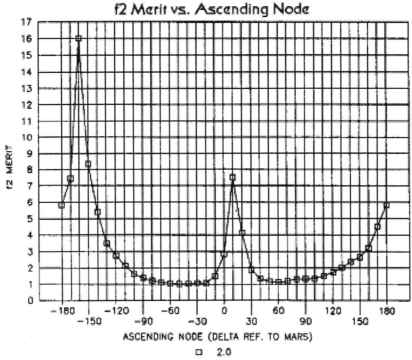

Figure 11

As long as the semimajor axis of the MARSSAT orbit is identical to that of Mars, its period is one Martian year. The other five orbital elements can be combined in a vast number of permutations to produce useful MARSSAT orbits. For a specified SUNMIN and $\Delta\Omega$, however, there is a unique combination of e, i, $\Delta\varpi$ and ΔL which produces an optimum F2MERIT, i.e., optimizes minimum solar angular separation, maximum range from Mars, and minimum orbital drift. This trade space can be characterized as a series of curves representing optimized F2MERIT for specified SUNMIN values plotted across the full range of $\Delta\Omega$ (Figure 11).

6.0 CONCLUSION

Selection of a specific MARSSAT orbit must be left to future system engineering trade studies based on the optimum Mars-to-MARSSAT range and the optimum Mars-to-MARSSAT solar separation angle, as well as ΔV requirements to insert a satellite into a given orbit. Human Mars mission studies should consider the minimum data rate that would be acceptable during the conjunction period. For instance, transmission to Earth of much of the science data recorded during the conjunction might be deferred until after the conjunction, while the minimum acceptable data rate might be primarily driven by voice links and engineering data relevant to mission safety. The minimum solar separation angle, and therefore the cost of the MARSSAT system, would be reduced accordingly.

Most space systems have one or more levels of redundancy designed into them, and even though these additional components increase the cost of a system, they are included in the design with the hope that they will never need to be used at all! In contrast, if assured communications with Mars throughout all phases of a human mission is a firm system requirement, the necessity of MARSSAT in providing that vital link with Earth will be a certainty.

MAR 98-052

MOBILITY OF LARGE MANNED ROVERS ON MARS

George William Herbert[*]

A study of a number of earth analogs for large, heavy, long range offroad vehicles generates useful inputs for rover design on Mars. An investigation into the types of suspensions available, wheeled and tracked, looks at their suitability for Mars use. Analysis of the mechanics and dynamics of power requirements indicates that motive power for a given speed should scale with the local gravity. Relative stability and mobility of vehicles are examined and the lesser stability of Mars vehicles quantified and shown to be within workable limits. Rover propulsion options are examined, investigating characteristics of possible fuel cell and internal combustion powertrains. Finally, three sample long range expeditionary rovers are described with estimated traverse ranges of over 5,000 km.

1. INTRODUCTION

Manned exploration of the surface of Mars will be critically dependent on the capabilities and safety of the surface transport vehicles available. Previous works have analyzed manned rovers [Clark 96; French 89, McCann 89, Zubrin 92, 96] from some high level assumptions to derive basic information. I feel it is important to analyze Mars rovers by starting with information known about similar vehicles used on Earth, their performance and drawbacks. Fortunately, a large class of heavy off road vehicles exists and is readily documented [Janes 97]: military vehicles, tracked and wheeled, exist in profusion and are good models in many ways for the types of mobility that explorers are likely to need on Mars. This paper will continue to build up performance and other requirements for rover design from low level analysis moving upwards, while assuming little about the end vehicles. It is hoped that this analysis will be useful for the design and tradeoff of any sized and mission manned Mars rover.

2. SUSPENSION CLASSES

There are two major types of vehicle support and suspension systems: tracked and wheeled. There are a few alternative systems (walkers, hoppers, and the like) which are not yet in viable use on Earth. Lacking any practical field experience with the alternatives, we will limit our choices to tracked and wheeled.

Tracked vehicles utilize a continuous, articulated belt made up of generally flat contact plates as the ground contact surface, providing steady support on shifting and weak ground surfaces and effectively distributing loads imposed by heavy vehicles. Tanks, some other military vehicles, and commercial bulldozers are examples.

A number of prior papers have examined the "loop wheel" concept [French 89, McCann 89]; based on practical considerations, this is a special case of the tracked vehicle type. The prior investigations rejected loop wheels as insufficiently robust for long term usage on Mars,

---

[*] Retro Aerospace. E-mail: gherbert@crl.com.

and their lack of field experience on Earth to disprove that conclusion leads this author to agree that they are not at this time worth consideration as a viable option.

Wheeled vehicles use multiple tires to contact the ground, giving less contact surface area, higher ground pressure and lower performance on very weak ground surfaces. Cars, trucks, and numerous other vehicles are examples.

Wheeled vehicles have additional detailed design choices in the type of tire used, with the usual options investigated in the past being pneumatic or wire cage tires. Solid tires utilizing elastomeric materials also are used in some applications on earth (such as armored cars) and are of interest in Mars applications due to ruggedness and reliability.

A fair rule of thumb from analysis of a number of vehicles is that a tracked suspension will weigh roughly 15% of overall vehicle weight and a wheeled suspension roughly 10% of overall vehicle weight at 1G. It is likely that this can scale with vehicle weight (i.e. with local gravitational level) rather than vehicle mass.

3. MOTIVE POWER AND SPEED

In most vehicle power calculations in earthbound environments, power is related to velocity squared, cubed, or to the fourth power, depending on the details of how drag interacts with the vehicle type and speed range.

For wheeled and tracked vehicles, examination of 53 representative tracked vehicles and 22 wheeled vehicles [Janes 97] showed that the listed road speed was well predicted by the formula:

$$\text{speed} = \sqrt{p/w} \cdot F_t$$

Where speed is the square root of the power to weight ratio (in kilowatts per ton) times a factor based on the type of suspension (wheeled or tracked). Derived suspension type factors are:

Tracked: $F_t = 15.7 \pm 3.0$
Wheeled: $F_t = 24.7 \pm 2.0$

Horsepower per ton are used by [Janes 97] and other publications, but the calculations are being performed here in kW/ton as units remain more consistent throughout performance calculations.

The power to weight ratios for the vehicles studied ranged from 12.5 to 34.0 hp/ton (9.3 to 25.4 kW/ton) for tracked, with the average of 21.16 hp/ton and standard deviation 4.75 (15.78 ± 3.54 kW/ton); and 14.6 to 34.6 hp/ton (10.9 to 25.8 kW/ton) for wheeled, with an average of 19.34 hp/ton and standard deviation 6.1 (14.42 ± 4.55 kW/ton). The question now becomes what is the significance of this general finding for Mars vehicles and design tradeoffs?

The Mars environment is characterized by a near total lack of atmosphere (5-8 millibars avg), lower gravity (3.8 m/s^2), and a surface that is apparently relatively smooth at medium and small topology scales but rough in fine detail (many rocks of scales 1-100 cm). Without examining the effects of rocks in detail, the other effects should serve to reduce the power used to traverse terrain. The question becomes, to what degree does reduced gravity and atmosphere reduce the power consumption?

One case of interest is hill climbing. In this movement, a vehicle is ascending a sloped surface, spending propulsive power to increase altitude primarily.

The definition of power (force · distance / time, in metric units n · m / s) indicates that power spent in hillclimbing should be proportional to the local gravity if distance and time are kept constant, as force (mass times gravity force) is proportional to local gravity. So 0.38 of the power required on Earth is needed for hillclimbing on Mars.

Power used in friction within the propulsive system also may scale with local gravity. The relevant analysis factors here, frictional coefficients, rotational velocity, and surface area of bearings and other friction surfaces, do not directly change due to a lower gravity environment. However, if we assume that suspension system mass is proportional to vehicle weight and not mass, then as the vehicle weight decreases in local gravity then the required mass, and thus area of bearings and other contacting surfaces, should proportionally decrease.

I believe that a complete and detailed analysis of the mechanics and dynamics of vehicle motions in reduced gravity is called for, but the preliminary indications are that performance scales inversely with local gravitational acceleration.

4. MOBILITY

A number of tradeoffs exist in designing vehicles for rough terrain operations.

One key factor is ground pressure, the average force imposed by the vehicle over the contact area between it and the ground (the tread footprint area, or the contact footprint of the tires for wheeled vehicles). Vehicles with higher ground pressure tend to be more efficient on flat, hard surfaces, but are easier to bog down and get stuck on soft surfaces such as sand or mud. Existing tracked vehicles have ground pressures ranging from 0.75 to 1.05 kg/cm^2 (this being the standard unit used in [Janes 97]) for tanks and 0.55 to 0.75 kg/cm^2 for other tracked vehicles. Unfortunately, ground pressure is not commonly listed for wheeled vehicles, so direct comparisons are not available.

It will be easier to achieve low ground pressure at Mars, if desired, due to the lower gravity. This will be counterbalanced by a desire to reduce suspension mass and vehicle mass overall.

Acceptable ground pressure ranges depend on a number of factors. Foremost are the average and extreme values for soil type the vehicle will have to operate over. Of secondary importance is the capability of the crew to extract the vehicle should it become mired in loose or weak soil. A merely embarrassing situation for a military vehicle on Earth, remedied by calling in a recovery vehicle or other equipment, might be unsalvageable on Mars due to lack of support or EVA gear mobility and dexterity limitations.

The type of suspension system, wheeled or tracked, also significantly affects the mobility of the vehicle. Tracked suspensions are slower and heavier, but give reduced ground pressure and improved rough terrain mobility. Wheeled suspensions are faster over good terrain, but slower over bad terrain. Both are vulnerable to sufficiently large rocks or other obstructions, which will be discussed in more detail in the following section. The NASA Reference Mission [Weaver & Duke 93] and prior rover papers [French 89, McCann 89] with pre-Pathfinder and MGS data concluded that tracks were not optimal choices, but the topology evident in publicly released images from those missions to date appears to be less conclusive. Until further data is available, both types of suspension should be considered as alternatives.

5. MOBILITY, RELIABILITY AND SAFETY

A number of environmental factors influence the reliability of Mars surface vehicles. At the micro scale, Mars has dust on the surface, which would be kicked up by vehicle motion and will cause wear in rotating mechanisms and other problems. At mid-scales, rocks may damage the suspension or vehicle proper should it bottom out or strike a sufficiently large one. At larger scales, the topology has sufficient vertical profile that vehicle overturn incidents are possible, which might well be catastrophic. There is no readily apparent non-life-support significant vehicle design impact of the atmosphere, the insolation, or radiation environment, except that for long duration rovers some protection in case of solar particle events is a wise precaution.

Surface Features

Rocks much smaller than the tire or tread width largely do not affect motion, but those of the same scale as the track or tire can impede progress or damage the vehicle if it hits them at much speed. Unfortunately, visual examination of Viking 1 and 2 and Mars Pathfinder imagery indicates that rocks of sufficient size to damage vehicles are plentiful, at least in those areas. Based on thermal analysis of the surface, a high proportion of the surface area of Mars is covered by rocks 10 cm in diameter or larger [Christensen 92], typically 6% but in a number of interesting areas (such as north and east of Valles Marineris) at a more significant 15-20% coverage. Determination of actual diameter distribution from thermal inertia data is difficult, as thermal inertia of rocks 10 cm and up is roughly constant. Numerous rocks of 25 cm and larger diameter are visible in surface imagery, with some 1m rocks and some suggestion of larger boulders or outcroppings of bedrock of multi-meter scale.

The size range distribution indicates that rocks and boulders on the Mars surface are going to be a significant hindrance and design constraint on any Mars rover. Rover tires considered for prior designs tend to be 1 m to 2 m in diameter, with width of 0.25 m to 0.5 m. Treads would have similar 0.65 m to 1.5 m height and 0.25 m to 0.65 m width. Roughly speaking, surface rocks with sizes between 0.1 and 0.3 of the width of the contact surface are a control hazard and significant maintenance wear, while larger rocks are a threat to the integrity of the suspension, possibly causing damage (wheel destruction, suspension arm damage, or "thrown" tread, coming loose from the tread wheels). A preliminary comparison of the surface topology with available suspensions indicates that even in average surface areas, vehicles are going to be operating in dangerous conditions with significant risk of damage due to surface rocks. Vehicle operating speed restrictions, structural design requirements, and maintainability requirements need to be investigated in more detail and will pose significant operational constraints.

Vehicle Stability

Vehicle overturn accidents happen due to a number of causes, but they all come down to one simple criterion: the vehicles' center of gravity (COG) is rotated past the edge of the vehicles' ground footprint and thus then provides an overturning moment. Once the COG has passed that point, overturn is unavoidable.

There are three main causes of such accidents: the vehicle tipping due to terrain, due to rotational impulse, or due to sideways motion. Most actual overturns are due to a combination of two or more factors occurring simultaneously, but a detailed analysis of the interactions is relatively difficult. We will consider them separately and then analyze the interactions.

Tipping due to terrain is relatively straightforward: the vehicle moves onto a slope of sufficient inclination that the COG passes outside the contact footprint, and the vehicle then overturns. The mechanics of such overturns are the same under fractional gravity as on Earth, though the dynamics are slower, and the risk and design impact are not affected.

Tipping due to rotational impulse is relatively rare on Earth. While most drivers, particularly those with off road or rough road experience, are familiar with the phenomena of running over a bump that caused one side of the vehicle to momentarily lift off the surface, such encounters rarely result in rollovers on Earth. On Mars, the situation is somewhat different. An analytical solution is difficult for this problem, but computational simulation provides some useful results. I simulated such overturns, keeping the wheel to COG distance constant at 2.0 meters, varying track width and COG height such that the angle between horizontal and the line from the tire to the COG varied from 0 to 40.5 degrees thus simulating varying initial slope and initial height of center of gravity, and found the following angular velocities needed to overturn the vehicle:

Table 1

Minimum initial upset angular velocity to overturn degrees per second

Initial Angle degrees	g=4.0	g=6.0	g=8.0	g=10.0 m/s
0	193	237	273	304
4.5	187	230	265	296
9.0	181	223	257	287
13.5	176	216	249	278
18.0	170	209	241	269
22.5	164	201	232	259
27.0	158	193	223	250
31.5	152	185	214	239
36.0	145	177	205	228
40.5	138	169	195	218
45.0	131	160	185	206

From this we deduce that the analytical solution is of the form $(1/G^{0.5})$. This indicates that Mars vehicles are 62% more vulnerable to this type of overturn than Earth vehicles of similar geometry.

Tipping due to sideways motion is similar to tipping due to rotational impulse. Rather than an external triggering event, a change in vehicle orientation relative to its direction of movement causes a destabilizing moment due to the height of the center of gravity and thus the center of inertial forces, and traction of the vehicle's tires or treads on the ground. This can be countered by returning the vehicle to pointing in the direction of motion, if it is under control. Other than doing so, the tipping once started is likely to continue unless the vehicle stops moving prior to overturning. The coefficient of friction between the vehicle and the ground is unlikely to decrease, and in fact may increase as tires dig in or encounter heavy rocks or other obstructions. Tipping will start if overturning moment exceeds the moment that gravity and the vehicle width inherently provide:

$$Vy \cdot Hg \cdot Cf > G \cdot tw/2$$

Vy is the sideways velocity, Hg is the height of the center of gravity, and Cf is the applicable coefficient of friction (static or sliding) between the surface and the vehicle. tw is the width of the track, or distance between the tracks or wheels, and G is the local gravity.

As can be seen from the equation, reduced local gravity decreases the vehicle stability resisting overturn due to sideways motion, as it does with rotational impulse overturn. Vehicles on Mars will be roughly 2.63 times more likely to overturn due to dynamic upset or accidents than vehicles with similar configurations (height of center of gravity and width of track) on Earth.

In summary, Mars rovers are significantly more likely to overturn than Earth vehicles in several scenarios; 1.62 times as likely ($1/G^{0.5}$) due to impulsive overturning impacts or bounces, and 2.63 times as likely ($1/G$) due to skidding sideways. As the overall forces due to both impacts and bounces and due to sideways skids are proportional to the velocity squared, the relative safe speeds for geometrically comparable vehicles are 0.78 if concerned with impulsive overturns and 0.61 as fast if concerned with sideways skids. These are workable fractions, reducing safe speeds below earth analog conditions but not by an unreasonable amount. Proper training and vehicle design can compensate.

6. MOTIVE POWER PRODUCTION

The power available to Mars rovers will likely be less than that available to Earth vehicles; lower speeds are acceptable, and tighter tradeoffs and technical problems lower performance. A number of prior works [Clark 96; Zubrin 92, 96] have argued effectively that required power cannot be provided by solar power systems. Nuclear reactors are too dangerous in close proximity with human crew, and RTG type power systems have very poor power to weight. We will therefore concentrate on chemically fueled rovers, utilizing internal combustion (piston or turbine), and fuel cell type power sources.

Internal Combustion

For internal combustion powered vehicles, if the motor and gearset are considered as a unit, power to weight ratios exceeding 1,000 W/kg are possible for short-duty lightweight motors. For rugged, high endurance, conventional motors, power to weight ratios are typically are much less than that: the Cummins VTA-903-T600 engine used in the M2 Bradley [Janes Upgrades 97] generates 493 kW and masses 1,190 kg. It is coupled with a 960 kg transmission. System mass is 2150 kg and specific power 230 W/kg. Gas turbines have specific power in the range of 1 to 6 kW/kg, but for vehicular applications gas turbines require significantly heavier transmissions (the M-1 tank requires a 1,960 kg transmission for 1,108 kW power). System specific power of gas turbine vehicle propulsion is 500 W/kg in the optimistic case, and can be worse. While substitution of aerospace materials and aggressive design and test can improve the specific power of internal combustion motor / transmission sets, there are practical limits to how far that process is likely to be able to go. We will assume that combustion based propulsion can generate 500 W/kg including the whole powertrain, though this may not be achievable in all cases.

Electrical Motors

For electrically powered vehicles, there are two segments to address, motors and power generation systems. For motors, industrial motor performance is 75 W/kg, with high performance lightweight motors around 150 W/kg. We will assume 150 W/kg motors can be rated to ground vehicle drive reliability requirements, but this assumption will need further test and justification.

Hydrogen Fuel Cells

Fuel cells providing electrical power can run at up to 200 to 300 W/kg, though most existing cells are lower performance. The Space Shuttle fuel cells are rated at 12 kWe continuous and 16 kWe peak performance and weigh 118 kg, a performance of 101 to 136 W/kg [Hamilton Standard 97]. Other references list specific power up to 370 W/kg [Fortescue & Stark 95] for other more modern fuel cell designs. These are hydrogen/oxygen fuel cells, which offer the greatest performance but low fuel density: conversion efficiency in these cells is 70% or greater from chemical energy to electrical output energy, with some high performance cells with topping cycle thermal turbines operating at 85%. We will assume 300 W/kg and 70% efficient.

Hydrocarbon Fuel Cells

Fuel cells using hydrocarbon reactants such as methanol or methane, either directly or pre-cracking into hydrogen also exist, but at lower specific efficiency and somewhat lower power. Current power figures for these cells appear to range up to 430 W/kg [Ballard 98]; details are hard to determine, though, as most of the programs are aimed at methanol or gasoline fuel cell cars, which have significant commercial potential and most details are being closely guarded. Some information is available on some cells; low power direct-methanol cells for non-auto applications [DTI 97] being developed now are working at 60 W/kg and 36% efficiency, though 45% efficiency is expected shortly. We will assume 400 W/kg and 40% efficiency though better may be available and announced reasonably shortly. We will also assume that the proton-exchange direct methanol cells can run on methane with no performance degradation, though this needs detailed analysis and test to verify.

Combining these into an integrated power system, we get the following projected system performance figures:

For hydrogen/oxygen fuel cells, power system specific mass of 100 W/kg at 70% efficiency is well justified at the current state of the art.

For methanol/oxygen fuel cells power system specific mass of 110 W/kg is reasonable with 40% efficiency.

7. FUEL CHOICES

The fuel consumption rate of an ideal power system is the useful power produced divided by a system efficiency and the energy content of the fuel.

The most promising fuel and oxidizer combination for Mars surface operations appears to be methane and oxygen; both are storable, easily produced with a little bring-along hydrogen, and well known. Methane can be used in fuel cells and internal combustion engines quite flexibly. The combination has 2,800 W-hr/kg specific power density.

Alternatively, we can consider hydrogen/oxygen fuel cells at 3,750 W-hk/kg, though much lower fuel density (1,312 W-hr/l versus 2,380 W-hr/l for methane/oxygen).

As was discussed in the previous section, hydrogen/oxygen cells can run at over 70% efficiency, and the best projected methane/oxygen cells at 45%.

Internal combustion engines have significantly lower overall efficiencies. Any combustive engine is a heat engine, in thermodynamic terms, and is first and foremost subject to the Carnot equation describing maximum efficiency for heat engines, $E = (T_h - T_c) / T_h$. No heat engine may exceed the Carnot efficiency, and typical overall values for practical engines are generally only 0.25 to 0.35.

8. SAMPLE VEHICLES

For purposes of demonstration, we will present some sample rover designs utilizing various propulsive options as described above. The following rover concepts are not intended as baseline vehicles for any particular mission; particularly, they are larger and more capable than needed by most proposed missions. These samples are done to be easy to analyze and compare. Actual rovers of half the proposed size or less are more likely; I picked 10 tons as a round number.

We will consider three sample 10 ton rovers: one that is internal combustion engine powered, one that is hydrogen fuel cell powered, and one that is methane fuel cell powered.

Power Requirement

Existing earth offroad vehicles have power to weight ratios of roughly 20 hp/ton (15 kW/t). We will use the equivalent rating at Mars.

Suspension

Designed for earth gravity, both vehicles would require about 1000 kg worth of tires and suspension. As they will be operating under 0.38 G Martian gravity, this is reduced to 380 kg (400 kg) each.

Motive Power

The target power to weight ratio equivalent is 20 hp/ton, or 15 kW/ton. We earlier argue that for Mars missions this should be multiplied by local gravity, so the desired power is 15 · 0.38 or 5.7 kW/t, or 57 kW total.

Our internal combustion vehicle with estimated power system specific mass of 500 W/kg will require 115 kg of motor and transmission.

The hydrogen fuel cell vehicle will require 570 kg of fuel cell and electric motors if the 100 W/kg specific power density is assumed.

The methane fuel cell vehicle will require 520 kg of powerplant and drive at 110 W/kg.

Mass Budgets

If we assume that both vehicles have a 500 kg structural frame on which other components are mounted, the total vehicle frame, suspension, and drivetrain systems are:
1,015 kg for internal combustion
1,450 kg for hydrogen fuel cells
1,400 kg for methane fuel cells.

We will assume this rover is designed to accommodate 4 persons for up to a month, requiring 100 m^3 of volume (a 4 m diameter cylinder 10 m long) massing about 2,000 kg and with 2,000 kg of life support and other equipment and 750 kg of expendables. Total mission payload is thus 4,750 kg. Remaining payload for the three sample vehicles are:
4,235 kg for internal combustion,
3,800 kg for hydrogen fuel cells, and
3,850 kg for methane fuel cells.

If we assume that fuel tankage mass is 75 kg/m^3, then allocating the entire remaining mass for fuel and tanks gives us useful fuel of:

4.63 m³ (3.85 t) for IC
7.78 m³ (3.22 t) for hydrogen fuel cell
4.21 m³ (3.50 t) for methane fuel cell

Basic Range and Performance

For a simplistic range estimate, we will divide the total fuel by 30 and assume 8 hours a day are spent driving.

The IC engine consumes 16.0 kg/hr. the hydrogen fuel cell 13.4 kg/hr, and the methane fuel cell 14.6 kg/hr. At their system efficiencies of 0.30, 0.70, and 0.40 respectively and fuel power densities of 2.80, 3.75, and 2.80 kW-Hr/kg of chemical energy, the system average output power values are:
13.44 kW for the IC engine
35.18 kW for the hydrogen fuel cell system
16.33 kW for the methane fuel cell system

Using these average output power levels, we can derive average speeds. The rovers mass 10,000 kg loaded and weigh 380,000 N each, equivalent at earth to 3.8 tons mass. The equivalent average active power to weight ratio is therefore:
3.5 kW/ton for the IC engine
9.26 kW/ton for the hydrogen fuel cell system
4.30 kW/ton for the methane fuel cell system

Recalling that the formula for road speed is $v = \sqrt{p/w} \cdot 24.7$ for wheeled vehicles, we get average speeds of:
46.2 kph for the IC engine
75.1 kph for the hydrogen fuel cell system
51.2 kph for the methane fuel cell system

If the equivalent cross country speed is half of the road speed, which is a normal assumption for earth vehicles, then the three Mars rovers can be expected to actually move at around 23 kph, 37 kph, and 26 kph respectively. Among other considerations, faster speeds may risk suspension and tire damage due to rock impacts. Over their 30 day endurance, this gives ranges of:
5,520 km for the IC engine rover
8,880 km for the hydrogen fuel cell system
6,240 km for the methane fuel cell system

If increasing performance as fuel is burned off is taken into account, the range increases proportionally.

Range with Hotel Loads

If the propulsive power must also power the life support and other electronics systems, then range decreases proportionally. Let us assume that the hotel power load for the rover is 3 kW. This corresponds to fuel consumption of:
IC engine: 3.57 kg/hr, 85.7 kg/day
H_2/O_2 FC: 1.14 kg/hr, 27.4 kg/day
Me/O_2 FC: 2.68 kg/hr, 64.3 kg/day

If the vehicle is driven 12 hrs/day at an average actual power load of 0.25 peak (57 · 0.25 or 14.25 kW), then rover speed will average 24 kph using the 0.5 assumed multiplier for offroad travel, daily movement will average 288 km, and propulsive fuel use will be:
IC engine: 205 kg/day
H2/O2 FC: 65 kg/day
Me/O2 FC: 153 kg/day

Combined usage is:
IC engine: 291 kg/day
H2/O2 FC: 93 kg/day
Me/O2 FC: 218 kg/day

0-margin endurance will be:
IC engine: 13.23 days
H2/O2 FC: 34.6 days
Me/O2 FC: 16.05 days

This gives effective straight line distances of:
IC engine: 3,810 km
H2/O2 FC: 9,965 km
Me/O2 FC: 4,622 km

In this case, the hydrogen/oxygen fuel cell vehicle shows more than a factor of two advantage in range and endurance over the competitive vehicles, and the methane/oxygen fuel cell has 21% range advantage over the internal combustion engine vehicle.

9. CONCLUSIONS

A number of conclusions spring from this research into Mars rover design.

Stability dictates vehicle configuration, and the reduced stability of Mars rovers will dictate that they be designed to be lower to the ground than Earth vehicles, not taller. Most detail designs for large Mars rovers have assumed a large cylindrical pressure hull that had considerable ground clearance. It is better to go low, oval pressure hull and lower ground clearance (earth off road vehicles get by with 0.5 meter, Mars probably can too, but rock size distribution and vehicle belly structural strengths need to be appropriately considered). This may be inconvenient in packaging for transport to Mars. Nobody said this was going to be easy.

The impact of surface rocks on rover operations and speed is significant: rocks of the right sizes to significantly damage rovers are plentiful near all lander sites and probably are present in lesser but still significant quantities over most of the Mars surface, particularly some of the more interesting traverse targets. The detailed impact on design and operations is not yet known, but may significantly decrease safe operating speeds below commonly considered values and increase maintenance and vehicle damage risks.

The best fuel combination, despite the low density, is hydrogen/oxygen fuel cells and electric motors for the powertrain. This gives at least 50% better range than competing systems. If large quantities of hydrogen are not available, or its storage proves to be overly difficult, then use of methane/oxygen internal combustion or fuel cell systems are roughly comparable in terms of rover performance, though the fuel cell vehicles enjoy an advantage of about 22% better range per fuel mass consumed and thus reduce consumables and ISRU loads.

Finally, there are literally hundreds of large, high performance off road vehicles available on Earth for comparison and analysis purposes, and with due attention to the scaling factors and areas of difference, much can be learned in detail design, configuration, and other areas from these many examples.

10. DIRECTIONS FOR RESEARCH

A number of areas in which research is needed for further progress exist.

Surface Detail Investigation

A greater level of understanding of the detail surface characteristics of Mars is required for rover design. Of particular interest are:

- soil characteristics and rock size distributions over the relevant areas of the surface that surface traverse expeditions are likely to cover.
- topology details such as slopes, gully widths, small and mid-scale surface disruptions.

At first analysis, we need the ability to count size distributions of 10 cm and larger rocks as input into rover detail design. We have total surface rock coverage percentage already from thermal data, but the size distribution from 10cm up is critical as the apparent distribution corresponds exactly with sizes likely to damage rover suspensions. Determining such distributions near landers is trivial, but hard from orbit, that resolution being roughly the same as achievable by Earth orbiting US military visual imaging satellites. While low weight optics and narrow swaths of view should be acceptable, it is still a difficult proposition. It may be necessary to fly a small field of view high resolution visual imaging camera, with 10cm or greater resolution, on a future orbiter to resolve the question.

Methane Fuel Cell Performance Determination

Methane is an attractive fuel on Mars, but not being examined for earth based fuel cell applications, losing out to methanol, gasoline, and hydrogen for a number of reasons. The chemistry of proton-exchange direct reactant methanol fuel cells appears to the author to be usable with methane as well. This assumption needs further analysis and practical testing.

Mechanics of Vehicle Maintenance in EVA Suits

A key question in designing rovers is what level of external repair and maintenance will be performable given the mobility and dexterity limits imposed by EVA suits. It is too much to hope that rovers on long traverses will be 100% reliable, thus rover design must follow experimental validation of what repairs will be possible in the field in suits. It is possible that repair of thrown or broken track-type suspensions will be impossible or impractical in EVA gear, in which case only wheeled rovers should be planned.

This question can be answered in some detail using fairly inexpensive testing. Using simulated EVA suits, pressurized but obviously not requiring actual full closed loop life support, we can have experienced vehicle crewmen of appropriately scaled wheeled and tracked vehicles perform various field maintenance procedures, recording the results for analysis. The test EVA suit simulators merely have to replicate dexterity and mobility limits with moderate to high fidelity, and this should be relatively inexpensive to develop and validate.

The tests should include replacement of tires, suspension elements, replacing thrown or damaged tracks, and recovering vehicles stuck in sand and/or other soil types which mire the vehicle.

Dynamics of Motion, Drag, and Power at Fractional Gravity

A detailed theoretical look at the influence of gravitational force on vehicle motion at the macroscale (not just light rovers) is required. Once that has been performed, then experimental verification, perhaps using testbeds with sub-scale driving tracks flown on fractional gravity parabolic arc test aircraft, should be performed to validate the detailed theory.

11. APPENDIX: EARTH VEHICLE DATA

Type	Susp	Hp/ton	kW/ton	tons	Rd Spd	Ft(hp)	Ft(kw)
Rooikat	8x8	20.1	15.0	28.0	120	26.77	31.00
LAV600	6x6	14.6	10.9	18.5	100	26.17	30.31
EE-9	6x6	15.8	11.8	13.4	100	25.16	29.13
Cntauro	8x8	20.8	15.5	25.0	105	23.02	26.66
VBC90	6x6	16.0	11.9	13.5	92	23.00	26.63
ERC1	6x6	17.5	13.0	8.3	95	22.71	26.30
BDRM2	4x4	20.0	14.9	7.0	100	22.36	25.89
Eagle	4x4	31.3	23.3	5.1	125	22.34	25.87
AML	4x4	16.4	12.2	5.5	90	22.22	25.74
Type87	6x6	20.3	15.1	15.0	100	22.19	25.70
EE-3	4x4	20.7	15.4	5.8	100	21.98	25.45
LAV25	8x8	21.5	16.0	12.8	100	21.57	24.97
M1114	4x4	34.6	25.8	5.5	125	21.25	24.61
AMX-10	6x6	16.5	12.3	15.9	85	20.93	24.23
Luchs	8x8	20.0	14.9	20.0	90	20.12	23.30
BDRM1	4x4	16.1	12.0	5.6	80	19.94	23.09
Scout	4x4	20.6	15.4	7.2	88	19.39	22.45
Fox	4x4	30.0	22.4	6.1	104	18.99	21.99
Saladin	6x6	14.7	11.0	11.6	72	18.78	21.75
VBL	4x4	26.7	19.9	3.6	95	18.39	21.29
T64B	Track	17.7	13.2	39.5	75	17.83	20.64
Warrior	Track	19.6	14.6	28.0	75	16.94	19.62
Ferret	4x4	30.6	22.8	4.2	93	16.81	19.47
Marder	Track	20.5	15.3	29.2	75	16.56	19.18
SK105	Track	18.1	13.5	17.7	70	16.45	19.05
TAM	Track	22.8	17.0	31.6	76	15.92	18.43
Eland	4x4	19.4	14.5	5.3	70	15.89	18.40
M41	Track	21.2	15.8	23.5	72	15.64	18.11
AMX10P	Track	17.9	13.3	14.5	65	15.36	17.79
LVTP7	Track	17.5	13.0	22.8	64	15.30	17.72
Type89	Track	22.2	16.6	27.0	70	14.86	17.20
Type73	Track	22.6	16.9	13.3	70	14.72	17.05
Dardo	Track	22.6	16.9	23.0	70	14.72	17.05
Leoprd1	Track	19.6	14.6	42.4	65	14.68	17.00
AMX13	Track	16.7	12.5	15.0	60	14.68	17.00
T90	Track	16.8	12.5	50.0	60	14.64	16.95
M2	Track	20.4	15.2	22.9	66	14.61	16.92
CV90	Track	24.1	18.0	22.8	70	14.26	16.51
OT-40	Track	18.2	13.6	43.1	60	14.06	16.29
WZ501	Track	22.0	16.4	13.3	65	13.86	16.05
Type90	Track	22.0	16.4	14.5	65	13.86	16.05
M1	Track	27.0	20.1	54.5	72	13.86	16.05
Leoprd2	Track	27.0	20.1	55.1	72	13.86	16.05
MTLB	Track	20.1	15.0	11.9	62	13.83	16.01
Type74	Track	18.9	14.1	38.0	60	13.80	15.98

T72S	Track	18.9	14.1	44.5	60	13.80	15.98
BMP1	Track	22.2	16.6	13.5	65	13.80	15.98
Type63	Track	21.7	16.2	18.4	64	13.74	15.91
M113	Track	19.3	14.4	11.3	60	13.66	15.82
BMP3	Track	26.7	19.9	18.7	70	13.55	15.69
Wiesel	Track	30.7	22.9	2.8	75	13.54	15.68
Leclerc	Track	27.5	20.5	54.5	71	13.54	15.68
CV90105	Track	27.0	20.1	22.5	70	13.47	15.60
T80U	Track	27.2	20.3	46.0	70	13.42	15.54
Type531	Track	23.5	17.5	13.6	65	13.41	15.53
K1	Track	23.5	17.5	51.1	65	13.41	15.53
Stingry	Track	25.9	19.3	21.2	67	13.17	15.25
T62	Track	14.5	10.8	40.0	50	13.13	15.21
XM8	Track	30.5	22.7	16.7	72	13.04	15.10
Chieft	Track	13.6	10.1	55.0	48	13.02	15.07
C-1	Track	25.0	18.6	52.0	65	13.00	15.05
Type90	Track	30.0	22.4	50.0	70	12.78	14.80
Chlngr2	Track	19.2	14.3	62.5	56	12.78	14.80
M60A3	Track	14.2	10.6	52.6	48	12.74	14.75
Chlngr1	Track	19.4	14.5	62.0	56	12.71	14.72
T55	Track	16.1	12.0	36.0	50	12.46	14.43
BMD3	Track	34.0	25.4	13.2	70	12.00	13.90
Merkava	Track	15.0	11.2	60.0	46	11.88	13.75
Strv103	Track	18.4	13.7	39.7	50	11.66	13.50
AMX-30	Track	20.0	14.9	36.0	50	11.18	12.95
Type61	Track	16.3	12.2	35.0	45	11.15	12.91
PT-76	Track	16.4	12.2	14.6	44	10.87	12.58
SU60	Track	18.6	13.9	11.8	45	10.43	12.08
Cent	Track	12.5	9.3	51.8	35	9.90	11.46
M48	Track	18.3	13.6	44.9	42	9.82	11.37

12. REFERENCES

anon, "Direct Liquid Feed Methanol Fuel Cell", DTI Press Release, http://137.79.70.19/success/fuelcell.html.

anon, "Space Shuttle Fuel Cells", Hamilton Standard Co., http://www.hamilton-standard.com/ifc-onsi/cell/alkaline.html.

Ballard Power Systems, Private communications, June 1998.

Philip R. Christensen and Henry J. Moore, "The Martian Surface Layer", in *Mars*, H.H Keiffer *et al.* eds., University of Arizona Press, Tucson, 1992, pp. 686-729.

Benton Clark, "Mars Rovers," AAS 95-490, in *Strategies for Mars*, Carol Stoker and Carter Emmart eds., Vol. 86, AAS *Science and Technology Series*, San Diego, 1996, pp. 445-462.

Peter Fortescue and John Stark ed., *Spacecraft Systems Engineering*, John Wiley & Sons, Chichester, 1995, pp. 324-325.

Christopher F Foss ed., *Janes Armor and Artillery 1997-98*, Janes Information Group, London, 1997.

Christopher F Foss ed., *Janes Armor and Artillery Upgrades 1997-98*, Janes Information Group, London, 1997.

James R French, "An Overview of Mars Surface Mobility Justification and Options," AAS 87-220, in *The Case for Mars III*, Carol Stoker ed., Vol. 74, AAS *Science and Technology Series*, San Diego, 1989, pp. 619-632.

J. Mark McCann, Mark J. Snaufer, and Robert J. Svenson, "Mars Global Exploration Vehicle," AAS 87-222, in *The Case for Mars III*, Carol Stoker ed., Vol. 74, AAS *Science and Technology Series*, San Diego, 1989, pp. 647-663.

David B. Weaver & Michael B. Duke, "Mars Exploration Strategies: A Reference Program and Comparison of Alternative Architectures," [NASA Reference Mission], AIAA 93-4212.

J. L. Meriam and L. G Kraige, *Engineering Mechanics* Volume 2: Dynamics, second edition SI Version, John Wiley & Sons, New York, 1987.

Robert Zubrin with Richard Wagner, *The Case for Mars*, Free Press, New York, 1996.

MAR 98-053

BOOSTERS FOR MANNED MISSIONS TO MARS, PAST AND PRESENT*

Scott Lowther

INTRODUCTION

A large number of space launch boosters have been proposed throughout the years, many of which were designed for or were applicable to launching components required for manned missions to Mars. In this paper, past and current designs for booster vehicles needed to perform manned Mars missions are examined and compared, with emphasis on current and very recent concepts. Included are such designs as: von Braun's Ferry Rocket, the Saturn V and various Saturn V derived vehicles, various Nova studies from the 1960's, the Soviet N-1, the Soviet Energia, various Shuttle Derived Vehicles (such as Shuttle-C, Ares and similar designs), the Evolved Expendable Launch Vehicle, VentureStar, StarBooster 400, StarBooster 1800, VentureStar and Magnum. Vehicle designs are described and shown and capabilities are compared, along with any manned Mars mission modes originally proposed for each booster. The utility of each booster for piloted missions and cargo missions at current technological levels are examined, as are launch site and infrastructure modification requirements.

Three baseline Mars missions based on modern technology are also briefly described: a Mars Direct architecture, a JSC baseline Mars Semi-Direct architecture and a stereotypical large on-orbit assembled single vehicle. Each booster is described in relation to these missions. The results are then compared and contrasted. Conclusions are drawn based upon these results, with certain boosters showing a higher level of ability and confidence.

Notes on Data Tables

Where possible, payloads for the examined boosters have been referenced for the same orbit (typically 407 kilometers and either 28.5 or 51.6 degrees inclination) to permit more direct comparisons. For some vehicles, however, this direct comparison is not possible.

For launchers with multiple strap-on boosters, the data for the strap-ons is the cumulative total for all, not for a single strap-on.

COMPARISON MISSIONS

On-Orbit Assembled

From the late 1940's until the late 1980's, the main line of thinking for mission modes for manned Mars missions involved some degree of on-orbit assembly. This has ranged everywhere from a few components simply docked together to far more involved orbital construction programs.

---

* Greatly expanded version will be made available at: http://www.webcreations.com/ptm/pubs.htm.

One essentially stereotypical on-orbit assembled Mars mission was produced by the NASA Lewis Research Center in 1991 for the Space Exploration Initiative. This concept utilized nuclear thermal rocket research then underway at Lewis for lunar logistics support. The main stage of the Mars vehicle used a derivative of the reusable lunar core stage, with additional liquid hydrogen tanks. The Mars vehicle included a space-only mission module and an Apollo capsule-shaped Mars Excursion Module similar to 1960's designs. Total vehicle length was 105.6 meters, with a mass in Low Earth Orbit of 668,000 kilograms. Heavy lift boosters derived from the Saturn V were the presumed launch system.

On-orbit assembled Mars vehicles were the standard NASA and industry design until the early to mid 1990's, when smaller single-launch vehicles (Mars Direct) and dual launch, on orbit docked (Design Reference Mission) vehicles utilizing Martian resources for propellant became preferred. While the design of such vehicles have been many and varied, on-orbit assembled vehicles that do not make use of Martian resources have tended to be extremely massive. A selection of such designs produced by Boeing for NASA-Marshall in the late 1980's and early 1990's showed initial masses on Low Earth Orbit of between 478,000 kilogram (for designs with advanced nuclear propulsion) to 828,000 kilograms (for all-chemical vehicles). Such vehicles have always required heavy lift boosters due to the weights of individual components, and have often had outsized components (fuel tanks, aeroshells, truss structures, etc.) that required large payload fairings.

Mars Direct

The idea of using Martian resources for rocket fuel (primarily the carbon dioxide in the atmosphere) has been around for several decades, but did not receive a great deal of study until 1990. A group of scientists and engineers at Martin-Marietta proposed a manned Mars mission mode that made maximum use of Martian resources in response to the Space Exploration Initiative. The mission mode became known as "Mars Direct," primarily due to the use of a single heavy lift launch vehicle to loft components of the mission directly to Mars with no stopover needed in Earth orbit.

Two launches of the "Ares" booster would be required for a single manned expedition. The first would send an unmanned Earth Return Vehicle directly to Mars, with its fuel tanks virtually dry. Once there, the Earth Return Vehicle would begin processing the local carbon dioxide with onboard chemical reactors and a small supply of liquid hydrogen. Sufficient fuel and oxidizer can be produced to allow the two-stage Earth Return Vehicle to launch directly back to Earth.

The piloted mission would be launched to Mars on the next available window, after the ERV has completed its fueling. This mission would contain a single habitation module with a pressurized rover and four crew. The habitation module would, it was planned, be landed within walking or driving distance of the ERV. The crew would spend in excess of one Earth year on the surface of Mars; at the end of which time they would board the ERV and return home. The mission would leave behind the habitation module and all its provisions, to serve as a basis for future explorations.

The ERV had a mass of 28,600 kilograms; the habitation module massed 25,200 kilograms. These masses do not include the masses for the required aeroshells, parachutes and lander stages.

While the Mars Direct concept is generally regarded as overly optimistic, some of the basic premises have been incorporated into current NASA thinking. The current Design Reference

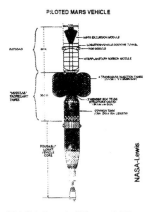

Figure 1: NASA-Lewis Manned Mars Vehicle.

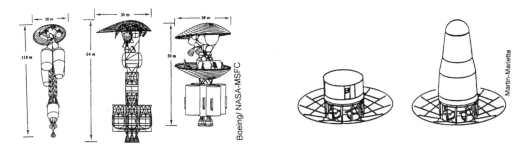

Figure 2 (left): Representative On-Orbit Assembled Concepts.
Figure 3 (right): Mars Direct - Hab and ERV with Aeroshells.

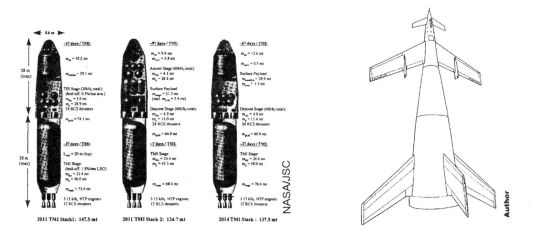

Figure 4 (left): Design Reference Mission Hardware.
Figure 5 (right): von Braun Ferry Rocket.

Mission uses In Situ propellant production to fuel the ascent stage used to boost the crew from the Martian surface to the orbiting Earth return vehicle. The habitation module will also be left behind.

Design Reference Mission

NASA's current studies for a manned mission to Mars revolve around the Design Reference Mission. This is an ongoing study to design low cost and reliable Mars mission using available or near term technology. As of this writing, the current NASA Design Reference Mission involves the use of six launches of a Magnum class booster. Each launch would orbit a Mars-bound spacecraft or its trans-Mars injection stage; the payload and stage would be docked in LEO, and then sent to Mars.. The current baseline for the trans-Mars injection stage is a nuclear thermal rocket (NTR) system with three engines using liquid hydrogen as the reaction mass. Trans Mars Injections stage has a dry mass of 27,000 kilograms and a propellant loading of 50,000 kilograms. The three payloads to be sent to Mars are the Earth return vehicle (71,400 kilograms, to be left in Mars orbit, with 147,500 kilograms launched from Earth), the ascent stage (66,000 kilograms, to be landed on Mars, with 134,700 kilograms launched from Earth) and the piloted TransHab (60,800 kilograms, to be landed on Mars, with 137,500 kilograms launched from Earth). Total payload to be launched into Earth orbit (407 kilometers by 51.6 degrees) is 419,700 kilograms.

The Design Reference Mission follows a mission similar to that proposed for Mars Direct, in that the Earth return vehicle and the ascent stage are sent to Mars unmanned and several years before the manned stage. The ascent stage would use an ISRU plant to produce the fuel required to make the ascent from the Martian surface to the orbiting Earth return vehicle; the Earth return vehicle would be fully fueled prior to launch from Earth.

WERNHER VON BRAUN'S "FERRY ROCKET"

The first truly large space launch vehicle ever designed at any level of practical detail was Wernher von Braun's "Ferry Rocket." This vehicle, which first appeared in von Braun's 1948 book "Das Marsprojekt," had it's origins with the A-4, A-9 and A-10 rockets of World War Two. The Ferry Rocket was designed to fulfill three main missions: launching components for a space station, a lunar mission and a Mars mission.

The Ferry Rocket's design had elements that, seen from a vantage point fifty years further on, seem both quaint and oddly futuristic. The Ferry Rocket was intentionally designed to not break new ground, technologically; everything about it was meant to be possible using early 1950's technology.

Propulsion for the giant three stage vehicle was to be provided by no less than 51 rocket engines in the first stage, 22 in the second and 5 in the winged third stage. These engines burned the propellant combination of hydrazine and nitric acid; while thoroughly toxic, this fuel combination did at least have the advantage of being hypergolic, thus reducing problems of ignition. These propellants were also non-cryogenic. The rocket engines on the first two stages were to be packed so closely together that instead of having circular cross-section bell nozzles they had hexagonal cross section nozzles, a practice virtually unknown in the world of rocketry.

The third stage of the ferry rocket was winged and manned, and used conventional tricycle landing gear for a return to a runway. In this respect it was similar to the current Space

Shuttle. However, it was to be made of primarily stainless steel, and had a small, hockey-puck shaped cargo bay. In this respect its growth capability was limited.

The Mars mission designed around the Ferry Rocket was stunningly large and complex. Each ferry rocket was expected to launch 39,000 kilograms into a 1,730 kilometer circular orbit. The Mars missions would be composed of ten ships carrying seventy men; three of these ships would be gigantic gliders for landing on Mars, while the other seven would be pure spacecraft. Total mass in Earth orbit of this flotilla would be 37,200,000 kilograms; to launch this, approximately 46 ferry rockets would fly 950 flights over a period of eight months... a flight rate about 200 times that of the Space Shuttle.

Once on orbit, the spacecraft would be assembled and fueled; the workers would use a co-orbiting 250-foot diameter space station as a base... a space station that would require many more launches of the ferry rocket.

Pros:
- No high-tech items.
- Performance would greatly benefit from a modernization program.

Cons:
- Exceedingly large.
- Not well defined.
- Without a massive modernization program, the propellants would present ecological problems.
- Difficulty in adapting large upper stages to the manned third stage.
- No infrastructure in place.
- Would require a massive development effort even without modernization.

Conclusions:
All in all, the Ferry Rocket was an interesting notion, a good first attempt and in many ways ahead of its time. But time and technology have passed it by, and it is not a realistic option.

SATURN V

The Saturn V is well known as the vehicle that launched the manned Apollo lunar missions. It had the payload capacity to become a serious contender for a manned Mars booster, and numerous proposals were made for exactly that.

The basic Saturn V was composed of three stages: the S-IC first stage, with five kerosene/oxygen burning F-1 engines; the S-II stage, with five hydrogen/oxygen burning J-2 engines, and the S-IVb stage, with a single hydrogen/oxygen burning J-2 engine. This booster was specifically designed to serve as the Apollo program booster, but many studies were performed to determine its ability to perform alternate missions. The closest the Saturn V came to performing a mission similar to a manned Mars mission was the launch of the Skylab orbital workshop, which was of similar size to many proposed Mars mission modules.

Proposals for using the Saturn V to boost manned Mars mission hardware have rarely used the stock Saturn V; augmented Saturns were usually the choice of the mission designers. At the time in which the Saturn V was in use, Mars missions were typically very massive, requiring numerous large hydrogen-filled stages and NERVA engines. These missions were an outgrowth of Apollo-era technology and the Apollo-era mindset, when a single "flags and footprints" mission was considered acceptable.

	Ferry Rocket	
Stage 1		
	Dry Mass (KG)	700,000
	Propellant Mass (KG)	4800000
	No. Engines	51
	Thrust Each (N)	2462118
	Total Thrust (N)	125568000
	Specific Impulse (Isp)	229
Stage 2		
	Dry Mass (KG)	70,000
	Propellant Mass (KG)	700,000
	No. Engines	22
	Thrust Each (N)	713455
	Total Thrust (N)	15696000
	Specific Impulse (sec)	285
Stage 3		
	Dry Mass (KG)	22,000
	Propellant Mass (KG)	83,000
	No. Engines	5
	Thrust Each (N)	3924000
	Total Thrust (N)	1962000
	Specific Impulse (sec)	285
Payload (kg)	1730 km X 0 deg.	39,400

Table 1 (left): Ferry Rocket Data.
Figure 6 (right): Saturn V.

	Saturn V	
Stage 1		
S-IC	Dry Mass (KG)	130,273
	Propellant Mass (KG)	2,084,319
	No. Engines (F-1)	5
	Thrust Each (N)	6920368
	Total Thrust (N)	34601840
	Specific Impulse (sec)	265
Stage 2		
S-II	Dry Mass (KG)	35796
	Propellant Mass (KG)	451229
	No. Engines (J-2)	5
	Thrust Each (N)	1025570
	Total Thrust (N)	5127850
	Specific Impulse (sec)	425
Stage 3		
S-IVB	Dry Mass (KG)	11325
	Propellant Mass (KG)	108256
	No. Engines (J-2)	1
	Thrust Each (N)	891800
	Total Thrust (N)	891800
	Specific Impulse (sec)	425
Payload (kg)	407 km X 28.5 deg.	110000
	407 km X 51.6 deg.	105500

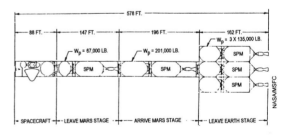

Table 2 (left): Saturn V Data.
Figure 7 (right): Saturn-Technology Mars Vehicle Configuration.

One representative Saturn V-launched Mars mission was presented by engineers of the Marshall Space Flight Center (based on prior work from several industry contractors) in 1966. The Saturn booster was actually the MLV-3 design, which utilized advanced F-1 and J-2 engines, and had extended S-IC and S-II stages. This Mars mission called for the use of five NERVA stages; each stage was identical, with the exception of second hydrogen fuel tanks of different sizes. Payload consisted of an Earth entry module for eight men, a Mars excursion module and a mission module. Total mass of the vehicle in Earth orbit at the beginning of the mission was 1,260,886 kilograms. A total of ten Saturn V/MLV-3 boosters were required to launch this vehicle.

Pros:
- Straightforward, proven booster.
- Adaptable to many different upper stages.
- Capable of heavy lift.
- Would benefit from modernization.

Cons:
- Infrastructure, including tooling, no longer exist.
- Fully expendable.

Conclusions:

The Saturn was the most capable booster of its day, rivaled only by the failed N-1 and cancelled Energia boosters. Had production continued, it certainly would play a major role in any American manned Mars mission. Its payload capacity was marginal for manned Mars missions of its day; but modern, lower mass mission certainly would have put the Saturn V to good use.

However, the fact remains that it was put out of production; while it could be returned to production status, enough advances have been made in the fields of propulsion and materials that it would only make sense to design a radically new Saturn V. This would provide for a more capable booster, but would be enough of a development effort that a completely new booster would make more sense.

SATURN V-23(L)

The design of the Saturn V was advanced for its time; but the pace of technological advancement in the mid to late 1960's was such that advanced and improved versions of the Saturn V were foreseen. Boeing in particular designed a large number of improved Saturn Vs, based primarily around an improved S-IC stage (built by Boeing) or the addition of strap-on auxiliary boosters.

One of the many studies for improving the Saturn V was Boeings Saturn V-23(L). This design called for the use of a modified Saturn V (lengthened S-IC and S-II stages with F-1A and J-2S engines) with the addition of four strap-on booster pods. These pods were each equipped with two F-1A engines, bringing the total launch thrust to more than two and a half times that of the standard Saturn V.

The strap-on booster pods were very nearly launch vehicles themselves. Prior to the design of the Saturn C-5 (renamed the Saturn V), the Saturn C-3 was examined by NASA and industry engineers. This vehicle used the S-IB first stage; smaller in diameter than the S-IC stage, it was to be equipped with only two F-1 engines. The design of the Saturn V-23(L) pods was similar to that of the S-IB stage.

The result of the lengthening of the Saturn S-IC and S-II stages and the addition of the four booster pods was more than doubling the payload capability of the Saturn.

Pros:
- Straightforward and mostly proven booster.
- Adaptable to many different upper stages.
- Capable of very heavy lift.
- Would benefit from modernization.

Cons:
- Infrastructure, including tooling, no longer exist for Saturn.
- Would require new launch facilities.
- Development would be needed for the new strap-on liquids.
- Mostly expendable.

Conclusions:

The Saturn V-23(L) design would have been a valuable booster for manned Mars missions. But for the present day, its technology was somewhat antiquated; the design was superceded by the somewhat more capable Saturn V-derived HLLV of the Space Exploration Initiative of the early 1990's.

SEI SATURN V-DERIVED HLLV

When the Space Exploration Initiative program was in existence in the early 1990's, a heavy lift booster was foreseen as needed to boost the massive payloads into orbit. Manned missions to Mars were to be part of the Space Exploration Initiative; the mission modes envisioned were essentially the same as had been proposed in the late 1960's, with on-orbit assembly of large nuclear powered vehicles that take the entire missions' fuel. Two types of boosters were proposed. A modified version of the unbuilt National Launch System used modified STS external tanks and SSMEs as the core stage, with a number of strap-on boosters. The other concept was to revive the Saturn V and update it with modern materials and equipment.

The SEI Saturn V was based largely around the Saturn V-23(L) design. The new vehicle also had four strap-on boosters each with two F-1As as well as five F-1A engines under the extended S-IC first stage. A highly modified S-II stage was equipped with six J-2S engines.

Two variants of the revived Saturn V were described, one for direct flights to the moon and the other to launch massive payloads into Earth orbit to support Mars missions. The lunar variant had a further hydrogen/oxygen stage atop the MS-II stage, equipped with a single J-2S engine. This stage would serve to inject the lunar payload onto a trans-lunar trajectory. The Mars variant did not have an upper stage above the MS-II stage. It did, however, have a greatly enlarge payload fairing.

The Mars mission envisioned for this booster was of the stereotypical on-orbit assembled variety. While a specific vehicle configuration was not chosen (nor was the booster vehicle designed in any great detail), the mission would have used NTR engines burning hydrogen, with the entire missions propellant mass having been carried from Earth orbit. Large aerobrakes were also proposed; this was one of the reasons why the payload shroud had such a large diameter - the aerobrakes were stronger and safer when launched as a single piece.

Pros:
- Straightforward booster, based on proven design.
- Adaptable to many different upper stages.
- Capable of very heavy lift.

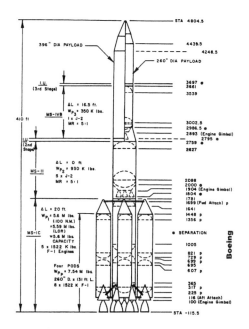

	Saturn V-23(L)	
Stage 1		
Pods	Dry Mass (KG)	273595
	Propellant Mass (KG)	3430409
	No. Engines (F-1)	8
	Thrust Each (N)	6786598
	Total Thrust (N)	54292784
	Specific Impulse (sec)	265
Stage 2		
MS-IC	Dry Mass (KG)	208982
	Propellant Mass (KG)	2545454
	No. Engines (F-1)	5
	Thrust Each (N)	6786598
	Total Thrust (N)	33932990
	Specific Impulse (sec)	265
Stage 3		
MS-II	Dry Mass (KG)	43191
	Propellant Mass (KG)	422727
	No. Engines (J-2)	5
	Thrust Each (N)	914095
	Total Thrust (N)	4570475
	Specific Impulse (sec)	425
Payload (kg)	185 km X 28.5 deg.	263296
	407 km X 28.5 deg.	232142

Figure 8 (left): Saturn V-23(L).
Table 3 (right): Saturn V-23(L) Data.

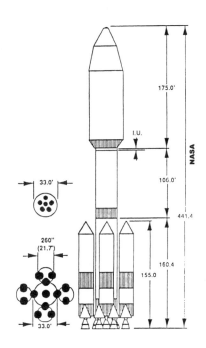

	SEI Saturn V	
Stage 1		
Pods	Dry Mass (KG)	303350
	Propellant Mass (KG)	3949776
	No. Engines (F-1A)	8
	Thrust Each (N)	8026200
	Total Thrust (N)	64209600
	Specific Impulse (sec)	270
Stage 2		
MS-IC	Dry Mass (KG)	209471
	Propellant Mass (KG)	2735520
	No. Engines (F-1A)	5
	Thrust Each (N)	8026200
	Total Thrust (N)	40131000
	Specific Impulse (sec)	270
Stage 3		
MS-II	Dry Mass (KG)	60895
	Propellant Mass (KG)	627730
	No. Engines (J-2S)	6
	Thrust Each (N)	1181635
	Total Thrust (N)	7089810
	Specific Impulse (sec)	436
Payload (kg)	407 km X 28.5 deg.	289000

Figure 9 (left): SEI Saturn V Derivative Mars Booster.
Table 4 (right): Saturn V Derived Mars Booster Data.

Cons:
- Infrastructure, including tooling, no longer exist for Saturn.
- Would require new launch facilities.
- Development would be needed for the new strap-on liquids and upper stages.
- Fully expendable.

Conclusions:

This vehicle would be a valuable booster for manned Mars missions. However, design definition on it is rather poor, so a complete development effort would be required to bring this booster into being.

More capable propulsion systems are available today than the F-1A engines proposed for this booster. The RD-170 engines provide comparable thrust and weight but with higher specific impulse. Utilizing these engines would increase performance, but would extend the development effort even further.

NOVA STUDIES

In the early days of the Apollo program, prior to the acceptance of Lunar Orbit Rendezvous (LOR), NASA expected to have to develop a very large booster capable of landing an Apollo Command and Service Module directly onto the lunar surface. The booster needed to boost this heavier mission was known as Nova.

When LOR was accepted in July of 1962, the design studies for Nova continued. There was no specific goal to be reached for Nova (soon renamed "Post Saturn"), so a wide variety of boosters of vastly differing design and capabilities were soon produced by NASA and its contractors. While NASA had no specific plans for Nova class vehicles, many studies were made of just what could be done with them. It was clear early on that a booster larger than the Saturn V would be of great value in a program of manned Mars exploration.

Wernher von Braun, head of the Marshall Space Flight Center, was a particular fan of the Nova/Post-Saturn vehicles and their applicability to Mars. Von Braun had always seen the Apollo program not as an end unto itself, but as an important first step towards a more important goal: manned exploration of Mars. As a result, Nova/Post-Saturn boosters did not die with the acceptance of LOR and the Saturn V, particularly at MSFC.

Several specific Post-Saturn vehicle designs are available to show the sort of capabilities envisioned and the possible Mars missions resulting from the acquisition of such a booster. Krafft Ehricke, then at Convair, designed in 1963 a giant Single Stage To Orbit booster known as Nexus. This hydrogen/oxygen behemoth would have had the ability to launch 455,000 kilograms (1 million pounds) into LEO; a somewhat larger version would have orbited twice that. The vehicle then would have re-entered and splashed down in the ocean, where it could be recovered and refurbished for reuse. The Nexus does not seem to have generated much interest at NASA, however.

The Martin corporation made a large number of studies of Nova/Post-Saturn boosters in the 1960's. One design, given the lackluster designation "T10RR-3," was a fairly typical design. This particular vehicle was the third major design in a series. The T10EE-1 was a similar but fully expendable design; the T10RE-1 was a very similar design with a reusable first stage; the T10RR-3 had fully reusable first and second stages. While this decreased performance somewhat (420,455 kilograms payload to LEO for the T10RR-3 as opposed to 463,280 kilograms to LEO for the T10EE-1), economy was served by having this vast vehicle fully recovered. Recovery was by means of ocean splashdown.

The Post-Saturn vehicle was expected to launch its payload into a 225 kilometer orbit around the Earth; an upper stage would boost the manned Mars mission components to a 568 kilometer orbit for assembly. Two launches of the T10RR-3 could carry a complete manned Mars mission vehicle to orbit; one vehicle carrying the modular cluster of three NERVA stages (used for trans-Mars injection), the other carrying the Mars-bound spacecraft, the Mars arrival stage and the Mars departure stage. This Mars vehicle was basically the standard US design in the 1960's.

The Post-Saturn class launch vehicle would, had it been built, have had a far greater payload capability than any existing launcher. The only paper launchers that have competed with the lift capability of the Post-Saturn class booster were the paper studies of Heavy Lift Launch Vehicles for the Solar Power Satellite in the mid to late 1970's, and even these would have had a hard time competing. In the early 1960's, before the collapse of the American space program, NASA and its industry contractors tended to dream big... and with Nova/Post-Saturn, they dreamed far bigger than anyone else, before or since.

Even under the best of circumstances, such as the Vietnam war not having happened and the American public remaining enthusiastic about the space program, it is doubtful that a Post-Saturn vehicle similar in potential to the T10RR-3 or the Nexus would have been built. Augmentation of the Saturn V would have been a more likely course.

Pros:
- Adaptable to many different upper stages.
- Capable of heavy to extremely heavy lift.
- Would benefit from modernization.
- Mostly reusable (depending upon design).

Cons:
- Infrastructure, including tooling, do not exist.
- Would require new launch facilities.
- Development of new stages and engines would be needed.

Conclusions:
While it would be of extreme advantage to have boosters with million-pound payloads, the Nova/Post Saturn vehicles were never fully designed. Components of these vehicles were built and tested (such as the M-1 hydrogen/oxygen engine), but no specific booster design was ever selected for development.

In order to build a Nova/Post-Saturn class booster today, a considerable national effort would be required. Shuttle derived vehicles such as the Ares and the Magnum are currently at the pinnacle of NASAs developmental interests; to build such a vast vehicle as the Nexus and the associated launch infrastructure would require a program more vast and nationally important than Apollo. The Solar Power Satellite studies of the 1970's showed that such giant boosters are not completely beyond contemplation, but can only be built for an appropriately vast space program.

SHUTTLE-C

The only Shuttle derived vehicle to approach production was the Shuttle-C. This vehicle used the ET and SRBs from the Shuttle, with a Shuttle engine module and an extended cargo bay. Payload was substantially increased by deleting the wings, crew compartment, re-entry and recovery systems. The project was geared specifically towards launching components of Space Station Freedom.

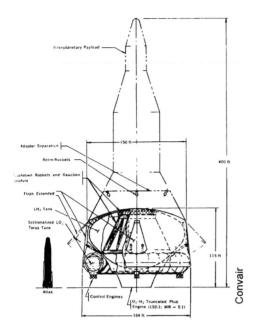
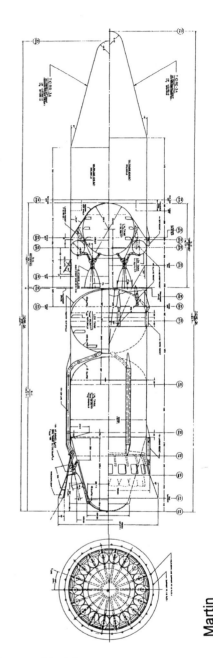

	NOVA - T10RR-3	
Stage 1		
	Dry Mass (KG)	627273
	Propellant Mass (KG)	5333780
	No. Engines (M-1)	18
	Thrust Each (N)	5043129
	Total Thrust (N)	90776322
	Specific Impulse (sec)	
Stage 2		
	Dry Mass (KG)	145455
	Propellant Mass (KG)	776218
	No. Engines (M-1)	2
	Thrust Each (N)	9809800
	Total Thrust (N)	19619600
	Specific Impulse (sec)	428
Payload (kg)	225 km X 28.5 deg.	420455

Figure 10 (top left): Nexus Compared to Atlas.
Figure 11 (right): T10RR-3 Post-Saturn Booster.
Table 5 (bottom left): T10RR-3 Data.

The payload shroud was derived from the cargo bay of the Shuttle. Except for length (81 feet instead of the Shuttle's 60), it was designed to be compatible with standard Shuttle payloads; therefore the diameter was only 15 feet. Many alternate proposals for the Shuttle-C were produced that used larger diameter payload shrouds (up to and including using a second and modified External Tank as a payload shroud), but the 15 foot version was selected as the baseline first generation Shuttle-C.

The lift capacity of the Shuttle-C was substantially greater than that of the Space Shuttle, but was only about 2/3 that of the older Saturn V. Also, the choice of a 15 foot diameter payload bay seriously hampered the possibility of carrying larger diameter payloads.

Two main variants of the same vehicle were offered; one used all three of the SSMEs from the Shuttle, while the second only used two. The three-SSME variant had greater lift capacity, but also expended one more of the expensive rocket engines on each launch.

Pros:
- Straightforward booster.
- Many components are in use and in active production.
- Would use current launch facilities.
- Capable of moderately heavy lift.

Cons:
- Limited ability to carry outsized payloads.
- Limited adaptability to different upper stages.
- Mostly expendable.

Conclusions:

The Shuttle-C was designed specifically for Space Station-based missions. As a result, it would be of somewhat limited value for manned Mars missions except for supporting roles. The cargo capacity is low for even small missions, meaning that on-orbit assembly would be a must; also, the cargo bay, while long, is narrow at only 15 feet. This would seriously hamper loading of many payloads. However, systems similar to the Shuttle-C could prove valuable; payload could be increased by eliminating the mass of the payload bay and using a much lighter payload shroud.

SHUTTLE-Z

Conceptually similar to the Shuttle-C, the Shuttle-Z (from Code-Z, the NASA Office of Policy and Plans) was developed at Martin Marietta as a vehicle capable of launching an entire Mars mission (Mars Direct class missions) with a single booster. A series of Shuttle-derived vehicle designs had been developed at Martin Marietta; these ranged in payloads from 120 tons to 666 tons. The design known as Shuttle Z was given the greatest study.

The narrow payload bay of the Shuttle-C was replace by a relatively giant shroud; this allowed for stowage of large items such as aerobrakes. The three SSME engine pod was replaced by a new engine pod containing four SSME engines. Also, an entirely new hydrogen/oxygen upper stage was to be developed and fitted within the payload shroud.

Due to the drag induced by the larger payload shroud, the SRB's would have to be replaced by the more powerful ASRM's then under development.

Pros:
- Straightforward booster.
- Many components are in use and in active production.

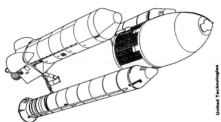

Figure 12: Shuttle-C.

	Shuttle-Z	
Stage 1		
ASRM	Dry Mass (KG)	126810
	Propellant Mass (KG)	1095552
	No. Engines (ASRM)	2
	Thrust Each (N)	15539615
	Total Thrust (N)	31079230
	Specific Impulse (sec)	269
Stage 2		
ET and P/A	Dry Mass (KG)	57000
	Propellant Mass (KG)	723000
	No. Engines (SSME)	4
	Thrust Each (N)	2095730
	Total Thrust (N)	8382920
	Specific Impulse (sec)	450
Stage 3		
	Dry Mass (KG)	13000
	Propellant Mass (KG)	127000
	No. Engines (SSME)	1
	Thrust Each (N)	2095730
	Total Thrust (N)	2095730
	Specific Impulse (sec)	450
Payload (kg)	407 km X 28.5 deg.	138700

Table 7: Shuttle-Z Data.

	Shuttle-C	
Stage 1		
SRBs	Dry Mass (KG)	136385
	Propellant Mass (KG)	1006036
	No. Engines (SRM)	2
	Thrust Each (N)	14759290
	Total Thrust (N)	29518580
	Specific Impulse (sec)	268
Stage 2		
	Dry Mass (KG)	~ 25570
	Propellant Mass (KG)	723000
	No. Engines (SSME)	3
	Thrust Each (N)	2176500
	Total Thrust (N)	6529500
	Specific Impulse (sec)	453
Payload (kg)	407 km X 28.5 deg.	68200

Table 6: Shuttle-C Data.

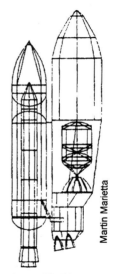

Figure 13: Shuttle-Z.

- Would use current launch facilities.
- Capable of heavy lift.

Cons:
- New upper stages and P/A module would require development.
- Mostly expendable.

Conclusions:

The Shuttle-Z would be a valuable and relatively straightforward booster for manned Mars missions. Payload is certainly sufficient to perform considerable mission launchings, being more than fifty percent greater than the Magnum launch vehicle, NASA's current baseline Mars booster. Outsized payloads such as aerobrakes and habitation modules could also be launched with relative ease. While a concerted development effort would be required for the upper stage and the main P/A module, these are relatively insignificant compared to the development required for a complete booster system and the production, assembly and launch infrastructure development that a new booster would require.

ARES

Derived from prior work on the Shuttle-Z, the Ares booster was designed specifically to perform the role of launching manned Mars missions, Mars Direct in particular. But instead of a side-mounted payload, the Ares had it's payload mounted at the fore end of the fuel ET-derived fuel tank. While this required that the fuel tank be modified, it also lowered drag considerably.

A Propulsion/Avionics module containing four SSMEs was to attached to the side of the main fuel tank in about the same location as the engine module of the Space Shuttle. This would have allowed the Ares to launch from existing Shuttle facilities with minimal modifications. The P/A module was given an ASSET-derived lifting body shape to allow for recovery.

A hammerhead payload shroud was to be used to allow for the transport of 27-foot/8.4 meter diameter Earth Return Vehicles or Habitation Modules with a folded aerobrake.

Standard Shuttle SRB's would have provided boost thrust, with the four SSME's also ignited at launch. The all-new hydrogen/oxygen upper stage would have been ignited while still sub-orbital; an initial 185 kilometer parking orbit would be attained prior to trans-Mars injection.

The Ares booster could be used to launch other payloads into Earth orbit. Without using the upper stage, the Ares could launch 67,000 kilograms into a 300 kilometer orbit. By using the standard upper stage, 106,000 kilograms could be launched; by doubling the upper stages thrust, 115,000 kilograms could be launched.

Pros:
- Straightforward booster.
- Many components are in use and in active production.
- Would use current launch facilities.
- Capable of heavy lift.

Cons:
- New upper stages and P/A module would require development.
- Mostly expendable.

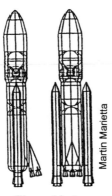

Figure 14: Ares.

	Ares	
Stage 1		
	Dry Mass (KG)	136385
	Propellant Mass (KG)	1006036
	No. Engines (SRB)	2
	Thrust Each (N)	14759290
	Total Thrust (N)	29518580
	Specific Impulse (sec)	268
Stage 2		
	Dry Mass (KG)	64229
	Propellant Mass (KG)	723492
	No. Engines (SSME, 104%)	4
	Thrust Each (N)	2176500
	Total Thrust (N)	870600
	Specific Impulse (sec)	453
Stage 3		
	Dry Mass (KG)	13245
	Propellant Mass (KG)	158760
	No. Engines	1
	Thrust Each (N)	1113000
	Total Thrust (N)	1113000
	Specific Impulse (sec)	450
	Fairing (KG)	20412
Payload (kg)	300 km X 28.5 deg.	106,000
	Trans-Mars	40,189

Table 8: Ares Data.

	N-1	
Stage 1		
	Dry Mass (KG)	125000
	Propellant Mass (KG)	1750000
	No. Engines (NK-33)	30
	Thrust Each (N)	1510000
	Total Thrust (N)	45300000
	Specific Impulse (sec)	308
Stage 2		
	Dry Mass (KG)	35000
	Propellant Mass (KG)	505000
	No. Engines (NK-43)	8
	Thrust Each (N)	1725000
	Total Thrust (N)	13800000
	Specific Impulse (sec)	346
Stage 3		
	Dry Mass (KG)	10000
	Propellant Mass (KG)	175000
	No. Engines (NK-39)	4
	Thrust Each (N)	401250
	Total Thrust (N)	1605000
	Specific Impulse (sec)	352
Payload (kg)	200 km X51.6 deg	100000

Table 9: N-1 Data.

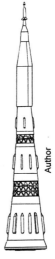

Figure 15: N-1 Booster.

Conclusions:

The Ares would be a useful and relatively straightforward booster for manned Mars missions. While the upper stage, modified fuel tank and the main P/A module would require development, this is a relatively minor problem.

As the Ares was designed specifically with the Mars Direct mission in mind, it is clear that it would be valuable for that mission at least.

N-1

To compete with the American Apollo/Saturn program to land men on the Moon, in the early 1960's the Soviet Korolev OKB-1 design bureau began design of the N-1 superbooster. This vehicle, while designed to perform the same mission and have virtually the same payload capability as the Saturn V, followed a very different design philosophy. All three of the main N-1 stages were kerosene/oxygen fueled, unlike the Saturn that used kerosene only on the first but hydrogen on the second and third stages. The N-1 also used a relatively vast number of smaller rocket engines to generate a similar amount of thrust; this reliance on large numbers of engines (30 NK-33s on the first stage alone) helped lead to the downfall of the project.

The three N-1 stages stacked up to form a cone, unlike the Saturn V's generally cylindrical shape. The fuel tanks also differed from the Saturn V in that they were spherical, unlike the nested and integrated cylindrical and oblate spheroidal tanks.

The first launch attempt, in February, 1969, failed a minute into the flight when the onboard computers shut down all of the first stage engines after one of the thirty engines tore loose. The second launch failed immediately after liftoff when one of the engines exploded; the onboard systems began shutting down engines until the N-1 no longer had the thrust needs to remain airborne. It fell back to the pad and destroyed it. The third vehicle was launched in June of 1970; exhaust interacting with the airstream imparted a roll the vehicle could not counteract. The fourth and final launch occurred in November of 1972; near the end of the first stage burn, one of the engines again exploded (due to fuel line hammering because of a sudden but preprogrammed shutdown of six engines to reduce G-loading). In May of 1974 the program was cancelled.

Even though the N-1 failed and the Soviet Union never landed a man on the moon, several Soviet proposals were made for using the N-1 to launch components of a manned Mars vehicle. Early designs called for a vehicle assembled in Earth orbit with a mass of 1,630,000 kilograms; 25 N-1s would be needed to launch it. A more realistic 1959 design called for the design and construction of a "Heavy Piloted Interplanetary Spacecraft," or TMK. This would not actually land on Mars, however; it would be only a flyby mission. It could be launched by a single N-1.

In 1969, OKB-1 proposed the Mars Expeditionary Complex (MEK). This vehicle would have been similar to concurrent NASA Mars proposals, with a nuclear reactor powering ions engines for thrust, and with a payload of an in-transit mission module, a Mars landing craft and an Earth return vehicle. Initial mass in Earth orbit was to be 150,000 kilograms; two launches of the more advanced N-1M boosters would have orbited the components.

Proposals for manned Mars vehicles using nuclear thermal rockets for propulsion were also made, but none made it into space.

Pros:
- Development mostly done.
- Capable of heavy lift.

Cons:
- Unreliable booster.
- Infrastructure, including tooling, no longer exist.
- Reliance upon an unstable foreign government.
- Fully expendable.

Conclusions:

In four launch attempts, the N-1 never succeeded to even attain orbit, much less send a payload to the Moon. The program was scrapped, as were the tools required to build the N-1. The idea of using the N-1 as a booster for manned Mars missions is at best dubious, and more likely foolhardy.

ENERGIA

In 1976, design began at NPO Energia on the Energia booster. This vehicle was a replacement to the failed N-1, and was vastly different in design. A core stage with four large hydrogen/oxygen engines was flanked by four liquid-fuel strap on boosters, each powered by a single kerosene/oxygen burning engine. The strap-on boosters were developed from the Zenit booster, capable vehicles in their own right.

The Energia booster bears a striking similarity to the SRB/ET stack used by the US Space Shuttle. The main difference is the difference in the number of strap-on boosters and the use of main engine on the core. This allows the Energia to have somewhat greater operational variability with its payload, unlike the US SRB/ET stack with would require separate hydrogen/oxygen burning engines.

The Energia is designed to NOT put its side-mounted payload into orbit. Rather, it puts the payload onto a high sub-orbital trajectory; it is up to the payload to either circularize its Earth orbit or boost directly onto an escape (or geostationary transfer) trajectory. To this end, several upper stages have been designed for use on the Energia, the two main being a kerosene/oxygen stage useful mainly for injecting into Earth orbit, and a hydrogen/oxygen stage for higher energy escape trajectories.

The Energia booster was launched only twice; in May of 1987 it launched the Polyus spacecraft (which failed to attain orbit due to upper stage failure) and in November of 1988 it launched the Buran shuttle. The Energia performed well on both launches; but shortly thereafter the Soviet Union collapsed and the Soviet space program began to crumble. The Energia was never launched again, and the assembly and launch infrastructures have remained unused. Reports indicate exceedingly poor conditions at the Energia launch facility.

Advanced versions of the Energia were proposed, all of which are fairly straightforward modifications. By increasing the number of strap-on boosters, payload capacity could be increased substantially. While the standard four booster Energia can loft 105,000 kilograms on a sub-orbital trajectory, a six booster version could loft 150,000 kilograms and an eight booster variant, which would require that the payload be relocated to the fore end of the core stage, could loft 200,000 kilograms.

NPO Energia made several proposals for using the Energia booster for manned Mars missions, many similar to concurrent US proposals. Designs using ion engines with power coming from vast solar panels were made (5 Energia launches required to launch the 355,000 kilograms of hardware into Earth orbit for assembly), as were designs using nuclear thermal engines. By the late 1980's, however, all hope of Soviet manned Mars missions were gone. However, interest in the Energia booster remained high; it has been proposed in NASA circles to launch components of a US space station, components of a conventional US on-orbit assembled manned Mars vehicle, and to launch Mars Direct hardware.

Pros:
- Proven booster.
- Capable of heavy lift.
- Adaptable to alternate payloads and upper stages.

Cons:
- Out of production.
- Infrastructure requires refurbishment.
- Reliance upon an unstable foreign government.
- Fully expendable.

Conclusions:

The Energia would be a capable booster for manned Mars missions. But given the effort that would be required to revive the program and its launch facilities, and the dangers posed by dealing with a government of doubtful stability, it seems likely that Shuttle derived launch vehicles of comparable lift capacity would be more advisable.

VENTURESTAR

Lockheed-Martin is currently engaged in an effort to develop Single Stage To Orbit technology. The VentureStar is a proposed operational follow-on to the X-33 technology demonstrator.

A lifting body based on early Space Shuttle design work done by Lockheed in the late 1960's, the VentureStar is to be equipped with seven linear aerospike engines at the blunt trailing edge of the vehicle. Burning hydrogen and oxygen, these engines, which form a single large linear aerospike, are to provide the power to boost the vehicle into Earth orbit carrying a Space Shuttle class payload in an internal bay. If all goes as planned, the VentureStar will be able to launch its payload for about $1000/pound ($2200/kilogram), as compared to the approximate $7000/pound for the Space Shuttle.

The main target markets for the VentureStar are the International Space Station and commercial telecommunications satellites. However, there has been great debate in the aerospace community regarding expendable HLV vs. reusable SSTO, and the VentureStar is the only SSTO currently under active development.

With a payload capacity to LEO of only 27 tons (in a payload bay 45 feet long and 15 feet in diameter), VentureStar would be hard pressed to provide a Mars mission in a single launch. For example, using a nuclear thermal upper stage and payload similar to those proposed for the Design Reference Mission (137.5 tons in LEO), of that 27 tons of payload delivered to LEO, only 5.7 tons would be habitation module - very cramped. In order to launch the entire Design Reference Mission as currently proposed, a total of 16 launches of the VentureStar would be needed. But as the Design Reference Mission calls for the equipment to be integrated on the ground, the equipment will weigh more when re-designed for on orbit assembly. Another concern will be in fueling the Mars bound craft. Each of the three vehicles will need 50 tons of liquid hydrogen propellant. Assuming the VentureStar payload tanks are extremely light weight (2000 kilograms), then two VentureStar flights can carry this quantity of hydrogen. But

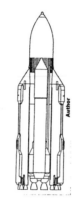

Figure 16: Energia.

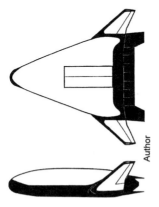

Figure 17: VentureStar.

	Energia	
Stage 1		
Strap-ons	Dry Mass (KG)	140000
	Propellant Mass (KG)	1280000
	No. Engines (RD-170)	4
	Thrust Each (N)	7268000
	Total Thrust (N)	29072000
	Specific Impulse (sec)	309
Stage 2		
Core	Dry Mass (KG)	85000
	Propellant Mass (KG)	820000
	No. Engines (RD-0120)	4
	Thrust Each (N)	1962000
	Total Thrust (N)	7848000
	Specific Impulse (sec)	452
Stage 3		
RCS	Dry Mass (KG)	2000
	Propellant Mass (KG)	15000
	No. Engines (1158DM)	1
	Thrust Each (N)	85000
	Total Thrust (N)	85000
	Specific Impulse (sec)	352
Payload (kg)	407 km X 51.6 deg.	83000
	trans Mars: LH2/LOX stage	28000

Table 10: Energia Data.

	VentureStar	
Stage 1		
	Dry Mass (KG)	89545
	Propellant Mass (KG)	876820
	No. Engines (RS-2200)	7
	Thrust Each (N)	1917370
	Total Thrust (N)	13421590
	Specific Impulse (sec)	361
Payload (kg)	185 km X 28.5 deg.	26820

Table 11: VentureStar Data.

Figure 18: Boeing EELV/Delta IV.

	EELV	
		LEO Payload (KG)
	Design	
Lockheed	MLV-D	3900
	HLV	7320
	HLV-A	18640
Boeing	Delta IV Small	5680
	Delta IV Medium	8640
	Delta IV Heavy	26100

Table 12: EELV Data.

with the losses likely to be encountered, at least three VentureStar tankers will be required for each vehicle. Three separate VentureStars will need to be on station at virtually the same time, as boiloff rates will be high in Earth orbit.

Pros:
- Reusable.
- Low cost.

Cons:
- Low payload.
- Development uncertain.
- Not adaptable to outsized payloads or upper stages.
- Unproven technology.
- New launch infrastructure required.

Conclusions:

The VentureStar, if built and successful, could serve a useful role in servicing Mars missions that required considerable on-orbit assembly. But given the low payload capacity (both in terms of mass and bay dimensions), the VentureStar is unlikely to serve much purpose in launching hardware. On orbit assembly would be the only mission mode available for VentureStar launched payloads, and payload bay size limitations would drive Mars vehicle designs. Mars Direct and Design Reference Mission modes would require a considerable redesign of mission elements to allow them to be transported by VentureStar.

EELV

Currently under development for the United States Air Force are two Evolved Expendable Launch Vehicles (EELVs). Both designs are largely conventional boosters, intended to reduce launch costs for light to heavy payloads. Complete data on these boosters is lacking.

Boeing is developing the Delta IV EELV. This vehicle has a hydrogen/oxygen core powered by a single RS-68 engine. The RS-68 is a high-thrust, medium specific impulse engine, designed for low cost, low parts count and complexity and to be used only once. Three basic variants are planned: a low payload version using only a single core and a small upper stage; a medium lift version using a single core and a large upper stage and a heavy lift version using three cores strapped together with a large upper stage.

Lockheed-Martin is following a similar design philosophy, but is using different propellants. Their core vehicle will be derived from the Atlas IIAR boosters, and will use kerosene/oxygen for propellants, with a single twin-chambered RD-180 engine. Storable propellant or Centaur upper stages can be used.

Pros:
- Development likely.
- Probably adaptable to outsized payloads and upper stages.

Cons:
- Low payload.
- New launch infrastructure required.
- Expendable.

Conclusions:

The Evolved Expendable Launch Vehicle could serve a useful if highly secondary role in servicing Mars missions that require considerable on-orbit assembly. However, payload capacity, even for the heavy lift variants, is low compared to most boosters examined, and the EELV is not planned to be man-rated.

MAGNUM

The current Design Reference Mission utilizes the proposed Magnum as the baseline launch vehicle. The Magnum is intended to use as much off the shelf hardware as possible to lower development time and cost. The core stage is derived from the external tanks from the Space Shuttle; the RS-68 hydrogen/oxygen engines are under development for the United States Air Forces EELV/Delta IV project. Solid Rocket Boosters from the Space Shuttle can be used to supply launch boost thrust, but it is hoped that eventually they can be replaced by Liquid Fly Back Boosters when and if they become available. A circularization stage based on the cryogenic stage of the Delta III is to be used for final orbital insertion into LEO.

Detailed design work does not seem to have been done as yet on the Magnum. However, basic costing models have been generated, showing a development cost on the order of $2.5 billion, with a payload launch cost around $1000 per pound (similar to the fully reusable SSTO VentureStar). Lower costs are expected from using existing designs and launch infrastructure (the ET and pads from the STS program), low cost expendable RS-68 engines and flyback boosters.

The Magnum is not currently under development, but it does represent the current thinking at NASA of what will be needed to fulfill manned Mars booster requirements.

Pros:
- Straightforward booster.
- Many components are in use and in active production.
- Capable of heavy lift.

Cons:
- New lower stages (first and LFBB) would require development.
- Partially expendable.
- Uncertain development.
- Would require new or modified launch facilities.

Conclusions:

As with the Ares booster, the Magnum is designed to fulfill a specific Mars mission required (the current Design Reference Mission) and is therefore admirably fitted to performing that task. Also, the Magnum seems likely to be a good all-around heavy lifter, thus capable of boosting alternate Mars mission payloads into Earth orbit for assembly. Development would likely be relatively straightforward due to the use of so many off the shelf or marginally modified components. However, actual interest shown for Magnum in NASA, industry and the public, at this point seems rather lukewarm, and development is uncertain.

STARBOOSTER 400

A private venture of StarCraft Boosters, Inc. is the StarBooster series of liquid fueled fly back boosters. The basic vehicle, the Star Booster 400, is designed as a relatively simple straight-winged aircraft equipped with turbofan engines in the nose for ferry and flyback. Main propulsion is to be provided by a single Russian Zenit booster which is inserted into the aircraft; in this way, the Zenit booster and its RD-170 rocket engine can be reused. While performance of the Zenit is of course reduced by the added mass of the aircraft, it is hoped that

Figure 19: Lockheed Martin EELV.

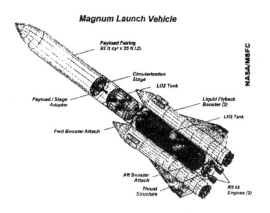

Figure 20: Magnum

Figure 21: StarBooster 400 with Zenit Upper Stages.

		Magnum	
Stage 1			
LFBB		Dry Mass (KG)	?
		Propellant Mass (KG)	?
		No. Engines (RD-180)	8
		Thrust Each (N)	3836523
		Total Thrust (N)	30692184
		Specific Impulse (sec)	311
Stage 2			
Core		Dry Mass (KG)	?
		Propellant Mass (KG)	767823
		No. Engines (RS-68)	2
		Thrust Each (N)	2898350
		Total Thrust (N)	5796700
		Specific Impulse (sec)	368
Stage 3			
Circularization	Dry Mass (KG)		2886
Delta III-derived	Propellant Mass (KG)		16830
	No. Engines (RLB10-2)		1
	Thrust Each (N)		
	Total Thrust (N)		
	Specific Impulse (sec)		466
Payload (kg)	407 km X 28.5 deg.		89300
	407 km X 51.6 deg.		83700

Table 13: Magnum Data.

StarBooster 400	
Dry Mass (KG)	65000
Propellant Mass (KG)	317000
No. Engines (RD-170)	1
Thrust Each (N)	7697000
Total Thrust (N)	7697000
Specific Impulse (sec)	309

Table 14: StarBooster 400 Data.

launch costs per pound of payload will be substantially lowered by reusing the expensive rocket propulsion system elements.

A large number of current (Titan cores, Ariane cores, Zenit upper stages) and planned (X-33, VentureStar, Kistler, EELV) boosters can be boosted or augmented by the addition of one or more StarBooster 400 LFBBs. This will be done to either improve performance of the booster or to lower the launch cost of the booster by replacing more costly expandable lower stages or strap-on boosters. While the StarBooster 400 has not been specifically proposed as a booster for Mars missions, the use of such low-cost thrust augmentation systems can make a marginal Mars booster more adequate, or can lower the launch cost of an otherwise well-performing but costly Mars booster.

Pros:
- Reusable.
- Based on proved boosters.
- Variable payload.
- Adaptable to outsized payloads or upper stages.

Cons:
- Development uncertain.
- Design and concept come from outside NASA and large contractors.
- New launch infrastructure required.

Conclusions:

The StarBooster 400 could be a useful component in a manned Mars mission, used to either increase the throw capacity of a heavy-lift booster or to decrease launch costs for supply vehicles for an on-orbit assembled Mars vehicle. However, while the concept is relatively low-tech (an aluminum aircraft wrapped around an existing booster), development is highly uncertain.

STARBOOSTER 1800

Intended as a launch vehicle for Solar Power Satellites, the StarBooster 1800 is a much larger cousin to the StarBooster 400. As opposed to the StarBooster 400, the fuel tanks and rocket engines are fully integrated into this larger design. With a total of five RD-170 rocket engines, the StarBooster 1800 produces more thrust than the Saturn V.

The basic StarBooster 1800 is designed to serve as a booster for an all-new hydrogen/oxygen powered spaceplane. This spaceplane, powered by five SSMEs, carries its liquid oxygen in internal, fully reusable tanks, but carries the liquid hydrogen externally in the form of a cluster of tanks that will serve as part of the structure of the proposed Solar Power Satellite.

The use of the StarBooster 1800 as a dedicated trans-Mars launcher has also undergone some investigation. By using a new expendable core stage (derived from the SPS spaceplane by removing wings and recovery systems) equipped with SSMEs and an upper stage also equipped with SSMEs, two StarBooster 1800s flanking the stage can provide a massive payload capability for direct launch to Mars.

Pros:
- Reusable.
- Based on proven propulsion systems.
- Variable payload.
- Adaptable to outsized payloads or upper stages.

Cons:
- Development uncertain.
- Design and concept come from outside NASA and large contractors.
- New launch infrastructure required.
- Would require considerable development.

Conclusions:

The StarBooster 1800 would provide a considerable payload capability for manned Mars missions, whether used as a modified SPS launcher to orbit large payloads for on-orbit assembly, or as a dedicated HLLV for direct throws of heavy payloads to Mars. If the concept can find support enough within NASA and the aerospace community to build it, the StarBooster 1800 or a similar vehicle could provide relatively vast trans-Mars payload capability.

FINAL CONCLUSIONS:

A wide variety of design options have been developed over the years and presented in this paper. However, all options are not equal; and some may be better for some missions than others.

In general terms, booster with low payloads are not good candidates. While VentureStar and EELV may promise low launch costs, the large amount of on-orbit assembly that would be required would effectively eliminate any cost savings.

Also, boosters designed in the relatively distant past are not the best candidates for these missions. Fabrication and launch infrastructures are largely gone for those vehicles that flew and have been withdrawn from service. To revive them would require considerable effort; effort that could be more productively spent on designing new systems using the latest technologies.

However, when positing manned Mars missions two or three generations further down the historical line, boosters with capabilities only equaled by past hypothetical boosters such as the Nova/Post-Saturn class of vehicles may be required. In this eventuality, it may prove most efficient to revive these older design studies and rethink them with modern materials. Worthy of study for these future missions are modern reusable heavy lifters such as the StarBooster 1800.

For on-orbit assembled Mars missions, heavy payload lift is not necessarily of paramount importance. In this case, lowest launch cost on a per pound to orbit basis is important, as is commonality with existing infrastructure. A Shuttle derived system (such as Shuttle-Z) with backup from smaller VentureStar-type RLV's would seem a logical choice.

For the current Design Reference Mission, the Magnum is obviously the booster of choice given that it is designed for that specific task. However, similar vehicles such as the Ares would clearly work as well, as would a revived Energia. The use of abandoned foreign boosters, however, is to be looked upon with skepticism.

For Mars Direct type missions, it is important that the booster have the ability to throw large payloads directly onto an interplanetary trajectory. This effectively eliminates most two-stage boosters; the first two stage expend their greatest energy in attaining orbit, using a partially emptied second stage to provide a high-energy trajectory is not efficient. Vehicles such as the Magnum are therefore dubious; but boosters such as the Shuttle-Z and Ares, each with a specially designed hydrogen/oxygen upper stage, would serve adequately.

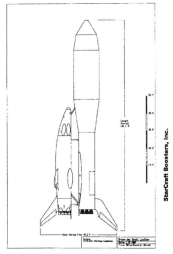

	StarBooster 1800	
Stage 1		
	Dry Mass (KG)	265000
	Propellant Mass (KG)	1594000
	No. Engines (RD-170)	5
	Thrust Each (N)	7697000
	Total Thrust (N)	38485000
	Specific Impulse (sec)	309
Stage 2		
	Dry Mass (KG)	126400
	Propellant Mass (KG)	1075000
	No. Engines (SSME)	5
	Thrust Each (N)	2095730
	Total Thrust (N)	10478650
	Specific Impulse (sec)	450
Payload (kg)	407 km X 28.5 deg.	138700

Figure 22 (left): StarBooster 1800 with SPS Payload.
Table 15 (right): StarBooster 1800 Data, SPS Payload.

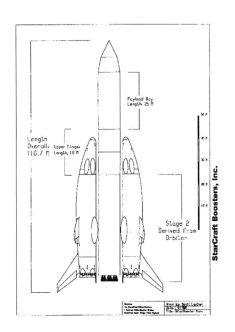

	Mars StarBooster 1800	
Stage 1		
	Dry Mass (KG)	530000
	Propellant Mass (KG)	3188000
	No. Engines (RD-170)	10
	Thrust Each (N)	15394000
	Total Thrust (N)	76970000
	Specific Impulse (sec)	309
Stage 2		
	Dry Mass (KG)	192380
	Propellant Mass (KG)	1603180
	No. Engines	
	Thrust Each (N)	
	Total Thrust (N)	17981788
	Specific Impulse (sec)	450
Stage 3		
	Dry Mass (KG)	41620
	Propellant Mass (KG)	346836
	No. Engines	
	Thrust Each (N)	
	Total Thrust (N)	2917650
	Specific Impulse (sec)	450
Payload (kg)	Trans-Mars	107238

Figure 23 (left): StarBooster 1800 as a Mars Booster.
Table 16 (right): Mars StarBooster 1800 Data

AN RLV / SHUTTLE COMPATIBLE HABITATION SYSTEM

Kurt Anthony Micheels[*]

The non-availability of Heavy Lift Launch Vehicles mandates use of the Shuttle or proposed RLV to transport material for Human Mars exploration to LEO. Current Mars exploration scenarios make use of landers and habitats compatible only with HLLVs. The size of the Habitat elements prevents economical delivery to polar test sites or practical use in other terrestrial extreme environments. A study of technologies was conducted regarding deployment of habitats capable of transport via LC-130H Antarctic logistics aircraft or integration with a robotic lander vehicle with launch to LEO via Shuttle or RLV. The result was a logistics support module capable of functioning as an inflatable deployment system. The module design enables Mars landing and surface transport via a robotic rover and facilitates establishment of a surface base derived from the current Mars reference mission.

INTRODUCTION

The recent decision to pursue testing and development of several re-usable launch vehicles seems to ensure that an abundance of transport to low Earth orbit (LEO) will exist in the early 21st century. A fundamental characteristic shared by these proposals is their apparent payload capability approximately 20-50MT (comparable to the shuttle). Attempts to design a new heavy lift vehicle, with the ability to place greater mass on orbit, are doubtful. Given this situation, it is not likely that the various 6 to 10m habitats, proposed for human Mars exploration, would be compatible with the next generation of launch vehicles.

Habitat Volume

The large cylindrical habitats proposed for the Mars Direct and Semi-direct mission plans typically contain a mass of between 40 and 60MT. This mass is contained in a relatively large volume yielding a low mass to volume ratio; i.e. an inefficient packing factor. Most of the mass is attributed to equipment and consumables, thus denying additional space to the human occupants.

Habitat Testing

The scale of the Mars Direct habitats is such that they cannot be transported in conventional aircraft. As a result, they are not easily moved to desirable terrestrial testing sites, such as Antarctica. Current plans call for testing of an inflatable transit habitat at the international space station. But, terrestrial testing would be difficult and is unlikely.

Mission Compatibility

The large scale habitats are designed for use within a specific mission plan and assumes the future existence of a heavy lift launch vehicle. They are designed as single mission units - a new habitat is required for every additional mission.

---

[*] Surface Extreme Environment Dwelling Systems, 357 Boardman St. #2, Auburn, California 95603.

As an alternative, a modular system can be designed for integration with most of the equipment necessary to support a Mars or lunar surface mission and be compatible with the shuttle / RLV payload bay. When reconfigured for terrestrial operations, it will be transportable via C-130, C-17 or C-141 aircraft. Its mass will be approximately 1/3 that of the Mars Direct habitats, be capable of terrestrial testing and be adapted to a wide variety of mission plans. It is based on the assumption that the crew members will make their trip in a separate habitat specifically designed for interplanetary transit. The modular system, when integrated with a lander may be impulsed on its trajectory individually or combined with other modules on an interplanetary vehicle similar to that used for crew transit, later in the mission. When launched individually, the system could arrive in advance of the crew and deploy telerobotically. This would also allow time for in-situ consumable production and crew adaptation upon arrival at Mars. It is this option that shall be discussed in detail in this paper.

METHOD

The issue of how to design an inflatable habitat contained within an RLV / Shuttle compatible module, was approached as an architectural design problem. This means, several relevant design parameters are identified, and the process evolves to a nearly convergent solution. For this project, a series of previous projects enabled the attaining of several conclusions. Three of the project conclusions developed heuristically, with a fourth demonstrating a failure in the design logic and termination of that line of investigation. The result was a redirection in overall design philosophy, and is the topic of this study.

Design Parameters

For each of the five projects, a series of design parameters was established. In that this study proposes design of human dwellings for environments completely alien to humans, it was decided an adaptation of the traditional approach to human dwelling design was required. An analysis of terrestrial dwelling requirements (this being an on going task, worthy of a separate study) revealed three primary constituents.

1. Function - what it will do?
2. Context - how it relates to its environment?
3. User needs - how it will be used by people?

These parameters were then applied to each design by first establishing what is to be accomplished. This results in a list of individual functions and equipment, with required area, volume, mass and power necessary to meet function requirements. In traditional architectural jargon, this list is the "program". Since planetary dwellings must function in a realm of great diversity, integrating spacecraft, surface vehicles, power production systems, etc., a better term for this list is "mission requirements".

The second parameter, context, refers to requirements for the environment the habitat is to occupy. It also refers to the overall mission the habitation system is to be integrated with. So, once it is established what is needed to accomplish the mission, it must be ascertained how all of the systems and equipment can be designed to fit into a mission compatible spacecraft, and function optimally when deployed at Mars.

The third parameter, user needs, is the most important, and most overlooked, especially in traditional applications. This deals with the overall design of the habitat for human use. It begins with habitat organization (both exterior and interior), to enable suitable way-finding, and proceeds to such apparently minimal issues as door knob design. Inadequate attention to user issues can lead to serious psychological and physiological dysfunction affecting mission completion.

How do these issues get organized to create the desired result? As an architectural exercise, it is accepted that the path to a finished result is not linear. Architect Louis Kahn has said in several articles and lectures that "architecture is the thoughtful making of space"(12). If this is accepted as dogma, the design path may be perceived to progress something like this:

1. Mission Requirements - determine how much space is required.
2. User Needs - what form should the space take to provide for optimal human use and mission accomplishment?
3. Context - with the given space and form, how can it be made to physically fit within mission guidelines and function in the desired environment?

This is an over simplification of the beginning of the design process adopted for this investigation. But, it represents a valid starting point. From here on the process continues to test each parameter at various stopping points (design reviews / critiques) and information added until the individual project is deemed complete. This process is comparable to Zeisel's "design development spiral", with the spiral becoming smaller as the designer approaches acceptable resolution (25).

The Projects - Antarctic High Station (October 1992 - May 1993)

The first project in the series was the "Antarctic High Station" design competition, sponsored by the National Science Foundation and the American Institute of Architecture Students. Mission requirements called for living and working space for sixteen research personnel in a space analog habitat, located on Dome A, approximately 2000 miles from McMurdo Sound. The design required the habitat be capable of sustaining itself for the duration of the Antarctic winter (24).

Mission requirements dictated a total habitat scale such that use of fixed modules, similar to a space station module, was undesirable due to the number of flights to deliver it to the site. Use of ski equipped LC-130H aircraft was preferred due to the remoteness of the site and the safety and economy relative to ground transport.

The design approach favored use of a module to contain an inflatable habitat and other necessary systems. The deployable module concept, in concert with aircraft, has been used successfully for military field hospitals and support facilities. A variety of self contained deployable habitats were postulated in the sixties by British architectural design team, Archigram.

The user requirements suggested a division of functions between public, private and work. This enabled clear physical delineation between social activity, sleep and mission related tasks (10). Special attention was granted the design of the private living spaces, due to the extreme amount of time winter over personnel spend in them (21).

Contextual demands of the site required a high degree of redundancy. The modules, after habitat deployment, would be used as emergency shelters (Crew Assured Survival Modules) in the event of life support failure. Consequently, each module would house emergency consumables and equipment, in addition to habitat systems.

The principle result of this project was the definition of an air transportable inflatable deployment system, fully integrated with a life support system, power and consumables. Established behavioral design principles were employed in establishing the extent of the interior space and its organization.

Projects - Mars SEEDS (January 1994 - December 1994)

The second project was the topic for a Master of Architecture thesis at the University of South Florida (17). The primary intent was to adapt the same module system developed for Antarctic High Station to the 1993 NASA Mars Reference Mission (8). The Reference Mission postulated the use of an HLLV to launch a cylindrical transit / surface habitat on a "direct ascent" to Trans Mars Injection (TMI), followed by aerobraking, direct descent and landing at Mars. Since the interior volume of the cylindrical habitat was fixed, it was decided use of a module contained inflatable could increase the interior volume of the habitat when on the surface, thus eliminating the need for landing of additional habitats.

The design approach utilized a combination, or "stack" of Antarctic modules to create a spacecraft with volume comparable to the Reference Mission habitats, when configured for transit. The modules at the bottom of the stack contained inflammables deployed after landing at Mars. Other modules were adapted to serve as crew compartments and a variety of equipment bays.

The mission requirements called for accommodation of six crew members during a 120-180 day transit and 600 day surface visit. The habitat would integrate with a variety of surface systems previously landed in vehicles derived from the same module.

A number of references were used to develop both habitat and surface systems program. These include *The Case for Mars* conference proceedings, *Mars 2008* (Cohen, 1993) and the International Space University *International Mars Mission* (16)(3)(15).

User requirements were similar to those for the previous project. The primary references for this project were *NASA Standard 3000* and *Space Station Crew Safety Alternatives Study* (Peercy, *et al.*, 1986) (14)(20).

The context required a fully pressurized envelope and some degree of radiation protection. This project assumed a solar storm shelter could be used both in transit and on the surface. This was designed to occur in the circulation space created by the junction of four modules, the shielding material being aluminum (20). The project assumed a closed LSS with 25% dedicated to CELSS.

This project demonstrated how the Antarctic module could be integrated into a comprehensive Mars mission plan. It also allowed extensive investigation into the design of a private living space adaptable to micro-gravity and Mars surface operations, compatible with the module.

Projects - Mars SEEDS II (May 1995 - June 1996)

SEEDS II was a post-graduate iteration of SEEDS 1, in preparation for Space 96 and ICES 96 conferences. Emphasis was on continued development of the private living space, derivation of an optimal model and its integration with the module. Architectural conventions, such as proportioning systems and standardized scale of furnishings were applied (19).

The analysis resulted in a private living space area of 5.76m² (2.4m x 2.4m), and 2.4m high. The space interior dimensions were based on a 20cm grid, this enabling a compromise between requirements for rack mounted equipment, standard furnishings and anthropometric constraints. The 2.4m cube could easily be manifest as a singular module, having overall exterior dimensions of no more than 2.54m, enabling air transport. It could be combined / stacked with other modules containing LSS, inflatables and other required equipment (18).

This project resulted in a highly evolved private living space, compatible with an air transportable module, derived from previous research regarding behavior in isolated environments. The same module could be used for other habitat logistics and systems functions.

Projects - Mars SEEDS III (June - September 1996)

The cube module, as well as the Antarctic module (an extruded octagon), packed efficiently, but were poor pressure vessels. SEEDS III sought to place the SEEDS II module in a cylindrical envelope compatible with the Shuttle payload bay, or when combined in groups of three to four, integrated with an HLLV payload shroud. Announcement of selection of a contractor to build the X-33 in Summer of '96 motivated this approach. The resulting paper was presented at the AIAA Space Programs and Technologies Conference 1996, in Huntsville.

Mission requirements were the same as SEEDS 1, with the exception that eight crew members would be accommodated, four in one of two vehicles. This was due to the 5.76m² floor plate (13.824m³ volume), when doubled, occupying most of the space inside the pressure vessel. The required pressure vessel was found to be 9m in length, 4m in diameter with 1:1 ratio end domes (domes at the end of the cylinder have a radius equal to the radius of the cylinder). Three pressure vessels would be combined with a supporting frame to create one transit / surface system; two vessels having crew accommodations, the third crew living / working space and end domes, the inflatables and LSS.

This study examined in some detail the number and type of surface systems necessary to support the mission. This enabled a volumetric analysis to determine what could be carried in the habitat vehicles and what must be landed separately. The volumetric analysis was followed by a mass analysis that found each of the pressure vessels mass to be approximately 23MT. Consequently, the mass of the entire vehicle (excluding the support system) is 69MT.

Since the Reference Mission habitat mass was given as +61.5MT, it was clear that SEEDS III was non-optimal (8). This, of course, was due to the use of separate pressure vessels and the need to provide independent systems for each.

The conclusion drawn from this line of investigation is that the habitat module must be designed for Shuttle / RLV or HLLV. Not for both.

The RLV / Shuttle Compatible Module

Since a module optimized for HLLV and Shuttle / RLV seemed unlikely, and development of an HLLV less likely, a redirection in research led to an approach focusing on a Shuttle / RLV compatible module.

Significant research regarding radiation protection for interplanetary spacecraft determined Galactic Cosmic Radiation (GCR) may pose a long term threat equal in magnitude to Solar Radiation (5). This condition requires consideration of shielding for the entire interplanetary spacecraft habitat, resulting in doubling the mass of the conceptual vehicle. This increase in mass would require the interplanetary vehicle to be placed at LEO as separate components and to be a system separate from the surface habitat. Establishment of separate surface and interplanetary habitats also enables optimization of each for its function; i.e. surface operations at 37% g versus micro-g (6). Consequently, it was clear the new module would house the habitat and its systems, while the crew would transit in a separate vehicle.

An additional aspect of this approach, enabled by the overall reduction of scale, is the consideration of the module for terrestrial applications. This module, like the original Antarctic Module, is designed to only contain an inflatable shelter and necessary equipment. It can be easily adapted to transport by logistics aircraft, allowing deployment in polar regions. Consequently, it could serve as an analog for planetary missions or be configured for a wide variety of scientific support or military missions in terrestrial environments.

DISCUSSION / PARAMETERS

Mission Requirements

The deployed habitat must provide an optimal living and working environment for eight crew members on the surface of Mars for 600 days, and when packed fit within a Shuttle or RLV payload bay. The packed module must be capable of integration with all spacecraft systems necessary for transit to Mars and landing. After Mars landing, it must be capable of telerobotic / autonomous deployment, allowing immediate occupation and use upon crew arrival. When not integrated with Mars surface systems, the habitat may be transported by conventional aircraft.

A detailed listing of the program requirements follows:

1. Air / Water - provide a physical / chemical life support system capable of supporting the crew for the duration of the mission, as well as follow-on crews for the entire mission cycle. The desired ECLSS must have 98% closure. A prototype system, including consumables, has a mass of 4MT and volume of $15m^3$ (9). When the habitat is deployed, it will be housed in the module.

2. Food - The habitat must provide sufficient internal volume dedicated to food storage. Food is assumed to be transported in a separate module. The internal volume will house relocated dehydrated food and fresh food, as it becomes available. The required habitat volume is $3m^3$ (9).

3. Shelter - The habitat envelope must meet or exceed existing volumes for cylindrical habitats. This requires a volume range from $265m^3$ to $1000m^3$ (4). For this study, an inflatable with internal volume of approximately $86.4m^3$ was chosen (23). To meet stated requirements a minimum of four are needed. To accurately assess habitat volume, it is necessary to itemize it by function:
A. Private living space and hygiene facility - $85m^3$ (18).
B. Work / Lab - $57m^3$ (9).
C. Social, dining - $45m^3$ (9).
D. Circulation / Stowage - $90m^3$ (17). This figure includes EVA storage and maintenance, and food storage. It also includes additional mechanical and plumbing space. The ECLSS is considered to be housed in the module, but external to the habitat.

The total habitat required volume is $277m^3$. The total allowable volume using the four CEISS inflammables is $345.6m^3$.

4. Egress - Two means of egress are required from each inflatable (7). For purposes of this dwelling, this requirement was amended to "two means of egress to another habitable volume". This creates a requirement for pressurized passageways between each inflatable. Egress to the exterior must be through one of two airlocks (personnel locks). Two equipment / sample locks must also be provided, as well as a pressure port to serve as a rover dock. The rover port, sample lock and personnel lock may be combined, so long as a minimum of two locks exist. Two combined pressure ports may have a combined mass of 1MT (5).

5. CELSS - A plant growth chamber capable of providing 25% of the atmospheric life support and some fresh food will occupy approximately $62m^3$ of one inflatable (27). Plant growth systems mass is 3MT (5).

6. Power - Approximately 100kw will be required for the entire habitat, when fully operational (8). For initial deployment and ECLSS operation, 50kw is sufficient. A proposed nuclear power system, combined with energy conversion systems, should be part of the entire habitat surface "package". The mode of deployment should enable remote placement of the power source prior to activation. Power source mass is 1.5MT (11).

7. Communications - S band and Ka/Ku band communications must be provided for direct and over the horizon communication. Communications mass is 1.5MT (5).

8. Active Thermal Control System Required for elimination of systems waste heat. Its mass is 2MT (9).

9. Crew accommodations - This includes habitat furnishings that must be installed after crew arrival. Some of these items may be landed in a separate module. Approximate mass is 2MT (9).

10. The module - Must contain packed inflammables, ECLSS and accommodations package. All other systems are external. They may be integrated directly with the module or delivered via the same lander vehicle and deployed separately; the power systems being a case in point. The module and its inflation system have a mass of 0.81MT.

11. To descend to the surface of Mars, the module and related systems require integration with a lander vehicle. Estimated mass of structure, RCS descent engine and related systems is 7.99MT (15).

The total habitat lander mass (less the inflatable) is 23.8MT.

User Requirements

Habitat organization, internally and externally is the primary user driven requirement not directly transferable from the previous projects. Given the data established regarding habitat living arrangements for isolated settings, it is possible to develop many alternatives. Since the entire crew will maintain a daily schedule based on a normal diurnal cycle, it is desirable to separate the private living spaces from working and group activity spaces. This approach was used in the Antarctic and SEEDS I projects; the three functions occupying their own discrete inflatable, enabling maximum acoustic privacy.

Another consideration is the long term "group dynamic" facilitated by the eight crew members. An arrangement allowing pairs of private compartments with central removable partitions, and remote locations, relative to one another may be the optimal situation for this mission.

A second area of concern regards the ability of the crew to immediately occupy the habitat upon arrival, with little preparatory activity. This is necessary as the crew will be suffering physiologically after an extended transit at micro-g. The reference mission cylindrical habitats have the advantage of being one vehicle, requiring no deployment or post-landing "set-up". This requirement mandates an advanced telerobotic deployment system, enabling full activation of all systems necessary to support human life, prior to crew arrival.

Contextual Requirements

The basic needs, air, water, food and shelter, as they relate to the Martian environment are considered in *Mission Requirements*. However, two other issues remain of specific importance to operations at Mars; dust and radiation.

It is assumed Martian regolith is fine grained and cohesive, as its origin is meteoric (2). However, constant wind and moisture exposure make it more akin to a "soil"(1). Nevertheless, contact with internal habitat equipment may cause damage. Prolonged inhalation may also prove unhealthful. Mitigation may be attained through use of an unpressurized vestibule at each airlock, not intended as an equipment or rover lock. This would allow initial dust removal prior to entering the lock, where a combination of negative pressure and vacuum systems could remove remaining dust.

The Martian atmosphere will provide some degree of protection from both GCR (Galactic Cosmic Radiation) and solar radiation. A shield of water, integral with the roof of each habitat inflatable, combined with site selection to make use of natural features, will mitigate a dangerous level of solar radiation during a worst case event (5). Water for the shield will be manufactured as part of the in-situ resource utilization process (ISRU). Seed hydrogen will be coupled with Martian CO_2 in a Sabatier reactor to produce methane and water. The ISRU plant would occupy a module separate from the habitat and be telerobotically located near the habitat site. Water transfer from the ISRU plant to the habitat may occur after crew arrival (26).

DESCRIPTION

Given the requirements previously outlined, it is possible to develop a prototype surface habitat and establish a module scale based on its packed dimensions.

The habitat envelope is assumed to be similar to the CEISS inflatable. Since four of these is required, the optimal organization requires two on either side of the module. An inflatable connector enables access to the ECLSS, located in the module. An additional connector between each pair of inflatables allows two means egress for the inflatables connected directly to the module. The "outlying" inflatables would house CELSS and lab functions, thus minimizing maximum occupancy and egress requirements. Consequently, the second means of egress for these inflatables would be to airlocks and / or logistics modules.

The internal organization selected favors separation of the private compartments into two groups of four, located in each inflatable, directly adjacent to the module. Half of each of these inflatables would serve as living and food preparation areas. The outlying inflatables contain lab, working and CELSS functions.

Description - The Module

Considering the habitat described above, it is possible to determine the necessary volume for packing the habitat. Since it is desirable to deploy as much of the habitat as possible, prior to crew arrival, it was assumed the interior partitions, conduits, ducts and plumbing would also be inflatable, or integrated with the packed inflatable prior to deployment. The packed mass of the CEISS inflatable, without any interior build-out, is given as 285kg (23). The mass of the inflatable described above is approximately 1500kg. The packed volume is estimated to be $3m^3$.

The addition of this inflatable to the habitat lander brings the total mass to 25.3MT.

Given the volume and mass for the ECLSS and packed inflatable, it is possible to determine the scale of the module. The total volume of the ECLSS and inflatable is $18m^3$. The inflation system and accommodations package are estimated to require $3.4m^3$. Consequently, the resultant module must be > 21.4m.

The gross volume of the SEEDS II-III modules is $13.83m^3$. This is the result of the use of a 20cm proportioning system (3 dimensional Cartesian grid, having 20cm intervals). The optimal scale for the module was found to be 2.4m per side. By manipulating the proportioning system, other modules can be created that are multiples of 2.4m and 20cm. For this phase of the study, a module of 1.2m per side was selected. Twelve modules would be combined to create a single module with a base 2.4m x 2.4m ($5.76m^2$) and 4.8m high. The gross volume of this construct is $27.65m^3$. This configuration allows an additional $6.25m^3$ for interior circulation and the structure of the module. It is assumed the ECLSS and other internal systems would be rack mounted in the modules.

The module, given its orthogonal nature, is a poor pressure vessel. It was decided the module would have no exterior envelope of its own, but would exist as a frame for equipment transport. For the Mars habitat mission, an inflatable of design similar to the CEISS inflatables, would be packed around the module. As previously described, the module, with its own inflatable would serve as a circulation node between the double inflatables at each side. Another option may be to pack a single large habitat inflatable around the module, simplifying the deployment process.

Description - Deployment Modes

Since the habitat must be fully deployed, with all support systems operational prior to crew arrival, some sort of autonomous/telerobotic system must be included with its design. Of the activities required for deployment, habitat movement is probably the most important. Mars Direct/Semi-Direct mission plans call for a 40-60MT habitat to be towed across the Martian surface by a rover to enable mating with a second habitat. Given the condition of the three known Mars surface sites, this may be extremely difficult, if not impossible.

For this proposal, a number of scenarios for habitat mobility were considered. An initial approach involved the habitat not moving at all, while required systems are brought to it via small wheeled telerobotic rovers. A second more ambitious approach favors adaptation of the Mars Global Surveyor "walking lander" as a habitat mobility platform (13). The MGS walking lander incorporates three non-moving jackstand legs and three fully articulated legs. This scenario allows the lander and its suite of scientific instruments to move from its landing site at the rate of 20m per hour. If scaled up to accommodate the habitat module and necessary support, the walking lander could relocate the habitat to an optimal location prior to deployment. It is also probable the walking lander could separate from the habitat after deployment and assist in the movement of other systems and equipment around the site.

Given the module configuration, the walking lander design would include deorbit/descent fuel and propulsion, RCS and descent guidance. It would aerobrake, descend with parachute assistance and utilize a terminal descent burn for a soft landing. After touch down and vehicle status verification, the lander would deploy an unactivated HPS type nuclear power system (11). It would then move away from the landing site to another location no less than 500m from the reactor. This location would, presumably provide some degree of natural shielding. The lander would have its own internal power source, such as a small HPS unit. After site selection and habitat placement, the primary reactor would be activated enabling deployment. Since it is assumed other systems required for complete habitat/base establishment will be landed separately, the walking lander could assist in their movement to the habitat site.

Description - Mission Modes

Two primary mission modes were investigated. The first assumes the habitat module/lander, along with other landers required for base establishment, would be combined with an interplanetary spacecraft in low Earth orbit (LEO). This vehicle would have been assembled on orbit and provide power for Trans Mars Injection (TMI). On arrival at Mars, the individual landers would be released to either aerobrake and "direct descend" to the site, or enter Mars orbit (the option to do one or the other allows for the possibility of dust storms at the site). The interplanetary spacecraft would continue on a free return trajectory to earth, for refurbishment and another mission.

The second option is the use of an individual TMI stage for each lander payload. In this case the habitat/lander would be docked to its TMI stage in LEO, a TMI burn initiated and arrival at Mars similar to that previously described.

The means of propulsion for each of the schemes is important in that it determines the number of shuttle/RLV flights to LEO to launch the habitat on its way to Mars. Two means of propulsion were investigated: Chemical (CH_4 and O_2) and Electric. The fuel and oxidizer choice was driven by the economic assumption that a common engine may be used, and it will have to be compatible with methane and oxygen, since these will be manufactured at Mars as part of the in-situ resource utilization (ISRU) process.

Given the specific impulse of CH_4 and O_2 and a delta-V of 3.8 kilometers/second, 66.7MT of propellant will be required for TMI, assuming a habitat/lander + TMI stage mass of 35.3MT (26). The total assembled mass will be 102MT. If the habitat/lander (m=25.3MT) is consider one shuttle/RLV equivalent flight (SEF), the TMI stage and fuel would be 4 SEFs (assumes launch to ISS orbital inclination). If chemical propulsion is used with the interplanetary spacecraft option, as much as 425.25MT of propellant are required, assuming a total vehicle and payload mass of 225MT. This is approximately 26 SEFs!

If electric propulsion is used, the propellant mass is considerably less. A specific impulse of at least 4900 seconds appears to be possible, with values as high as 7200 attainable (22). Using the same delta-V and the combined vehicle mass of 35.3MT, the fuel mass (fuel/reactant is xenon) is 2.9MT. The interplanetary spacecraft would require 18.5MT. Consequently, two SEF and ten SEFs respectively are required for each mission type. Its is clear that the optimal configuration favors use of the electric engine for TMI sequence.

Description - Site Integration

If the single TMI stage is used with a solar electric rocket (SER), two shuttle or RLV missions would place the entire system in LEO. The two elements could be telerobically docked in close proximity to the shuttle or ISS. Due to the low thrust of the electric rocket, the TMI burn would involve a lengthy "spiral" out of LEO (15). After a nine month transit to Mars, the Habitat module/lander would aerobrake and soft land at the designated landing site. The lander's initial activity would be to deploy its S-band antenna for receipt of commands necessary for deployment. Next it would place the HPS reactor and begin its motion away from it. When > 500m from the reactor, it would be activated and habitat deployment activity started.

The inflatables packed in the module would be jettisoned and Martian atmosphere used to pressurize the integral structural system. Once it was verified that the inflatables were properly in place, the walking lander would detach itself from the habitat and conduct a walk-around inspection to ensure proper deployment. Mars landing to full inflation of the habitat would demand a minimum of one week.

While this activity was underway, other landers would be touching down at predetermined landing zones within two kilometers of the habitat. They would include two seed hydrogen ISRU landers, two logistics landers (containing additional food, parts, accommodations, and extra reactor, scientific equipment and an unpressurized rover), a Mars Ascent Vehicle with integral in-situ propellant manufacturing capability and one pressurized rover for long duration surface exploration. The ISRU landers would move to the habitat first, connect to the power supply, with the help of a walking rover, and begin manufacturing methane, water and buffer gas (nitrogen). As oxygen and nitrogen were produced, they would be provided to the habitat ECLSS. Another week would be necessary for the ISRU landers to arrive at the habitat. Next would come the logistics landers, moving to a location adjacent to the pressure ports at the outlying inflatables. There would be little need for them to connect to the habitat before crew arrival. Finally, the pressurized rover would obtain a position one kilometer away form the designated landing site of the first crewed mission - an event that would not happen for approximately two years.

CONCLUSION

This paper has shown how the use of architectural research and design methods, conducted over a six year period, were used to develop a conceptual module capable of containing all systems necessary for deployment of an inflatable habitat on the surface of Mars. The module, when packed may be transported to LEO in a Shuttle payload bay or RLV. Once on orbit it may be docked to a TMI propulsion stage using Solar Electric Propulsion.

This scheme negates the requirement for HLLV launch capability and the need for the common transit vehicle/habitat. It makes use of existing, as well as evolving, technologies as opposed to being strictly "off the shelf".

Finally, it is adaptable to the philosophy of the Direct/Semi-Direct missions by providing means to include ISRU/ISPP systems, enabling an economical initial Mars exploration proposal.

REFERENCES

1. Banin, A., B. C. Clark and H. Wanke. "Surface Chemistry and Mineralogy", in Kieffer, H. H., *et al.*, ed. *Mars*. University of Arizona Press, Tucson, 1992. P. 595.

2. Christensen, Philip R. and Henry J. Moore. "The Martian Surface Layer", in Kieffer, H. H., *et al.*, ed. *Mars*. University of Arizona Press, Tucson, 1992. p. 695.
3. Cohen, Marc. *Mars 2008: Surface Habitation Study*. Advanced Space Technology Office, NASA-Ames Research Center, April 22, 1994.
4. Cohen, Marc. *Two Planetary Habitat Analyses in Parallel: Habitat Diameter as a Determinant of Heavy Lift Launch Vehicle (HLLV) Shroud Diameter: A Parametric Analysis*. Advanced Projects Branch, NASA Ames Research Center, Moffett Field, CA, September 14, 1995.
5. Cohen, Marc. *Design of a Planetary Habitat Versus an Interplanetary Habitat*. 26th International Conference on Environmental Systems, July 8-11, 1996, Monterey, CA.
6. Cohen, Marc. *Design Research Issues for an Interplanetary Habitat*. 27th International Conference on Environmental Systems, Lake Tahoe, Nevada, July 14-17, 1997.
7. Cote, Ron, Ed. *Life Safety Code Handbook*. National Fire Protection Association. 1997.
8. Duke, Michael B. and Nancy Ann Budden. *Mars Exploration Study: Work Shop II*. NASA Conference Publication 3243, May 24-25 1993.
9. Eagle Engineering. *Mars Interplanetary Mission Modules: A Conceptual Design Study*. Houston, TX. July, 1989.
10. Harrison, Albert A., Barrett Caldwell, Nancy Struthers and Yvonne Clearwater. *Incorporation of Privacy Elements in Space Station Design*. Aerospace Human Factors Research Division, NASA-Ames Research Center, May 20, 1988.
11. Houts, M., D. Posten, D. Berry and G. Polansky. *Near-Term, Low Cost Fission Power and Propulsion Systems for Space Applications*. AIAA Space Programs Conference, Huntsville, AL, September 24-26, 1996.
12. Latour, Alessandra, ed. *Louis I. Kahn: Writing, Lectures and Interviews*. Rizzoli International, New York, 1991. pp. 102-108.
13. Lockheed Martin. *Mars Mobile Lander Concept Study Results*, November 10, 1997.
14. *Man-Systems Integration Standards, NASA Standard 3000*, v.I. Revision A, 1989.
15. Mendell, Wendell, ed. *International Mars Mission*. International Space University, Toulouse, France, 1991.
16. McKay, Christopher P., ed. *The Case for Mars II*. American Astronautical Society, *Science and Technology Series*, Vol. 62 - Univelt, San Diego, CA, 1985.
17. Micheels, Kurt A. *The Surface Extreme Environment Dwelling System for Mars*. Master of Architecture Thesis, University of South Florida, December 1994.
18. Micheels, Kurt A. "The Surface Extreme Environment Dwelling System for Mars", in Johnson, Stewart, ed. *Space V: Proceedings of the Fifth International Conference on Space 96*. Albuquerque, NM, June 1-6, 1996.
19. Micheels, Kurt A. *Use of Architectural Proportioning Systems in the Design of a Mars Surface Habitat*. 26th International Conference on Environmental Systems, July 8-11, 1996, Monterey, CA.
20. Peercy, R. L., Jr., R. F. Raasch and L. A. Rockoff. *Space Station Crew Alternatives Study Final Report: Volume I-V, Final Summary Report*. NASA Contractor Report 3854, Contract NAS-17424, June 1985.
21. Raybeck, Douglas. "Proxemics and Privacy", in Harrison, Albert A., *et. al.*, ed. *From Antarctica to Outer Space: Life in Isolation and Confinement*. Springer-Verlag, New York, 1991.
22. Routier, D. "A Solar-Ion Propelled Mission to the Local Interstellar Medium", in *Journal of the British Interplanetary Society: Electric Propulsion (Part II)*. Vol. 49, No. 5, May 1996, pp. 193-200.
23. Sadeh, Willy Z., Jenine Abarbanel, Ted Bateman and Marvin Criswell. "A Framing System for a Lunar Martian Inflatable Structure", in Johnson, Stewart, ed., *Space V: Proceedings of the Fifth International Conference on Space 96*. June 16, 1996, pp. 1069-1075.
24. Tilley, Ray Don. "South Pole Prototype", *Architecture*. August 1993, pp. 105-107.
25. Zeisel, John. *Inquiry by Design: Tools for Environment Behavior - Research*. Cambridge University Press, 1981. pp. 3-17.
26. Zubrin, Robert. *The Case for Mars: The Plan to Settle the Red Planet and Why we Must*. The Free Press, New York, 1996.
27. Based on information obtained during a visit to the Kennedy Space Center, CELSS facility, January 8, 1998.

ARESAM:
STUDENT CONCEPT OF FUTURE MARS SPACE STATION[*]

Jonathon Smith[†] and Jim Bishop[‡]

A space station orbiting the planet Mars with the ability to house 18,000 full time inhabitants, equipped with such technology as Fractal Shape Changing Robots and a space elevator. Sound like fiction? Well, it is ... for now.

Aresam (the name of the station) was designed by a team of sophomores and freshman from Sanilac County, Michigan. The design is explained and detailed in a forty page report, including about thirty pages of typed print as well as drawings and diagrams. It was submitted for competition with other student designs from around the world in the "International Space Settlement Design Competition" (ISSDC). This design placed as one of eight finalists in the competition.

Our report was structured around the ISSDC's RFP, or Request for proposal. It basically required that every team consider the full implications of what a space station required, and to provide for and facilitate those needs. We had to consider many scientific factors such as how to provide food/air/water/fuel for a population of 18,000 people so far from earth, how to power the station, and how to repair it. But we also had to consider many social factors such as how to keep a population of 18,000 people mentally sane 48,000,000 miles from earth, and how to organize the judicial and law enforcement systems for this independent society.

In our design we greatly elaborate on the above, and include a limited price estimate, schedule for completion once construction has started, and plan for operation of Aresam. It goes over many major processes and contingencies that this station will use to survive emergencies in space, as well as addressing in depth many of the key technologies that make it possible.

1. EXECUTIVE SUMMARY

1.0 A glint of light flashes in the distance, largely obscured by the glow of the planet Mars. You feel the deck vibrate slightly under your feet as the engines of your craft push you forward, accelerating toward Aresam, your destination. Anticipation builds as you near the end of your six month journey to your new home.

After a while you notice that the glint you saw from afar has now grown into a huge network of interwoven solar panels, sucking up solar energy to be used by the machinery inside.

---

[*] On the following pages are excerpts from our report. If you would like to obtain a full copy of our report, send a check or money order for 10.00 (to cover cost of printing and shipping) to: Jonathon Smith, 4571 South Lakeshore Road, Lexington, MI 48450.

[†] Jonathon Smith is a 16 year old high school junior. He lives in Croswell, Michigan and attends the Sanilac County Science and Math Center. Upon graduation he hopes to enroll at Embry-Riddle Aeronautical University and major in aerospace engineering, eventually becoming part of a manned mission to Mars. E-mail: duesouth@greatlakes.net.

[‡] James Bishop is a 16 year old high school junior. He lives in Marlette, Michigan and attends the Sanilac County Math Science Center. He hopes to attend the U.S. Naval Academy in Annapolis, Maryland, and major in Engineering. E-mail: strangescience@hotmail.com.

Something catches your eye, and you are stunned as the entire inside portion of, a large torus, which has recently materialized from the blackness of space, melts away and becomes transparent. Small lights start appearing everywhere on the interior of the ring, and it dawns on you that you've just witnessed your first "sunset", or crossover from day to night on Aresam. A shudder rocks you in your seat as your ship makes its final adjustments for docking. You notice a protrusion extending from the torus, or rather from a large sphere filling the doughnut hole of the torus. As the ship keeps nudging forward you find that this protrusion connects to Aresam's docking facility; a giddy feeling races through your system as you realize that soon you'll be riding through that tunnel on one of Aresam's high speed maglevs.

Wow. This is the only word left to you as you view the docking facility, with the glowing red surface of Mars in the background. "Please remain in your seat until the ship has come to a full and complete stop," a voice says over the intercom. You hear a slight ticking sound and then a muffled THUD, as the ship mates with a docking airlock. The lights of your cabin come alive, bringing you back to the moment. You pick up your backpack and walk out into the hallway, already full with grinning faces and a low excited flux of conversation. Welcome to Aresam!

As required by the Foundation Societies Request for Proposal, the Northdonning Heedwell design of Aresam, if chosen, will be an independent community capable of supporting a population of 18,000 full time inhabitants, has deciphered and prepared for all elements of basic infrastructure, etc. ... However, the philosophy of our design, which will be exemplified throughout, was not to make a settlement that fit all the requirements of the RFP, or even fit them well. Rather, we sought to design the best station that could be designed; the most innovative, the most attractive, the cheapest, the best. This design philosophy was present through every meeting, discussion and decision made about the settlement; every dispute, every synergy, every last minute change, holding us to our purpose like a guiding star. We at Northdonning Heedwell now yield to you, the Foundation Society, the fruits of our labor; our response to the Aresam Space Settlement Request for Proposal.

2. STRUCTURAL ENGINEERING

2.1 The major structural material used in the construction of Aresam will be woven carbon nano-tube blocks. These blocks will be made by machines which link carbon atoms into linear molecule, molecules into rope and rope into blocks. These blocks are used in the construction of Aresam's main torus, central axis, two spherical hubs, and mounts/supports for solar cell array. In addition to being used on the exterior of the station, most of the interior structures such as houses, office buildings, restaurants, malls, tennis courts, etc...., are constructed of carbon-nano tube blocks.

Major studies have been done into the effects of the Zero-G environment on the human body. Most have shown a decrease in muscular strength, and an overall degenerative effect on human bones. Therefore it was determined early off that any area in the station housing large amounts of people for a substantial time would have to have artificial gravity. The main torus of Aresam is the only part of the station that truly fit the above description, therefore it is the only area of Aresam that is provided with artificial gravity.

There are several theoretical ways of producing an artificial gravity field (i.e. harnessing a small black hole, creating a gravity well with huge energy generator, etc....) but for all practical purposes, these will not be feasible until many years in the future. There is really only one manageable way to create artificial gravity in the space environment, and that is using centrifugal forces.

According to Einstein's special relativity, an accelerating body (space ship, etc ...) induces on its contents forces that equal a gravitational field. The greater the magnitude of acceleration, the greater the artificial gravitational effect. Therefore, by spinning a circular torus shaped structure, you create acceleration (and therefore gravity) on any surfaces facing the center of the circle. On Aresam, the intensity of the gravitational field will be 1 g, which will be produced by spinning the torus at approximately 0.6 rpm.

3. OPERATIONS ENGINEERING

3.1 The orbital location of Aresam will be a geo-synchronous orbit around the equator of Mars. There are several reasons for this. First, the space elevator is to be attached to Mars at Middle Point, a location on Pavonis Mons that lies directly on the equator. By keeping Aresam a set distance from the elevator, shuttles can easily make the orbit transfers necessary to transfer goods to and from the elevator. This also provides unlimited access to the Martian surface by inhabitants, as the two structures (Aresam and the elevator) will always be aligned in the most convenient orbital locations. Secondly, in case of an emergency evacuation, Aresam will always be above its emergency bunkers on the surface. This would prevent the stranding of the residents in the rugged Martian terrain that might occur if Aresam were on the opposite side of the planet when the emergency occurred.

During the construction phase of Aresam, all the materials used in construction will be lifted off the surface of Mars *via* SSTO RLV's (single stage to orbit re-usable launch vehicles) capable of lifting 100 tons to the construction site. The Lockheed Martin Corporation created a prototype model of an SSTO RLV, the VentureStar, in the 1990's, and since that time they have found wide-spread use on Earth. These ships will launch from quickly erected bases on Mars, and will land at the very same bases. They will be serviced and fueled on the surface by a skeleton crew of base operators.

Transportation in Aresam will have two overall dimensions, local and long distance travel. The local level deals with transportation around the homes of the people and to the local malls and restaurants. For these purposes, and ones similar, the residents will have two means of transportation available to them, sidewalks and escalators.

The side walks are ten feet wide and provide a specific walkway for residents to the nearest escalator. The escalators are located every half mile and run down to the train stations located in the middle of the torus. Residents can ride these escalators down into the "downtown" section of Aresam, where the theaters, office buildings and expensive restaurants are.

On the long distance scale, there will be really only one means of transportation for the residents, maglev trains. There are seven maglev trains servicing the population of Aresam, six in the main torus and one in the central axis. The six in the main torus will be stacked in two groups of three, each stack traveling the opposite direction of the other. Because the trains are propelled using magnets, they are just as safe suspended above the ground as on the ground, and this allows the "stacking" of several trains. The trains run two main circuits. The top train in each stack stops only at designated train stations and transports passengers to the agricultural section of the torus, The bottom two trains stop at every train station in the economic area, and don't venture out to the agricultural section of the torus. The function of the two bottom trains is similar to that of New York's or Toronto's subway system, i.e. to transport people to different parts of the city.

The central axis, as well as the main torus of Aresam, will spin, but the two spheres will remain stationary. Therefore, "buffer" sections will be built to allow passengers to go from the main sections of Aresam. into the spheres. The buffer zones are themselves only cylindrical shaped connectors between the central axis and the spheres that are capable of being accelerated and then slowed down. The procedure for transfer of passenger/materials from the central axis to one of the spheres by use of the buffer zone is as follows. A maglev train inside the spinning central axis approaches the stationary "buffer" zone. The passengers/materials on the train are off loaded onto a small station located very close to the "buffer" zone. Next, the buffer zone is accelerated using magnets to mach the spin of the central axis. Once it has achieved this acceleration, the passengers/materials are transferred into the zone. Once everyone/everything has been transferred into the "buffer" zone, it is then de-accelerated to a complete stop. The initial train's passengers/materials are transferred into the stationary sphere.

3.4 Independent Facilities Aresam, as a whole, will be a single integrated community. However, certain structures of Aresam exist and operate independent of the rest. These are Aresam's independent facilities. There are three major independent facilities, the space elevator, the Mars base and the eco-parks.

The space elevator is an innovative and efficient solution to one of space developments greatest obstacles; how to provide economical transportation to planetary orbit. Standard means of transporting goods to low planetary orbit requires a launch vehicle loaded with millions of dollars of fuel which accelerates its payload to a speed capable of breaking gravity's grip. This excessive and constant use of fuel drives up the price of planetary exports beyond that of being affordable. Using a space elevator, the scene is drastically altered. The space elevator literally pulls itself out of a planet's gravity well without the use of any fuel *per se*. Thus the price of extra-terrestrial exports delivered to low planetary orbit by a space elevator is barely increased, making the exports not only affordable, but also *profitable* to the one exporting. This is important because it encourages economic investment in space and space related fields. These are facts which in general show the superiority of the space elevator over rocket propulsion for transporting goods into LPO; however, the specific advantages that will be provided to Aresam by this important piece of infrastructure far surpass these.

First of all, Aresam, as an independent but limited community, will depend largely on imports of natural resources to drive its industry. These imports have to be supplied from someplace by some means. However, if these imports need to be transported over large distances, such as the asteroid belt, they may cost as much or more than the products to be manufactured out of them. The same may be true if the imports needed to be launched by chemical propulsion out of a gravity well, and this is clearly undesirable for a community wishing to export its goods for sale. However, using the space elevator, Aresam can have access to an entire planet of resources with none of the two problems previously stated. As already explained, the cost of transporting the raw material into planetary orbit with the elevator are relatively nil, and the materials need only be shipped from a low Mars orbit to the high Mars orbit Aresam will be occupying. Thus it is possible and profitable for Aresam's industry to produce goods for sale to the Earth system and the asteroid colonies.

5. AUTOMATION

5.1 Several thousand temperature, humidity, and atmospheric gas sensors will monitor the atmosphere of all of Aresam's living space. The information collected by these probes will be wired to Control Computer 2, a super computer of the same type as Control Computer 1, located also in the control building. This computer will have no other network ports and will not be accessible by any other computer system on Aresam. The collected information from the probes will be contrasted to set standards. The super computer will then send out commands to the atmospheric scrubbers as needed, adjusting the local atmosphere of the probes.

Several thousand pressure sensors will be wired to the same computer system. They will notify the computer that if the pressure for any one given area begins to rapidly decrease, which would indicate a hull breach. The computer will then issue orders to robots on the exterior of the station, instructing them with the location of the hull breach and instructions on how to fix it.

6. SCHEDULE AND COST

6.1 June 2040: Crew and materials for Mars base are sent out. This consists of a crew hab, along with a large number of FSCR's and the SSTO RLV ships. Also included will be chemical factories to begin mining of materials from the Martian soil, and factories mining carbon from the atmosphere, and factories to construct carbon blocks out of the carbon. The FSCR's will begin constructing the Mars bunkers, and launch pads. At the same time this is sent, two separate convoys will also be sent out, one to Diemos and the other to the Martian North Pole.

The convoy to Deimos will consist of FSCR's equipped for mining, and a factory capable of producing solar cells. This convoy will begin manufacturing solar cells for Aresam and the space elevator.

The convoy to the Martian north pole will consist of FSCR's equipped for mining ice, and a number of robotic rovers. Mining of the ice at the north pole will commence immediately, and the ice will start to be shipped to the Mars base construction site. This water will start to be stored, and part will be used to manufacture rocket fuel.

6.2.2. The main expense in the annual operation of Aresam will be paying its employees. The total annual expenditure on employees will be $180,000,000. This is assuming a work force of 2,000 people getting an average income of $90,000 (some of the less technical employees receive salaries of $50,000, while the technical employees receive $125,000 a year.)

ACKNOWLEDGMENTS

Special Thanks out to my mother, Joan Smith, for editing and often re-writing large sections of this presentation and report. Also to all the other team members who helped design the station, and to Robert Zubrin for helping me to believe that we will one day go to Mars.
—**Jonathon Smith**

Thanks to the whole team for helping out with the project; we wouldn't be here without you.
—**Jim**

"Too Low they Build,
Who build Beneath the Stars."

—*Edward Young*

MAR 98-056

CURRENT PROGRESS IN WATER RECLAMATION TECHNOLOGY*

Bradley S. Tice[†]

A review of the current status on research conducted at Ames-NASA Research Center in the area of water reclamation technologies. The following areas will be explored:

1. Advanced oxidant delivery screening
2. Catalyst screening
3. Development of special catalysts.

INTRODUCTION

The program operating in the Regenerative Life Sciences Department at the Ames-NASA Research Center in Mountain View, California U.S.A. is currently focusing on new catalysts and oxidant delivery systems. This is divided into three separate tasks; advanced oxidant delivery systems, catalyst screening, and development of specialty catalysts.

Advanced Oxidant Delivery Systems

The use of a non-traditional oxidant delivery systems, such as ozone and hydrogen peroxide, with the use of heterogeneous catalysts has been observed to facilitate NH_3 and organics oxidations. The program will evaluate these agents as a post-treatment technology that will be supported by a Phase II ARC SBIR grant to develop an *in situ* hydrogen peroxide generator-catalyst system. Current modifications to the ARC Catalyst Evaluation Test-stand (CET) for the use of ozone as an oxidant is being undertaken [2].

Activated carbon-supported, noble metal catalysts exhibit the highest activity of any catalysts for a wide variety of aerospace and chemical process applications. This is due to their superior surface area and adsorptive characteristics. The drawback of such catalysts agents is that they suffer degradation of the carbon support, resulting in inadequate operational life for long-term space applications. New surface treatments and impregnation agents have been demonstrated to be effective at inhibiting oxidation of the activated carbon support over short-term testing conducted at this time [4].

Catalyst Screening

The identification of over 20 potentially useful catalyst materials that have shown the properties of facilitating oxidation reactions, of which most of work was performed in the vapor phase. This is a re-evaluation of a group of highest potential catalysts for use in an aqueous

---

* I would like to thank Michael Flynn or Ames-NASA Research Center at Moffett Field, California U.S.A. for providing an intellectually stimulating work environment in his lab during my stay in 1997-1998.

† Dr. Tice is Director and Institute Professor of Chemistry, Advanced Human Design, Medical Research & Development, P.O. Box 2214, Cupertino, California 95015-2214.

phase reaction. The resulting catalysts will be used in the existing VRA/APCO post treatment processor to enhance that units ability to treat organics and NH3 [2].

Development of Speciality Catalysts

The development of novel catalysts and catalyst supports that is supported by a Phase II ARC SBIR grant that covers the development of oxidation and physical deterioration resistant activated carbon supports. Unique vapor and aqueous phase catalysts tailored to meet specific mission requirements [2].

Another area of interest is Direct Osmosis (DO) that is a two-step concentration process that has successfully concentrated many wastewaters and food/juice products while recovering very clean water. Direct Osmosis has recovered over 97% of the water from actual composite wastewater similar to those that would be generated in space [1].

SUMMARY

The potentials and justification of the advanced catalyst evaluation program at Ames-NASA Research center is that such aqueous phase catalytic oxidation reactors are either currently in use or proposed for use on Mir, ISS, and in the JSC EHTI Phase II and III water treatment systems. This class of technology represents one of the most effective fully regenerative technologies for both water and air purification. Industrial use in commercial applications is also a growing area of interest for such catalyst systems. Some of the potential benefits are as follows:

1. The development of a highly active aqueous phase ammonia oxidation catalyst would facilitate the development of biological treatment methods.

2. Reducing the reactor temperature of existing APCO reactors from 150°C to 60°C, by increasing the activity of catalysts, could reduce the power consumption of these processors by approximately forty percent.

3. By increasing the quantity of contaminates which can be oxidized by the conventional APCO processors could save around $2.5 million per year in resupply costs, using the ISSA WRS as a baseline, by reducing the amount of activated carbon used in the multifilteration beds.

The duration and cost of future human space-flight programs will be influenced by NASA's ability to provide purified water. In an open loop system, water accounts for 87% (mass) of the total life support resupply requirements. The impact of this requirement upon mission costs is substantial. If it is assumed that an astronaut is allowed about 27 kg of water per day for drinking, showering, and washing both clothes and dishes and that it cost about $18 thousand to move a kilogram of mass into low Earth-orbit, it would cost about $177 million dollars per person per year to supply water to an astronaut in an open-loop system.

For long range missions, such as a lunar or Martian base and transport vehicles, resupply of metabolic supplies and processor equipment will be infrequent or nonexistent [3].

The greatest importance of these programs is the potential for 'order of magnitude' improvements possible in the future for long-term manned space flights.

REFERENCES

1. Beaudry, E. G., Herron, J. R. (1997) "Direct osmosis for concentrating wastewater," Paper presented at the 27th International Conference on Environmental Systems, Lake Tahoe, Nevada July 14-17, 1997.
2. Flynn, M. "Improvements to existing water reclamation technologies," Unpublished notes. Ames-NASA Research Center, Moffett Field, California U.S.A.
3. Flynn, M. T., and Borchers, B. (1996) "An evaluation of the vapor phase catalytic ammonia removal process for use in space flight," Paper presented at the 26th International Conference on Environmental Systems, Monterey, California July 8-11, 1996.
4. Flynn, M., and Jolly, C. (1997) "Advanced oxidation catalysts for water and air treatment," Paper presented at the 27th International Conference on Environmental Systems, Lake Tahoe, Nevada July 14-17, 1997.

DESIGN OF A NUCLEAR-POWERED ROVER FOR LUNAR OR MARTIAN EXPLORATION

Holly R. Trellue, Rachelle Trautner, Michael G. Houts,
David I. Poston and Kenji Giovig*

J. A. Baca and R. J. Lipinski†

To perform more advanced studies on the surface of the moon or Mars, a rover must provide long-term power (≥ 10 kW$_e$). However, a majority of rovers in the past have been designed for much lower power levels (i.e., on the order of watts) or for shorter operating periods using stored power. Thus, more advanced systems are required to generate additional power. One possible design for a more highly powered rover involves using a nuclear reactor to supply energy to the rover and material from the surface of the moon or Mars to shield the electronics from high neutron fluxes and gamma doses. Typically, one of the main disadvantages of using a nuclear-powered rover is that the required shielding would be heavy and expensive to include as part of the payload on a mission. Obtaining most of the required shielding material from the surface of the moon or Mars would reduce the cost of the mission and still provide the necessary power. This paper describes the basic design of a rover that uses the Heatpipe Power System (HPS) as an energy source, including the shielding and reactor control issues associated with the design. It also discusses briefly the amount of power that can produced by other power methods (solar/photovoltaic cells, radioisotope power supplies, dynamic radioisotope power systems, and the production of methane or acetylene fuel from the surface of Mars) as a comparison to the HPS.

1. INTRODUCTION

As the number of explorations that take place in space increases, the purpose and extent of these missions will also increase. In particular, missions to explore the surface of Mars or the moon also will expand and increase in complexity. To perform such missions, a larger, more efficient power source will be required. Currently, there are several options available for supplying power to a rover system, including radioisotope power supplies (RPSs), Dynamic Radioisotope Power Systems (DIPSs), photovoltaic/solar cells, wind power, laser techniques, conversion of carbon dioxide (or CO_2) in the atmosphere to methane and/or acetylene fuel, and nuclear power. Nuclear power has the advantage in that a small amount of fuel can provide large amounts of energy if it is properly designed to ensure that the system remains safe. Unfortunately, one of the disadvantages of a nuclear-powered system is that it can produce large amounts of radiation and must be shielded appropriately. Transporting shielding from Earth on a space mission can become expensive; thus, it is more economical to obtain the material from the surface itself. There are several options for using surface material; the ones studied here include using lunar regolith (on the moon) and either Mars soil or carbon dioxide converted to dry ice on Mars. This paper presents some background information regarding the various power

---

* Los Alamos National Laboratory (TSA-10).
† Sandia National Laboratories.

methods for rovers and then describes proposed shielding options and operational safety issues associated with a nuclear-powered rover.

2. BACKGROUND/VARIOUS POWER SOURCES FOR ROVERS

Although a wide variety of power sources can be used for a rover system, most of the sources provide fairly low amounts of thermal power. Radioisotope power sources are radioisotope power systems that comprise a nuclear heat source and appropriate power conversion equipment.[1] The most common radioisotope used for this is ^{238}Pu. Approximately 0.56 W of power is produced from every gram of ^{238}Pu [see Eq. (1)], but every gram of a ^{238}Pu heat source is also fairly expensive to fabricate. Thus, the costs involved in the fuel requirements for a large-scale mission would not make the RPS option economical. An RPS also would require a power storage unit to satisfy peak power demands (an RPS simply provides a constant heat source that cannot be increased or decreased to suit power needs). Such storage requirements would require a fuel cell mass of up to 2000 kg for a 10-kW$_e$ system,[2] thus reducing the overall power density of this source. The efficiency of an RPS, using a silicon-germanium unicouple as a thermoelectric converter, is ~6%.

$$\text{Power } (W_t) = m * sa * E * 1.6*10^{-13} \text{ J/MeV} \qquad (1)$$
$$= 0.56 \text{ W/g} * m$$

where

m = mass (g)
sa = specific activity (Bq/g = disintegrations/s-g) = 6.36e+11 for ^{238}Pu, and
E = energy per disintegration (MeV) = 5.45 for ^{238}Pu.

Another power source related to this is a DIPS. This technique combines an RPS with a highly efficient dynamic thermal-to-electric conversion device to achieve higher electric-power conversion efficiencies than other types of conversion systems (i.e., it has an 18% efficiency compared to other systems with 5 to 10%) and can be used to increase the power from an RPS by up to three times. The conversion device uses an organic working fluid in a Rankine cycle. Although this can help reduce the mass of the RPS required for a given power level, it has the limitation that it is best suited for power demands between 1 and 5 kW$_e$.[1]

Another possible source of power involves producing methane and/or acetylene fuel from atmospheric carbon dioxide on Mars.[3] By supplying (i.e., transporting) liquid hydrogen from Earth to a processing module on the surface of Mars, carbon dioxide from the atmosphere (where it is 95% abundant) can be converted into methane (or CH_4) with the aide of a nickel/graphite catalyst. The methane then could be converted to acetylene fuel (or C_2H_2). The equations below describe the conversion reactions.

$$4H_2 + CO_2 \quad 2H_2O + CH_4 \qquad (2)$$

$$6CH_4 + O_2 \quad 2C_2H_2 + 2CO + 10H_2 \qquad (3)$$

One of the disadvantages of this option is that energy is still needed to convert the carbon dioxide into fuel; this would require a separate power source to be transported from Earth. Even more energy is needed to convert methane to acetylene fuel; however, using acetylene fuel would decrease the payload mass of liquid hydrogen (or H_2) by a factor of four because it produces liquid hydrogen that can be reused in the production of methane. In addition, methane produces water (or H_2O), which also can be converted back to liquid hydrogen for further use; however, this requires additional processing (e.g., electrolysis). To produce 100 kW of power, methane with a energy density of 49.3 MJ/kg needs a flow rate of 2.04 g/s produced from 1.02 g/s of liquid hydrogen. Acetylene, with a energy density of 30.81 MJ/kg, needs a flow rate of

3.25 g/s produced from 3.0 g/s of liquid hydrogen. The efficiency of using either methane or acetylene fuel is ~40%.[4] Although using chemical fuels such as methane or acetylene makes sense for power generation over short periods of time (i.e., on the order of a day), over a longer duration (i.e., a year), so much material is required that it becomes less economical. For example, the power density decreases from 228 and 143 W/kg for 1 day of operation for methane and acetylene fuel, respectively, to 0.625 and 0.391 W/kg, respectively, for 1 yr. of operation [power densities are obtained by dividing the energy density by the length of operation (in seconds)].

The final option discussed here is photovoltaic cells (i.e., solar power). Three options for using solar power[5] include (1) a deployable high-efficiency flat plate array, (2) an array with a thin layer of photovoltaic material placed onto flexible substrate, and (3) a concentration system. The high-efficiency flat-plate array typically uses either crystalline silicon (or Si) or gallium arsenide (or GaAs) solar cells and is the most widely used with an efficiency of 14.5%. An array with a thin layer of photovoltaic material provides a higher specific power but has a lower efficiency and has never been demonstrated in space. The third option, using a concentration system, focuses sunlight on highly efficient solar cells to yield a conversion efficiency >30%. However, this system is not practical in the Mars environment because it focuses on direct, not diffuse sunlight, of which the majority of the light on Mars is composed. Additionally, the Martian environmental conditions make the use of solar power difficult as compared to other power sources. There are ~10 h of sunlight per day during optimum conditions on Mars (this varies with distance from the equator), resulting in the need for a large power storage area for the 14.5 or more hours of darkness. A storage unit required to provide a power of 10 kW_e would weigh between 1000 and 5000 kg, depending on the type of fuel cell used.[2] The typical power production at the top of Mars' atmosphere is 590 W/m^2 (Ref. 5). Even if power conversion technology is optimized, a relatively large solar array still would be required to generate 10 kW_e of power. Building and transporting a rover with such an array would require many supports, large conversion equipment, and additional insulation to keep parts from freezing overnight. By adding the storage unit and other necessary equipment to the total weight of the system, the power density(s) would decrease by one-half to one-fifth, assuming a 1-ton solar array. In addition, dust storms would decrease the power intensity and the effectiveness of any solar array significantly.

In contrast, the nuclear-powered system presented in this paper produces ~100 kW_t of power and weighs only ~250 kg (excluding the shielding and power conversion system). Depending on whether the reactor were actually placed on the rover itself or used to power the rover remotely, shielding costs and weights could become relatively large. Such shielding options will be discussed further in Section 3.1.

3. NUCLEAR-POWERED OPTIONS

Nuclear power is an inexpensive and efficient method of producing energy for supporting human life and scientific research while on Mars. The reactor needed to produce this mission-critical energy would be relatively small and lightweight, but the shielding necessary to protect the crew of a manned rover mission or the electronics on an unmanned one could be large and heavy. An alternative to bringing a heavy radiation shield from Earth, which would be very expensive and inefficient, is to make a shield from the materials available on Mars.

One method of shielding the crew and habitat from radiation is connecting a stationary reactor to a rover using a long power cable, powering the rover for long excursions, and using the atmosphere and diffusion as the shield. This method is risky because it relies on the durability of the connection between the reactor and the rover, which could present difficulties. In addition, the power cable required for such a job would be heavy and take up valuable payload space. Instead, it was assumed for this analysis that the reactor is located on the rover itself. However, this in itself presents certain difficulties, including the large amount of shielding needed to protect the electronics for an unmanned mission (and/or human operators for a manned mission) and operational safety. These issues for an unmanned mission will be discussed in the following sections.

3.1. Shielding

As mentioned previously, one of the main concerns of using a nuclear-powered system is preventing the radiation from damaging the electronics. Although several different types of nuclear systems can be analyzed, a Heatpipe Power System (HPS) was modeled in this analysis.[6] In the HPS, heat pipes are surrounded by uranium-fueled pins (which avoids the political issues associated with using plutonium) and transfer energy to a conversion system on one end. The particular HPS model used in this analysis produced a power of 100 kW$_t$ and an average flux of $1.4*10^{12}$ neutrons/cm^2-s. To prevent damage to the electronics, the primary objective was to keep the gamma dose rate outside the shielding to $<10^5$ rads/yr. (11.4 rads/h) and the neutron flux to $<10^{13}$ neutrons/cm^2-yr ($3.17*10^5$ neutrons/cm^2-s). These shielding studies included the use of lithium hydride (or LiH), an effective neutron absorber, and tungsten (or W), an effective gamma absorber, as well as lunar/Martian soil or carbon dioxide condensed from dry ice.

3.1.1. Lunar Regolith/Martian Soil

To obtain the desired neutron flux and gamma dose rate, the HPS was modeled using the Monte Carlo N-Particle (MCNP^{TM}) transport code[7]; tally cards were used to perform dose calculations. The electronics were assumed to comprise pure silicon; thus, the gamma dose rate in silicon was desired for radiation calculations. This dose rate was obtained by using energy-dependent dose factors in silicon[8] in MCNP.

The rover modeled in MCNP was broken into four sections: the power conversion system and radiator, reactor, shielding, and electronics (see Figure 1). The reactor section of the rover was in the shape of a cylinder, and the shielding and electronics sections branched off this cylinder in a cone configuration. The cone was surrounded by a thin layer of aluminum, and the various sections were separated by a few centimeters of aluminum to prevent intermixing of the sections. The rover itself originally was contained in an aluminum "box" 5 cm thick. However, because this produced a great deal of scattering, the model was simplified to assume that a thin layer of low-density material holds the reactor in place instead. Calculations also showed that the presence of regolith underneath the rover also contributed to an increased dose rate and neutron flux caused by scattering of the particles. Thus, it was determined that a layer of shielding material must surround the bottom 120° of the reactor to reduce the number of particles that reach the regolith and scatter. Approximately 10 to 15 cm of lithium hydride was required to reduce the neutron flux, and 2 cm of tungsten placed in the middle of the lithium hydride reduced the gamma dose rate to the electronics to acceptable levels. The placement of the two materials relative to each other was extremely important. Although tungsten is a high-density material that can reduce the gamma dose rate, it also has a high probability of inelastic scatter for neutrons and is subject to (n,γ) reactions. This means that it captures neutrons and produces gamma rays with energies of ~7 to 8 MeV, which hinders rather than aids gamma shielding. When the tungsten was located in front of the lithium hydride, the neutron flux to the tungsten was relatively high, thus producing a large number of gamma rays. By placing the tungsten behind the lithium hydride, fewer gamma rays were produced because a lower neutron flux entered the tungsten. In addition, having the tungsten in front of the lithium hydride helps to reduce the total neutron flux as well. Thus, the "sandwich" configuration shown in Figure 1 was found to be the best in this assessment.

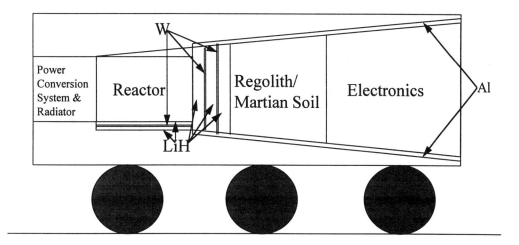

Figure 1 Rover configuration.

The required amount of shielding needed between the reactor and the electronics was determined to be ~85 cm of lithium hydride and 10 cm of tungsten. If material from the surface of the moon or Mars is used as shielding, the amount of lithium hydride and/or tungsten can be decreased and less material must be shipped as part of the payload on a mission. Approximately seven times as much regolith or Martian soil is needed as lithium hydride to decrease neutron fluxes, and approximately 20 times as much regolith or Martian soil is needed in place of tungsten to reduce the gamma dose rate. However, some amount of lithium hydride and/or tungsten still is needed to keep the size of the rover reasonable. Also, by increasing the thickness of material in the shield at the bottom of the reactor or by having the shield encompass 240° of the reactor core instead of 120°, dose rates and/or additional shielding requirements can be reduced. However, more design is necessary before detailed volume and mass requirements for the system can be made.

3.1.2. Carbon Dioxide-Dry-Ice Radiation Shield for a Nuclear Reactor on Mars

The atmosphere on Mars, although short on the oxygen needed to support human life, has an abundant supply of carbon dioxide. When in gas form, carbon dioxide is a poor radiation shield; however, in a solid, denser form, it can serve as an effective shield. In this scenario, the carbon dioxide would be obtained from the atmosphere of the planet and condensed into dry ice using a cooling fin. A shield's effectiveness is directly dependent on how much matter is contained in the shield between the source of the radiation and that which is being shielded. Thus, the effectiveness is dependent on the density and thickness of the shield (which is also why tungsten, which is a dense material, was used to reduce the gamma dose rate in the previous section). By using thermal conductivity equations,[9,10] the thickness of dry ice that can be created as a function of time can be obtained and is plotted in Figure 2.

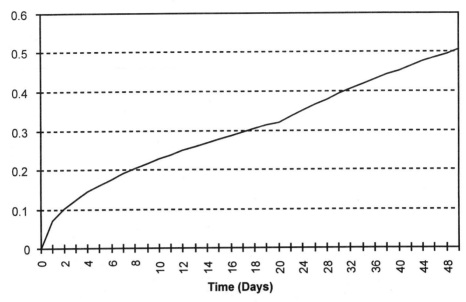

Figure 2   Thickness of dry ice as a function of time.

This figure was created under the assumption that the coolant will remain at a constant temperature. The coolant must be moving fast enough to remove all heat gained through the condensation of carbon dioxide; if it moves too slowly, it could heat the portions of the dry ice that are in contact with the cooling surface. The speed at which the dry ice can grow is limited by its thermal conductivity. The dry ice grows more slowly as it gets further from the cooling surface because the heat gained by the condensing carbon dioxide must travel a progressively greater distance to the cooling surface.

By combining the dry ice with lithium hydride and tungsten shielding, 50 to 100 cm of dry ice would be required to obtain the desired gamma dose rates and neutron fluxes. Figure 2 shows that 50 cm of carbon dioxide can be formed within 50 days; additional material can be formed relatively quickly thereafter.

In addition, the heat flux, which is gained by condensing the carbon dioxide, must be radiated into the atmosphere to maintain an effective coolant loop, which adds more power requirements to the system. The heat flux is plotted as a function of time in Figure 3. This analysis does not account for the heating rate of neutrons and gamma rays absorbed in the shield but rather is solely representative of the system itself.

Such heat could be removed from the system with a reasonable energy investment but would still require a small amount of power from the reactor itself.

3.2. Operations Safety

A nuclear reactor normally is kept at safe operating conditions through the use of control rods or drums. These devices contain a strong neutron absorber that captures neutrons and prevents them from fissioning. At a constant operating power, a reactor is said to be critical, which means either that its effective multiplication factor (k_{eff}) is equal to one or that one neutron effectively is produced per each neutron destroyed. Power production ultimately results from fission reactions within the reactor, which create neutrons as well as a large amount of energy. At the beginning of operation, a large number of fissile atoms (those capable of fission with any neutron) exist per absorbing atom, and a relatively large amount of absorbing material is needed to keep the reactor critical. As the reactor runs, more absorbing fission products are produced and the effective multiplication factor of the reactor decreases; thus, less absorbing material is needed to keep the reactor critical. Control rods or drums can be adjusted to provide more absorbing material when the reactor is first started and less as the reactor runs and the amount of fissile material in the system decreases. The HPS model follows this pattern, and control drums regulate the amount of absorbing material that is present at various steps.

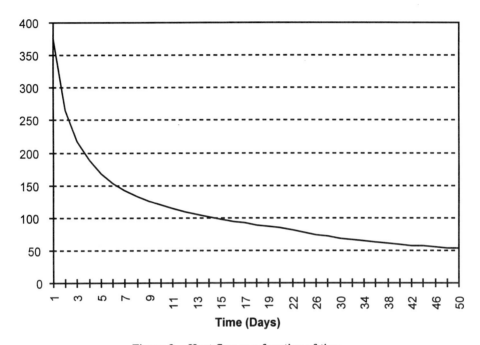

Figure 3 Heat flux as a function of time.

When a rover explores the surface of Mars, it will encounter rough terrain, many rocks, and other obstacles. One of the problems of having a nuclear reactor on a rover is that the control drums could get bumped and/or dislocated during reactor operation, thus causing the reactor to go either slightly above or below critical. Thus, the reactor must be designed to adjust back to a critical condition automatically. Fortunately, the amount of burnup in the HPS model is small, which means that the quantity of fissile material decreases slowly over time (once operation begins) and relatively few control drum adjustments are necessary. One option to avoid problems with unplanned control drum movement caused by this low burnup is to design sturdy control drums that can be locked into one position at the beginning of each excursion and then remain there throughout the excursion. Another option is to rely on the fact that the reactor is designed to have a negative temperature coefficient, which means that if the reactor goes supercritical (i.e., $k_{eff} > 1$), the temperature of the system will increase, materials will expand, and the value of k_{eff} will decrease as a result. In contrast, if the reactor goes subcritical (i.e., $k_{eff} < 1$), the temperature will decrease and materials will shrink, which will cause the k_{eff} to increase slightly. In either case, the reactor is designed so that it can remain at a critical condition and still allow steady-state operation to continue.

4. CONCLUSIONS

There are many methods by which a rover can be powered for excavations on the moon and/or Mars. Nuclear power is one of the most economical and efficient methods; however, certain considerations, such as shielding necessary for the electronics and safe operating conditions, must be taken into account in its design. Although a relatively small amount of shielding is needed to produce acceptable gamma dose rates and neutron fluxes for a nuclear reactor in a vacuum situation, much more is needed in a realistic situation for a rover on the moon or Mars because of scattering of particles off various surfaces. Lunar regolith, Martian soil, and/or carbon dioxide can be used to help reduce the large costs involved with the shielding (both in payload mass and obtaining the material), and the reactor can be designed so that the rugged terrain will not affect the safe operation of the system. A better option to reduce the necessary shielding may be to power the rover from a stationary nuclear reactor using either rechargeable fuel cells on the rover, a long cable connected to the rover, or a laser to beam power to the rover from a location in orbit. All options must be studied further before an absolute decision can be made.

5. REFERENCES

1. Joseph A. Angelo, Jr., Ph.D., and David Buden, *Space Nuclear Power* (Orbit Book Company, Inc., Malabar, Florida, 1985), pp. 126-157.
2. P. M. O'Donnell *et al.*, "Energy Storage Considerations for a Robotic Mars Surface Sampler," in *Proceedings of the 23rd Intersociety Energy Conversion Engineering Conference* (The American Society of Mechanical Engineers, New York),Vol. 2, pp. 385-389.
3. Geoffrey A. Landis and Diane L. Linne, "Acetylene Fuel from Atmospheric CO_2 on Mars," *Journal of Spacecraft and Rockets* 29:2, 294-295 (1991).
4. "Gas Industry Online: Gas Technology Spring '98," http://www. infoinc.com/ newgio/gio/gt/field.html.
5. Geoffrey A. Landis and Joseph Appelbaum, "Photovoltaic Power Options for Mars," *Space Power* 10:2 225-237 (1991).
6. Michael G. Houts and David I. Poston, "Heatpipe Power System and Heatpipe Bimodal System Development Status," in *Proceedings of the Space Technology and Applications International Forum* (Woodbury Publishing, New York, 1998), pp. 1189-1195.
7. Judith F. Breismeister et al., "MCNP™—A General Monte Carlo N-Particle Transport Code, Version 4B," Los Alamos National Laboratory report LA-12625-M, pp. 3-65, H-6 (March 1997).
8. Norman J. Rudie, *Principles and Techniques of Radiation Hardening* (Western Periodicals Co., North Hollywood, California, 1980), Vol. 1, p. 5-11.
9. Warren M. Rohsenow *et al.*, *Handbook of Heat Transfer Fundamentals* (McGraw-Hill, New York, 1985), pp. 3-88, 3-99.
10. Dwight E. Gray, *American Institute of Physics Handbook*, 3rd ed. (McGraw-Hill, New York, 1972), p. 4-226.

Chapter 12
POWER ON MARS

A technical point is explained.

MAR 98-058

SURVIVING ON MARS WITHOUT NUCLEAR ENERGY

George James, Gregory Chamitoff and Donald Barker

Current strategies for early missions to Mars are highly dependent on the assumption of nuclear power as the primary near-term energy supply. Since political considerations may prohibit the launch of nuclear systems, this paper investigates the potential for utilizing in situ energy sources on Mars to either supplement or replace nuclear power. The current knowledge of solar, wind, and areothermal energy resources on Mars is discussed, and the studies required to identify these resources and further characterize their distribution and abundance is given. A non-nuclear power system for the first Mars mission could be based on a combination of solar and wind energy coupled with a liquid fuel storage system. Indications are that such a system, using the latest solar cell technology, could be cost competitive with nuclear power in terms of kiloWatts per kilogram delivered to the Martian surface. Areothermal energy has significant potential, but the development of this resource will require longer term space and surface based exploration. The work reported herein advances the current knowledge base with the following accomplishments: (1) collecting models to jointly estimate solar and wind power production; (2) utilizing terrestrial geothermal exploration techniques coupled to remote sensing data from Mars to refine potential regions of exploitable planetary heat sources; and (3) suggesting that an early Mars mission that relies heavily on in-situ energy sources will require additional precursor information on energy resource location, extent, and accessibility. It is proposed that the production of energy on Mars, solely from local resources, may be practical enough to render a small outpost completely self-sufficient. Moreover, the addition of in-situ energy resource development to that for life support and transportation may advance the development of larger permanent self-sufficient human colonies on Mars.

INTRODUCTION

The development of economical strategies for the first human missions to Mars has been increasingly focused on the utilization of in-situ resources for providing required supplies for life support, surface mobility, and return-to-Earth capability. It has been demonstrated that mission robustness and affordability can been drastically improved by "living off the land." Recently, the Mars Direct plan (Zubrin 1996) and elements of the NASA Design Reference Mission (NASA 98) illustrate the importance of this concept. A common feature to most mission plans, however, is the transportation of a nuclear energy source to the Martian surface. While this may be the simplest short-term solution for meeting the energy requirements of a human base on Mars, it also has the potential to be the show-stopper due to the current political climate regarding the safety of launching nuclear materials and/or polluting another planet with nuclear waste.

The objective of this paper is to present alternative means for providing energy on Mars through the development of local resources. Available indigenous energy alternatives include solar power (photovoltaic, solar dynamic, or solar satellites coupled to power beaming), areothermal energy (Martian geothermal), and wind energy. The potential for exploiting photovoltaic solar energy on Mars is well established. However, the cost is high and the implementation involves certain obstacles, such as maintenance and restrictions in output performance due to aeolian deposition and dust storms. Solar dynamic power is even more susceptible to atmospheric dust than photovoltaic systems. Solar power satellites with power beaming to the surface has not yet been demonstrated on a large scale, although, in-situ construction of the rectenna may be quite feasible on Mars. Areothermal energy is a potential longer-term resource that could be plentiful in certain regions, but the utilization of this resource requires further remote sensing data as well as subsurface drilling and other surface based exploration. Wind energy has surprising potential for Mars due to the magnitude of geological features and temperature extremes that can produce highly reliable and localized winds. Therefore, the near-term energy system proposed herein couples a wind energy conversion system to photovoltaic power production. These two technologies are complimentary since wind energy tends to be active during times when solar energy is reduced or unavailable; and some locations are subjected to stronger winds during dust storms. Beyond the initial outpost on Mars, once areothermal resources are located, this form of energy may complete the energy equation for permanent, self-sufficient, and productive bases on Mars.

ENERGY REQUIREMENTS FOR A MARTIAN OUTPOST

Energy is a dominating factor in the design of a human mission to Mars. Long flight times and an extended surface interval requires an optimum combination of energy resources for redundancy, storage, and eventual indefinite self-sufficiency of a Martian base. Solar power is a reliable source of energy for meeting daily demands on the outbound and return flights. On the surface of Mars, however, a combination of nuclear or isotope generators, solar power, wind energy, or areothermal energy can be considered. The Mars Direct plan proposed by (Zubrin 1996), which is a low cost strategy for achieving the goal of putting humans on Mars and working toward permanent inhabitation, has become the guiding philosophy for a near-term mission. This plan, like others, assumes that the first manned mission to Mars will include at least one 100 kW nuclear generator. In addition to life support, energy production would be required to produce some 108 tonnes of methane and oxygen to fuel a return rocket as well as to provide fuel for local surface exploration in a Martian rover. The fundamental concept of the Mars Direct plan is to "Live off the Land" using in-situ resources to the greatest extent possible. This is of primary importance as a means of reducing the required payload delivered to Mars to support initial exploration initiatives. As such, fuel, oxidizer, water, and breathable air would be produced on Mars from local resources. Only a small supply of hydrogen is required until a local resource for this element can be developed. Eventually, many other materials could be processed on Mars, including cement, glass, iron, steel, even fertilized soil. Other than the exploration required to locate resources, and the refinement of resource utilization and extraction techniques, the only other major requirement is that of energy production.

The energy required for early missions to Mars is highly dependent on the mission design and objectives. Life support energy requirements per crew member is a function of living volume, mobility, quality of food provided, activity level, and so on. To this must be added requirements for scientific and resource exploration and research. The power to operate Biosphere 2, a closed ecological system that supported a crew of 8 for two years (Nelson 1996), was approximately 100 kW per person. However, this did not include the power required for materials production and pressurization which would raise the energy requirements. Nor did the program include the ability to extract new resources from the environment or to exchange wastes with the outside, which would lower the anticipated energy needs (Meyer 1996). Moreover, the habitable space in Biosphere 2 was 1335 cubic meters per person, perhaps 50 to 100 times that which is likely for a first human Mars mission. A Japanese study placed a value of 20 to 50 kW per person as the value needed for a 150 person Mars settlement. The current

NASA Mars Reference Mission suggests a value of 60kW to support the first crew of 6 astronauts on Mars, including the life support cache, ascent fuel propellant production, and surface exploration (NASA 1998). By the time the third crew arrives, this is expanded to a total of 160kW, which supports increases in habitable volume, life support capability, science and exploration. Based on this study, a power supply system on the order of 60 to 200 kW appears to encompass the reasonable range of current estimates for an initial Martian outpost.

ASSESSMENT OF APPLICABLE ENERGY SOURCES

Nuclear Energy

Nuclear power is typically assumed as the baseline power supply for Mars missions (Zubrin 1996; Meyer 1996; Duke 1985; NASA 1998). From the late 50's to 1972, a series of analytical and experimental projects were undertaken to produce viable space nuclear power systems (Colston 1985). One high point of this work was the 1965 flight of the 0.5 kW SNAP 10A reactor which operated in space for 40 days. The NASA sponsored SP-100 project was initiated in 1981 and targeted the production of a 100 kW system with a seven year lifetime. Estimates of the power output of the SP-100 class reactors as a function of mass range between 37 W/kg and 17 W/kg (Zubrin 1996; Haslach 1989). Although these projects remain uncompleted, the technological foundation has been laid and no stumbling blocks are seen. The Russians however, continued development of a space nuclear reactor to produce the 6 kW TOPAZ II with a three year life (Voss 1994).

Nuclear power is a compact method for generating power in the 100 to 500 kW range. The technology is well understood and at least one space qualified system has been produced. However, most power plants envisioned would produce radioactive wastes after a 7 to 10 year lifetime that would require disposal. The power units could not be recharged or repaired in-situ without a great deal of high technology and mineralogical support. Also, there is a great deal of political and public resistance to building and using nuclear reactors at this time. In view of the political, legal, and technological hurdles involved, for a near term Mars mission, the longest lead time item could be the ability to build and launch a significant nuclear energy source.

Solar Energy

Solar power is readily available on Mars, but dependent upon latitude, seasonal variation (due to orbital eccentricity and inclination), daily variation, day/night cycles, suspended aerosols, and dust storms (Zubrin 1996; Meyer 1996; Geels 1989; Meyer 1989). Orbital eccentricity causes a variation in the surface solar constant from 718 to 493 Wm^{-2}. On average, the Martian surface receives about 50% less solar flux on the surface when compared to Earth (mainly due to its distance from the sun). On Earth, the average daily insolation is 75 to 200 Wm^{-2} at the surface. On Mars, though, global dust storms occur one to two times a year, roughly at perihelion, and can last for months. Local dust storms may also last for several days (Meyer 1996). Martian dust storms do not render solar power ineffective, however, since the dust has a scattering effect rather than a blocking effect. Hence an all-sky, scattered light collector would continue to produce power at a reduced percentage of its maximum output.

Potential Solar Energy Extraction Systems

There are three extraction mechanisms for solar power: solar dynamic, photovoltaic, and space-based. Solar dynamic systems utilize collected light to heat a working fluid which drives a turbine. Such systems would convert 15 to 25 percent of the incident energy into electricity.

The major components of these systems (pipes, boilers, and collectors) are "low tech" and would be amenable to repair and eventual manufacturing on Mars (Zubrin 1996). However, solar collectors generally require a point source of light. Hence dust storms are estimated to drastically reduce the output of these systems - by as much as 95% (Geels 1989).

Space-based solar energy collection and microwave beaming for baseload power has only recently been studied for Martian applications (Mankins 1997), however, it has received a great deal of attention for Earth orbiting (Collins 1996) and Lunar surface installations (Criswell 1996). Space-based collection systems would not be subject to the power reductions due to atmosphere absorption and dust storms. Assuming a .25% conversion efficiency as provided by (Criswell 1996) for a lunar installation providing terrestrial power with 1980's technology, an equivalent Martian power density would be 0.0025 x 590 W/m^2 or 1.48 W/m^2. Hence such a system would require a solar array area of 68000 m^2 to produce 100 kW. This translates into a very large planar array of solar cells arranged in a 259 m x 259 m square. The Japanese are currently considering producing a technology demonstrator satellite of these dimensions to produce solar energy in Earth orbit and beam the power to Earth (Collins 1996). The footprint of the ground-based rectenna for the Earth orbiting Solar Power Satellite (SPS) is 1 km square. Although the rectenna is "low tech" and could be produced from Martian resources, it represents a longer term investment than this project is considering.

Photovoltaic systems do not require a single point source of light and would be less affected by global dust storms. These systems operate at about 13-15% efficiency for silicon and around 20% for advanced GaAs models and produce no excess heat. The specific power of state-of-the-art solar cells has now been demonstrated to exceed that expected from a nuclear power source. Space qualified silicon solar arrays producing 66 W/kg have been flown on the Space Shuttle. Arrays producing 130 W/kg have been manufactured, and the latest thin film solar cell technology promises specific powers from 1 kW/kg to 15 kW/kg. Comparing solar and nuclear energy for an average 100 kW power supply at the Viking 1 landing site, it has been estimated that a solar array of 1850 square meters (41 x 45 m - about half a football field) could provide the equivalent energy of a nuclear source (Haberle 1993). This assumes average daily insolation based on solar flux models accounting for latitude, absorption, season, and dust, but not storms. Conversion efficiency of 20% was assumed, specific mass was 0.9 kg/m^2, and the time interval for this study was 155 days during the Spring and Summer. The weight of the solar cell assembly is estimated to be around 1.67 metric tons, as compared to 3.96 tons for the nuclear generator. Accounting for the effect of dust storms, the irradiance at the surface is expected to diminish by a maximum of about 60%. For the Martian atmosphere, it has been shown that while the direct irradiance drops off quickly with increased optical depth, the light is largely scattered into diffuse radiation even up to high optical depths. For a solar flux of 200 W/m^2 at a typical optical depth of 1.0, this drops to about 100 W/m^2 for an optical depth of about 4.0, and to about 80 W/m^2 at an optical depth of 5.0. Typical values for optical depth during a global dust storm would range from 2 to 5, and nominal values for Mars are expected to be about 1.0 or less. It has been shown that for optical depths of 1.0 or greater, the efficiency of solar cells collecting diffuse radiation exceeds that for directed light (Haberle 1993). Therefore, a reasonable approximation for the area and mass required for a photovoltaic array to reliably produce the equivalent average energy of a 100 kW nuclear generator, despite a continuous dust storm, would be about 4000 square meters (slightly larger than the size of a football field) and 3.5 metric tons. This is comparable to the payload mass required for nuclear power.

Model of Available Energy

Several studies have been performed which assess the available solar energy on Mars (Haberle 1993; Appelbaum 1989 a, b; Geels 1989; Landis 1990; Appelbaum 1990). The following work utilizes the model of available solar energy which is common to many of the studies listed above. The primary equation for modeling the solar radiation S, at the top of the atmosphere is provided by the following (Haberle 1993):

$$S = S_0 R^2 \cos(z) \tag{1}$$

where
S_0 is the average solar insolation at Mars (590 Wm^{-2});
R is a nondimensional Sun-Mars distance parameter: and
z is the solar zenith angle.

$$R = \frac{1 + e\cos(L_s - L_s^p)}{1 - e^2}; \tag{2}$$

where
e is the orbital eccentricity (.0934);
L_s is the areocentric longitutde; and
L_s^p is the areocentric longitude at perihelion (250°)

$$\cos(z) = \sin(\theta)\sin(\delta) + \cos(\theta)\cos(\delta)\cos(h); \tag{3}$$

where
θ is the Martian latitude;
δ is the solar declination; and
h is the hour measured in angular units.

$$\delta = \sin^{-1}[\sin(\varepsilon)\sin(L_s)]; \tag{4}$$

where
ε is the current Martian obliquity (25.2°)

$$h = 2\pi \frac{t}{D}; \tag{5}$$

where
t is the time in seconds measured from noon; and
D is the length of the Martian day (88775 s).

The most important additional factor in modeling the available solar power on Mars is the effect of dust in the atmosphere. One approach adopted in (Haberle 1993) includes calculating the total downward irradiance which includes the incident solar radiation seen by a flat plate collector due to both direct illumination and scattered light:

$$I_{TD} = \frac{SI_{NN}}{1 - A}; \tag{6}$$

where
I_{NN} is the normalized net irradiance; and
A is the surface albedo (typically assumed to be .25).

The normalized net irradiance is a function which quantifies the effects due to surface albedo, solar zenith angle, and optical depth (or opacity of the atmosphere). The function is provided in tabular form in (Haberle 1993). The power produced by a photovoltaic array at any given time is then obtained by multiplying by the area of the collector and the efficiency of the solar cells (note that this model does not take into account losses due to dust loading on the array):

$$P_s = \textit{eff}*Area*I_{TD}. \qquad (7)$$

Figure 1 provides graphical information covering several parameters relating to the estimation of available solar power. The Viking 2 lander data was used in this example. The upper left hand plot provides the estimated solar insolation calculated above the Viking 2 lander site. Although, this calculation did not depend on measured data, the drop-out regions correspond to times during which the Viking 2 lander was not providing data. The upper right hand plot provides the optical depth data as measured by the Viking 2 lander and interpolated to match other meteorological measurements. The higher optical depth values denote times of a dust laden atmosphere. The normalized net irradiance function estimated using the Viking 2 data and the tables provided in (Haberle 1993) is provided in the lower left hand plot. The instantaneous power which would have been produced by an 1850 m2 photovoltaic array with 20% efficiency is provided in the lower right hand plot of Figure 1. Note the wide variation in power produced which is due to the orbital eccentricity, the dust laden atmosphere, and the seasonal solar zenith angle effects.

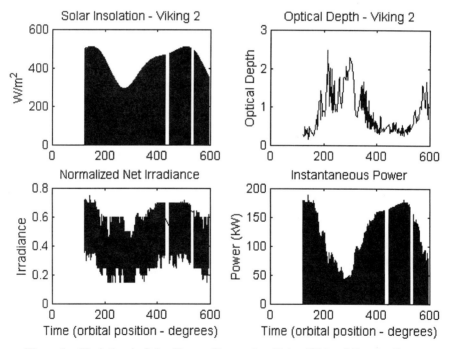

Figure 1   Variation in Solar Energy Parameters Using Viking 2 Lander Data.

Implementation Issues

Surface-based solar power offers several advantages for a Martian base. First of all, much terrestrial and space experience exists for solar energy production systems. Once sized correctly, solar power production would be sustainable indefinitely and would not produce any wastes. While the output would be degraded by dust accumulation on the collectors surface, this degradation is graceful, not catastrophic, and can likely be easily removed by manual or automated systems. Furthermore, a solar cell energy system could be designed to be robust enough to sustain substantial localized damage without significant effect on the total power production. One potential advantage due to the new thin film solar cell technology is the possibility of a mobile energy platform comprised of a balloon or other inflatable structures covered with solar cells. The feasibility of this concept has been demonstrated by (Ramohalli 1998). Other advantages include safety, global access to the resource, reliability, and expandability. It has also been suggested that in the future, solar cell sheets could be manufactured directly on Mars. However, since such production facilities represent a "high tech" endeavor, a mature base or colony would need to be in place.

Models of the distribution of solar radiation on Mars are well established and the availability and distribution of this resource is already well understood. Improvements in the fidelity of these models will come from new knowledge about the specific composition and size distribution of aerosols and dust particles in the Martian atmosphere. Nevertheless, enough is presently know about the Martian atmosphere and its effect on the incident radiation at the surface to design a robust solar power system for worst-case conditions.

An important consideration for solar energy on Mars, however, is the storage of energy for night time loads. Ultimately, development of a robust locally derived power system, would require the integration of multiple alternative sources, such as wind or areothermal energy, both of which supplement solar power, and reduce the cyclic storage requirements. Solar systems could function in a hybrid mode with other systems to create a redundant base-load system. Although assessing or examining energy storage techniques is beyond the scope of this paper, the potential for combined solar/wind energy systems will discussed in following sections.

Wind Energy

Although the atmospheric density is about 100 times less than the Earth, Mars has several advantages for successful wind power applications: less gravity (less massive components), large temperature and pressure swings (producing high winds), and tremendous surface relief and low atmospheric thermal inertia (produces consistent wind patterns). Unfortunately, the most direct observations of wind speed on Mars are limited to the Viking landers and from Pathfinder. Wind speeds at these locations were observed to average about 5 m/s, with a peak of 25 to 30 m/s recorded at the Viking Lander 1 site. A local dust storm over Chryse Planitia accompanied these peak readings which was also observed from orbit (Greeley 1982). The Viking landing sites, however, were selected on the basis of mission safety, which precluded complex or steep terrain that is more likely to harbor the high winds of interest. Computer models of the Martian atmosphere based on longer range observations and extrapolations from the wind blown sand streaks on Mars have predicted significantly larger values for surface wind speeds (Haslach 1989). It has been estimated that a well chosen site could harbor sustained speeds approaching 14 m s^{-1}. Possible sites include the horseshoe vortices around raised rim craters (as seen by dark streaks), and natural wind channels due to the topography of hills and valleys (such sites have been used successfully on Earth). Also, regions such as Hellas basin (the lowest region on Mars) have up to a 44% denser atmosphere (and hence a 44% increase in power). These regions would be favorable sites if high local wind speeds can be identified. Long low angle slopes (as seen on the shield volcanoes or slopes of large basins) may produce winds of 25 to 33 m s^{-1} at approximately 25 meters above the surface. It should be noted that the wind patterns at the Viking 1 landing site were believed to be dominated by this type of slope wind pattern (Zurek 1992). Recent measurements made at the nearby Pathfinder landing site further support this conjecture (Schofield 1997).

Martian Wind Exploration Techniques

A significant amount of observational evidence exists for the location and regularity of Martian winds. Sand and dust deposits, wind streaks, and erosional features are good indications of potential regions with significant winds. Wind streaks are present in the Tharsis region near the large volcanoes and in the Elysium region around Elysium Mons. Such streaks are the most numerous of the eolian markings in these regions and are accurate recorders of the direction of near-surface winds (Lee 1982). Observations of dark streaks in regions without visible points of origin have been identified as occurring in regions with extended slopes of 1 to 10 degrees. Similar concentrations of dark streaks on long slopes have also been found in the planet's southern hemisphere. A model of Martian slope winds, developed by (Magalhaes 1982), has predicted that slope winds will occur for slopes greater than 0.02 degrees, late at night and during the early morning, with maximum speeds being attained immediately before sunrise when the ground is at a minimum temperature. For slopes of more than 0.1 degrees, fully developed wind profiles would take less than 200 km to be achieved. For the central volcanoes, these conditions would occur along most of the flanks. It was also found that regions with low thermal inertia and high surface roughness tend to favor intense slope winds.

The majority of sediment deposits on Mars are composed of either sand or dust sized particles. The functional difference being that sand is transported by saltation and dust by suspension. In (Thomas 1982) the relationship between Martian winds and these deposits is investigated by comparing global wind streak data to the major deposit locations. Dunes tend to form in the large north polar erg, in craters, canyons, and similar topological traps. The largest concentration of sand on Mars lies in the north polar erg, mostly on erosional troughs in and around the polar layered deposits. Dunes in the low-latitude canyons are subject to complex wind patterns because flow up and down the canyon competes with north-south polar circulation. Dark streaks are found in the Juventae Chasma and Coprates Chasma indicating strong winds in these areas. Dunes in the south polar region are constrained by local topography. In the north polar region, winds show a strong outward flow (i.e., radially away from the topographic north pole). At low latitudes, wind streaks and eolian deposits seem to be a result of Hadley cell circulation (Thomas 1982).

While eolian processes on Mars provide clear evidence of wind on Mars, the transport of visible deposits can occur over long periods of time and do not necessarily imply recent high velocity winds. This evidence can be used, however, to identify likely locations for significant wind energy resources. Other than direct wind measurements taken at specific locations by robotic landers, the only other mechanism for global measurement of surface winds is via spaced-based remote sensing techniques. A recent technology called Coherent Doppler Lidar will soon be tested as an instrument on the Shuttle to perform space-based wind vector measurements (Kavaya 1998). Such an instrument could be incorporated in a near-term Mars orbiting spacecraft to identify and map ideal landing sites for maximizing wind energy resources.

Coherent Doppler Lidar (CDL), which has recently become a viable option for space-based wind vector measurements on a global scale, has the potential to drastically improve current knowledge of atmospheric processes on Earth and Mars. A precursor lidar mission for mapping winds on Mars could contribute significantly to the feasibility of a self-sufficient human mission to Mars. The concept of Doppler lidar is about 20 years old, and has been applied to the detection of aircraft wave vortices, pollution studies and high stack emissions, as well as the study of weather phenomenon such as tornado's and wind shear in thunderstorms. Developments in laser and signal processing technology have been instrumental in producing systems that enable the consideration of space-based lidar applications to the mapping of global winds from an orbiting platform. The technical obstacles for developing such a system were mainly due to power requirements and efficiency of the detection of the backscattered signal over long ranges. Presently, a NASA mission named SPARCLE (SPAce Readiness Coherent Lidar Experiment) is under development and is scheduled to fly on the Space Shuttle in 2001 (Kavaya 1998).

A space-based coherent Doppler lidar instrument for observations in the Martian atmosphere would serve dual purposes in terms of achieving scientific objectives and preparing for human missions to Mars. With respect to wind measurements, a CDL system could specifically determine global circulation patterns, closely track the full development of Martian dust storms, measure the annual variation in global and local winds, and locate target base locations with optimal wind characteristics for supporting wind energy generation.

An important issue in the design of a Martian CDL system would be the determination of required laser power for successful coherent backscatter measurements. The efficiency with which aerosol particles would scatter the laser light back to the detector is unknown for Mars, and this value is critical in determining the minimum energy required to obtain the minimal number of coherent photons for successful wind vector determination. Designs for a Mars mission would necessarily be conservative, thus affecting solar panel sizing and overall spacecraft design as well. The low atmospheric density and pressure at Mars is another related factor in the design of a Martian CDL instrument. The particle size distribution of suspended aerosols can be expected to be quite different from Earth. Optimal lasing frequencies for observing Martian winds would require models of the size distribution for these particles. In order to calibrate a CDL lidar system in Martian orbit, one concept could employ a series of ground-based observational stations. These stations could provide ground truth data that are matched simultaneously with space-based measurements to refine atmospheric model parameters, and improve the quality of lidar observations. Such a system is described by (Crisp 1994). The Mars Environmental Survey (MESUR) Program is envisioned as a network of micro weather stations for making in-situ meteorological measurements in the Martian boundary layer. This system would use micro sensors for measuring pressure, temperature, winds, humidity, and dust/ice optical depth. To maximize the scientific value of this program and provide ground truth data for remote sensing satellites, a network of at least 20 of these stations is required with sampling rates of at least 1 Hz. Wind measurements for these stations would be accomplished by micro-machined pitot-static ports or sonic anemometers. Each micro station would consume less than 0.1 Watt and be capable of communicating directly with a space-based instrument.

Given a CDL equipped spacecraft orbiting Mars, the objective of locating optimal sites for wind energy generation is not difficult to accomplish. The primary question is "Where on Mars can one expect high enough average wind speeds to justify an energy management approach partly based on wind energy ?" Solar energy considerations would also play a role in defining the optimal combination of the two resources. From the wind mapping perspective, however, a global assessment of boundary layer winds for a complete solar cycle would provide the necessary information. In particular, space-based observations could be guided to focus more on high probability regions, such as the north polar erg, low latitude canyons, long slopes in the Tharsis and Elysium regions, around large crater rims, and so on. Targeting specific sites with multiple pulses can be used to obtain better resolution. An important long term question that might be answered by a Mars CDL system is the relationship between local winds, dust storms, and solar flux attenuation at specific sites. An ideal location for cooperative solar and wind energy generation would be one where wind speeds increase significantly at night and during dust storms, at the same time that incident solar flux is absent or reduced. A CDL system is one promising approach to answering these fundamental questions, which in turn would enable the system trade studies required to design a self-sufficient, locally powered, human base.

Implementation Issues

Fortunately, for Martian utilization of wind energy, the power from a wind turbine is more a function of the wind velocity than the atmospheric density. The power available from a wind turbine is given by the following (Haslach 1989):

$$P = \tfrac{1}{2} c \rho v^3 A; \tag{8}$$

where
    P is the power produced by the turbine;
    A is the swept area of the wind turbine;
    v is the wind velocity
    ρ is the density of the atmosphere; and
    c is the power coefficient.

It can be seen that although the atmospheric density is 100 times less on Mars, the dominate term is the wind speed. Hence, assuming a ρ of .01665 kg/m^3 for Mars, a 30 m/s Martian wind will provide the same power a 6 m/s wind on the Earth (Zubrin 1996). Using an efficiency value of $c = 0.4$, as is common on terrestrial turbines, a 200 m^2 turbine could produce 2 kW in a 14 m/s wind and 12 kW in a 25 m/s wind (Haslach 1989). Haslach produced a concept which called for a 17.25 meter tall giromill turbine situated atop a 21.5 meter landing vehicle with a weight of 175 kg.

Wind power produces no wastes and is totally sustainable. Also, it requires "low tech" extraction technology. Hence, it would be easily maintainable and potentially expandable. In fact, most of the mass and volume of a wind turbine are in the blades and tower. These are components that could be manufactured early in a Martian outpost's life from native metallic or composite materials. Terrestrial experience has shown that wind/solar and wind/combustion hybrid systems are extremely effective and can provide near continuous power. Also the generator systems could be common to solar dynamic, combustion, and some nuclear systems. The power to mass ratio for wind power generated using a design suggested by (Haslach 1989) ranges from 7.64 W/kg for 14 m/s winds to 44.1 W/kg for 25 m/s winds. However, wind power is a variable and limited resource when used alone. The extraction system also would have to be sited appropriately to make full use of the resource. Current models suggest that the most effective winds could be as much as 25 meters above the surface.

Model of Available Energy

Currently wind speed measurements exist at only three sites on the Martian surface. This information can be utilized to assess the potential wind energy resource available on Mars. The model suggested by (Zurek 1992) was used to estimate the potential wind speeds that would have existed above the Viking 1 and Viking 2 lander measurement sites. This model takes the following form:

$$U = \frac{u_*}{k}\left[\ln\left(\frac{z}{z_0}\right) - \psi_u\right]; \tag{9}$$

where

    U is the wind speed estimated at altitude;
    u_* is the friction velocity which varies with time;
    k is von Kármán's constant
    z is the altitude desired;
    z_0 is the surface roughness height; and
    ψ_u is a function modeling atmospheric stability.

The stability function is based on Earth derived boundary layer results and is expected to be universally applicable (Sutton 1978):

$$\psi_u = \frac{-4.7}{L}(z - z_0); \text{ for stable conditions (when } L>0);$$

$$\psi_u = 0; \text{ for neutral conditions; (when } \frac{z}{L}=0); \qquad (10)$$

$$\psi_u = 4 \int_{u_0}^{u} \frac{u^2}{(1+u^2)(1+u)} du; \text{ for unstable conditions (when } L<0);$$

where

    L is the Monin-Obukhov height which varies over time; and

$$u = \left(1 - 15\frac{z}{L}\right)^{1/4}.$$

Values for L and u_* for the first part of each Viking lander mission are provided in Zurek (1992) and Sutton (1978). These parameters were estimated using wind measurements at an altitude of 1.6 m. Estimates of wind speeds at other altitudes over the Viking landing sites can be then be produced. Figure 2 provides the wind speed estimates produced by this model for ten days at the Viking 2 Lander site at 25 meters above the surface. The total wind speed peaks at 19 m/s as seen in the upper left plot. The neutral component of this model has a maximum speed of over 7 m/s as seen in the upper right plot. The stable component, shown in the lower right, tends to add up to 18 m/s to the neutral component during the night. The unstable component (lower left) reduces the wind speed by almost 3.5 m/s during the day. Equation (8) can then be utilized to estimate the power produced by a wind turbine of a specific design. Utilizing the parameters listed above for the wind turbine concept provided in (Haslach 1989), the power produced by the Viking 2 winds would peak over 4.5 kW and produce an average power of slightly under .5 kW per day.

Figure 3 provides the wind speed estimates 25 meters above the Viking 1 lander. The model predicts that the wind speeds peak at 60 m/s. The stable part of the winds are the largest component of these wind speeds with peaks of 55 m/s which is related to a nocturnal jet occurring after sunset. A second peak occurs pre-dawn and is likely related to the slope wind phenomena. The neutral component peaks at 15 m/s. The unstable component reduces the total wind speed by up to 10 m/s. This data represents a ten day window out of the 15 day window over which the model parameters are valid.

The expected power produced by the turbine design parameters listed above is shown in the lower graph of Figure 4. This figure provides estimates of the instantaneous power expected at 25 sample times per day. The associated estimates of the solar power (using the solar array size and efficiency provided earlier) for the same period of time are shown in the upper graph of Figure 4. The wind power is seen to peak at almost 150 kW. However, this power level is not maintained for an extended period of time. Daily averages are 58 kW per day for solar power and slightly less than 7 kW per day for wind power. Since these wind power production is extremely sensitive to the peak wind speed estimate, uncertainties in the model could significantly affect these estimates. Ongoing efforts include validating and expanding the wind speed estimation model (to other seasons, altitudes, and locations) as well as performing energy system trade-off studies.

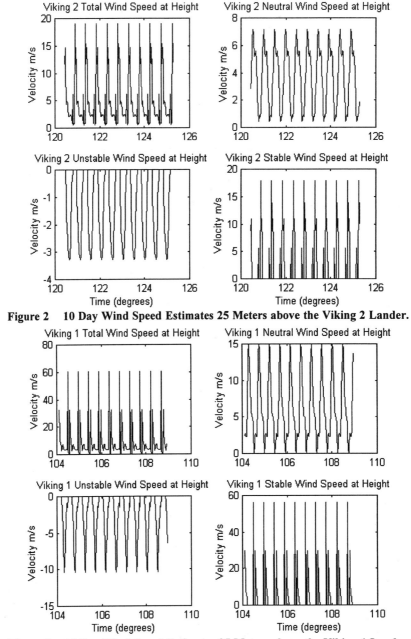

Figure 2   **10 Day Wind Speed Estimates 25 Meters above the Viking 2 Lander.**

Figure 3   **10 Day Wind Speed Estimates 25 Meters above the Viking 1 Lander.**

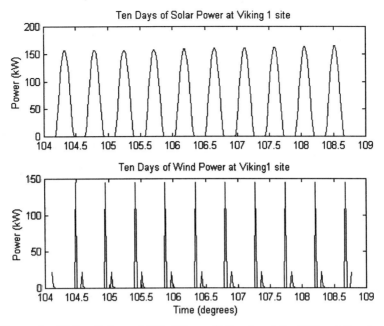

Figure 4   10 Day Solar and Wind Power Estimates at the Viking 1 Lander.

Areothermal Energy

This section discusses the availability of planetary heat sources on Mars. This resource will be labeled as Areothermal, in order to distinguish the Martian expression from the terrestrial resource which is labeled as Geothermal. Geological measurements are required to determine the heat flux on Mars. However, a current estimate suggests that the average value is close to 35 mW/m^2. This is less than the average terrestrial value of 80 mW/m^2. On the Earth, the process of plate tectonics typically serve to concentrate this energy (Meyer 1996). This process is not currently seen or believed to be possible on Mars. However, young volcanic features are also indicative of underground geothermal sources (Zubrin 1996). Seven percent of the Martian surface was geologically emplaced during the Upper Amazonian period (this dating comes from an impact crater count of less than 40 craters larger than 2 km per 10^6 km^2). Of this, about 3.1% is covered with formations resulting from young igneous intrusions or various fluvial processes which are indicative of near surface volcanic heating. Hence, 4.5 million km^2 of the Martian surface area is likely to have experienced volcanism in the last 700 to 250 million years. In fact, it is possible to have had active volcanism until recent times or even ongoing today. These regions are more likely candidates to harbor near surface reservoirs of areothermal energy (Fogg 1996). It should be noted that significant near-surface areothermal energy may not have a surface manifestation. With this in mind, (Fogg 1996) points out that most Amazonian volcanism (which is less than 2 billion years old) lies on the 28% of the surface (40 million km^2) contained between 20° and 220° W and 50° N to 15° S. This area may contain many regions of such cryptovolcanic or subsurface volcanism, which may be a result of a huge mantle plume of ascending magma (Fogg 1996). As such, there seem to be many potential regions on Mars which could provide areothermal resources.

The general model of an exploitable geothermal resource includes a heat source of magma or a cooling intrusion of volcanic material. In general, a supply of water in a permeable reservoir is heated by the source and contained by an impermeable cap rock layer. Therefore most terrestrial exploration techniques assume that heated water is available and is assisting in the modification of the environment surrounding the heat source. There are terrestrial applications where hot dry rock can be used for energy production assuming a ready supply of fluid is available. Also, there are postulated heat sources which are driven by concentrated radioactive materials as opposed to magma sources (Fogg 1996; Wright 1985).

Geothermal Exploration Techniques

The current techniques available for geothermal exploration are based on thermal, electrical, gravitational, magnetic, seismic, and radiometric methods, geophysical well logging, and studies of borehole geophysics. The thermal techniques measure elevated temperatures of rock or any nearby fluids. Hence, these techniques represent the most direct indication of geothermal reserves. Thermal studies include thermal gradient and heat flow, shallow-temperature surveys, snow-melt photography, and thermal-infrared imagery. An additional technique which has received little terrestrial attention is microwave emission mapping. The electrical techniques are the second most important procedures (after thermal) for locating geothermal resources when heated water is present. High temperatures, increased porosity, increased salinity, and the presence of certain clays and zeolites increase the ionic mobility (and hence conductivity) of the fluid-rock volume. All of these conditions may be present in Martian hydrothermal systems. Also, partial melts and magma can become very conductive, especially when water is present. The electrical exploration techniques include a variety of surface-based techniques in which an electromagnetic field or current is produced in the rock volume under test and the modifications due to the surrounding rock are interpreted to perform the exploration (Wright 1985).

Local gravity is modified by volcanic intrusions and changes in geologic structures. Typically, positive gravity anomalies are sought which indicate the presence metamorphic rock, alteration of rocks by hot water, and granitic intrusions. Negative gravity anomalies imply the presence of hot silicate magmas. Magnetic methods are useful for mapping regions where water or heat has modified the magnetic properties of the rock. Typically, these measurements indicate where magnetite has been modified to form nonmagnetic pyrite by interacting with hydrogen sulfide in the presence of heat. This creates a magnetic low (which will remain even after the source becomes extinct). Another utility of magnetic surveys is the ability to determine Curie point depth. This is the depth at which the temperature is high enough to destroy magnetite. Seismic events are usually associated with hydrothermal systems. There is some indication that these water/heat systems radiate seismic energy in the range from 1 to 100 Hz. There are also modifications which result when seismic waves pass through magma reservoirs that can be used to map the resource (Wright 1985).

Radiometric mapping is used to determine the extent of systems that have surface manifestations of hot springs containing radioactive elements. Geophysical well logging records physical properties of the lithosphere, which can be used to determine properties related to geothermal production issues. Finally, borehole geophysics utilizes surface and subsurface instruments to map electrical or seismic phenomena with greater depth resolution (Wright 1985).

Martian Thermal Mapping

In thermal mapping studies of Mars, the temperatures just before dawn are of particular interest. This is the coldest time of the day and the results are less dependent on albedo or terrain slope. The results are then most indicative of surface properties. Mariner 9 acquired the first predawn temperature measurements during the first successful orbital Martian orbital mission in 1972. The Mariner 9 results showed that there was a warm region along the northern edge of Hellas Planitia and high local temperatures in Valles Marineris. The area around the south of Tharsis Ridge was, in contrast, found to be cooler than the surrounding area. The Viking results showed that warmer regions were associated with Valles Marineris and extending into the Chryse Basin as well as the southern border of Isidis Planitia (Kieffer 1977). Upon comparison with a Viking thermal model, the Valles Marineris anomaly showed the largest temperature differences T_r (>16K). The floors of Ophir Chasma and Ganges Chasma also show T_r values comparable to Valles Marineris. Additionally, a region to the east of the Valles Marineris canyon system (-17°, 28° W) showed similar residual temperature values. Since none of these regions were found to be especially dark, the original authors assumed that high thermal inertia material was present (Kieffer 1977).

High residual temperatures (>10K) were also found in the Isidis Planitia region which had high actual temperatures. This region was on the boundary between Isidis Planitia, Syrtis Planitia, and the southern highlands (+5°, 270° W). Two adjacent regions (+30°, 255° W and +10°, 220° W) were indicated to contain high to moderate slopes. However, this is not expected to have had a serious effect on the residual temperature calculations. No other explanation was provided for these anomalies (Kieffer 1977).

The Viking data showed several areas with T_r values of 2 to 8 K. The most extensive of these is the Hellas Basin with a 4 K value. There were localized areas to the west and northeast of the basin which also showed similar values. Again, the Mariner 9 data suggests that these areas might contain higher thermal inertia material. Argyre Basin showed the same trend as Hellas but with a smaller magnitude. A few other localized areas such as the crater Huygens and the region of Solis Lacus (-20°, 90° W) also exhibited some residual temperature anomalies (Kieffer 1977).

Limitations in Areothermal Interpretation of Thermal Mapping Results

The primary limitation in using the thermal mapping studies performed to date for areothermal exploration is in spatial resolution. The level of detection of extremely localized thermal phenomena is dependent on size and temperature of the anomaly. As an example, a lava field or lava lake with a brightness temperature of 500K would need to be at least 200m in diameter to produce a 4K change in the 7 μm band. The Viking or Mariner 9 data did not show any regions with a 4K increase over adjacent measurements (Kieffer 1977).

Variations in albedo and thermal inertia can also have an effect on the measurements. However, the Viking instruments allowed the albedo to be measured using the solar reflectance channel. Thermal inertia then remains as one of the primary unknowns in understanding the residual temperature anomalies. Surface roughness and slope of the terrain are other important modifiers of the temperature profile (Kieffer 1977). Currently, the Mars Global Surveyor mission is carrying a thermal emission spectrometer which will function as an infrared spectrometer and radiometer. One study will be to determine the extent of the surface covered by rocks and boulders and to determine grain size (Smith 1996).

The atmosphere can also have a significant effect on the measured temperatures. In fact, the Viking analysis showed variations in residual temperature of up to 10K within a region between Olympus Mons and the Tharsis ridge. These variations were believed to be due to winds. It is expected that this region will contain some of the strongest winds on the planet. Modifications in the thermal profile due to ground ice can also be expected but as yet have not been measured (Kieffer 1977).

Photographic Analysis

Although not all areothermal energy sources can be expected to have a surface manifestation, those that do can be expected to have produced young terrain. The late Amazonian regions of Mars include the youngest surfaces, assumed to be less than 0.7 to 0.25 billion years old. There are a few localized areas which are largely uncratered and may be the youngest surfaces on the planet (Tanaka 1986). One of these regions is the Ceberus Plains in SE Elysium (5° N, 190° W). This region appears to be a flood basalt extrusion of significant depth (Fogg 1996). Another region is the western slope of Hecates Tholus in the Elysium province which is largely devoid of impact structures. This construct appears to be an ash-fall which would suggest a heated source of water (Moginis-Mark 1982).

The Medusa Fosse formation south of Amazonis Planitia appears to be a large pyroclastic emplacement which has been built up over extended periods of time (Scott 1982). The Tharsis region and especially the large Olympus Mons are covered with young lava flows. Some constructs in northwest Tharsis (16° N, 129° W) appear to be cut by water as opposed to lava and appear to postdate the lava flows (Mouginis-Mark 1990). The floor of Valles Marineris including the Ophir, Candor, and Coprates Chasmas show dark splotches which may be pyroclastically emplaced materials that have not been extensively weathered. These deposits appear to follow the fault line which forms the canyon system (Luchitta 1990).

Microwave Emission Mapping

Microwave emission mapping holds some promise as it can extract information from a few meters below the surface (Bowen 1979). Unfortunately, no current data exists for Mars.

Gravity Mapping

An analysis of the Doppler shift in the radio frequencies of the Viking Orbiters has allowed a determination of a gravity model for Mars. This model has gravity highs associated with Olympus Mons, Elysium Mons, Arisa Mons, Pavonis Mons, Ascraeus Mons, Alba Patera, and Utopia Planitia. Gravity lows are associated with Hellas Planitia, and Valles Marineris. Isidis Planitia has a gravity high in the center and a gravity low on the edges (Esposito 1992). The Mars Global Surveyor will be providing more detailed measurements of the gravity field (Smith 1996).

Local Magnetic Mapping

Regional magnetic mapping of Mars has been performed. In fact, the Martian Global Surveyor has begun the measurement of global Martian magnetic properties (Smith 1996). Although the results are not complete at the time of this writing, there are several localized crustal magnetic anomalies which have been reported (Acuna 1998).

Radioisotope Mapping

Radioisotope mapping of the Martian surface has not yet occurred. However, a gamma-ray spectrometer is planned for the 2001 Mars Surveyor Orbiter (Covault 1996).

Water Vapor Sources

Although there has not been a formal study of the relationship between water vapor sources and areothermal heat sources, the connection is plausible. In fact, it is analogous to the relation between terrestrial fumaroles and geothermal sources (Wright 1985). One study identified potential Martian water vapor sources based on terrestrial, Mariner 9, and Viking observations. The most probable source regions include Solis Lacus (25° S, 85° W) which has shown activity throughout the observational record. In fact, this is the probable source region for the clouds which form in the Tharsis and Syria Planum regions. Another strong source region for water vapor is the Noachis-Hellespontes region (30° S, 310° W). The winter clouds which form in Hellas are most likely derived from this source. The clouds which form around the Elysium construct are likely derived from a source on the border regions of Syrtis Major which is adjacent to Isidis Planitia. Other potential source regions of lesser importance include Argyre Planitia, Arcadia Planitia, Tempe Fosse, Candor Chasma, and Lunae Planum (Huguenin 1982).

Terrestrial Analogs

The most successful geothermal source on Earth is at the Mid-Atlantic Rift as it crosses Iceland. Since Mars is a one plate planet, such sources would not have analogs. However, rift valleys and large graben have produced exploitable energy sources on the Earth. This means that the Valles Marineris and many other smaller tectonic features associated with the volcanic regions represent potential sources of areothermal energy. Also, good geothermal sources exist in turbidite areas which are formed when intrusions are trapped under thick plastic sediments. This is believed to be especially true on Mars where buoyancy forces are less. Also, since water erosion and eolian deposition are postulated to have occurred over long periods of Martian history, such sedimentary layers can be postulated to exist (Wright 1985; Fogg 1996).

Correlation of Areothermal Indicators

The analysis of the Viking thermal mapping data required the use of a fairly crude model. Additional information will be available in the future concerning surface roughness, local slope, ground ice, atmospheric effects, and even more detailed soil information. When factored into the model along with more detailed experimental spatial resolution, the model could be much more precise. Additionally, several ground truth measurements should be available from the unmanned landers over the next few years. Hence, the thermal imagery results can be expected to improve in terms of data return and enhanced understanding. Nonetheless, it is possible to augment current understanding of the data by correlating the thermal imagery with other indicators of areothermal energy sources. Figure 5 collects this information on a single map. Potential areothermal regions dictate exploratory sites which stand out as the locations where several indicators converge.

The chasmas of Valles Marineris and the associated outflow regions have emerged as the most likely candidates for providing areothermal resources. Thermal imagery, visual imagery, gravity mapping, water vapor production, and terrestrial analogs all point to the large canyon system as a candidate. The regions around the Hellas basin especially to the west and northeast are also strong candidates based on thermal, gravity, and water vapor sources. Also, if the basin is filled with a thick sediment load, it could be a candidate area for turbidite deposits. The border area between Syrtis Major and Isidis Planitia is also a candidate given the thermal, gravity, and water vapor indicators. Again if the Isidis basin is sediment filled, then it could be trapping a rising magma source. The residual temperature anomaly and strong water vapor production source in Solis Lacus also provide indications that further study is needed in that region. Other regions such as the Ceberus Plains, Medusa Fosse, Hecates Tholus, and northwest Tharsis are potential candidates based on the photographic analyses.

New information from additional thermal infrared studies, gravity studies, magnetic mapping, and radioisotope mapping should be available in the next few years. This database will add significant knowledge to that which is discussed in this paper. The ability to perform microwave emission mapping could also add significant detail to the areothermal energy exploration efforts. However, surface studies will be needed to finally answer the question of energy source exploitability once remote sensing has defined the search areas.

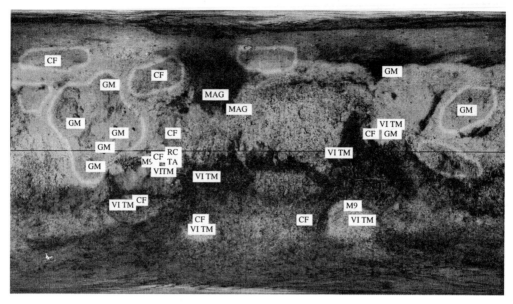

Key:

   Shaded Regions or **RC** - Photography (Recent Construct)
   M9 - Mariner 9 Thermal Imaging
   VITM - Viking Infrared Thermal Mapper
   GM - Viking Gravity Mapping
   MAG - Global Surveyor Magnetic Mapping
   CF - Cloud Formation
   TA - Terrestrial Analogs

Figure 5   Map of Potential Areothermal Sites.

The question of how deep and where these resources are will only be answered, conclusively, by drilling missions. However, some estimations are possible. Table 1 is provided in (Zubrin 1996) as a guide to the depth needed to reach areothermal resources given different geological ages. As a point of reference, a single-well areothermal source of 150° would produce 10 MW of power. This assumes that heated fluids are available to directly drive the energy production turbines. Fogg (1996) also discusses extraction techniques when lower temperatures are available and when areothermal fluids are not available. Electricity production represents an indirect use of geothermal energy and has a maximum conversion efficiency of about 20%. On the other hand, direct use of geothermal energy for heating has been demonstrated to attain efficiencies of close to 90% (Fogg 1996).

Fogg (1996) also suggests a 10 km limit of the depth of available resources. This maximum depth is set for two reasons. First at 10 km most pore spaces in the crust will be closed by compaction and heated fluids would not be available. Also, 10 km is an accepted maximum limit for terrestrial drilling technology. It can be seen from Table 1 that the capability to drill from 2 to 3 km holds the potential for utilizing significant areothermal resources. However, the early capabilities of a Mars outpost will only allow drilling to a few hundred meters for water exploration or extraction purposes. It should be noted that the current Mars reference mission is only assuming a 10 meter drill (NASA 1998). Unless near surface sources are found, full scale (i.e., deep drilling operations) use of areothermal energy will not be feasible until the capability exists to produce material supplies such as pipe and drill rod directly on Mars.

Table 1
DEPTH OF MARS AREOTHERMAL RESERVES AS
A FUNCTION OF YEARS SINCE LAST VOLCANIC ACTIVITY (Zubrin 1996)

Time since activity (Myr)	Depth to 0°C (km)	Depth to 60°C (km)	Depth to 100°C (km)	Depth to 200°C (km)	Depth to 300°C (km)
.5	.29	.62	.84	1.38	1.92
5	.65	1.38	1.87	3.09	4.30
10	.91	1.95	2.64	4.36	6.09
20	1.29	2.76	3.73	6.17	8.61
50	2.04	4.35	5.88	9.73	13.
>150	3.53	7.53	10.	17.	24.

Unfortunately, the true extent and location of areothermal energy is not known at this time. Hence, the initial mission planning could not rely on an areothermal source and successful exploitation will require a long term build-up of mass intensive components or a Martian manufacturing capability (Fogg 1996). Areothermal prospecting, in the long term, will be an important activity, however, to ensure future growth and prosperity of a permanent Mars colony.

INTEGRATED POWER SYSTEMS

While the preceding discussions have shown the potential for the utilization of in situ energy resources on Mars, a reliable power system for a first mission must necessarily depend upon known capability supplemented by auxiliary resources as they become available. Hence a strategy of "living off the land" while conducting resource exploration is the key to development of a substantial self-supporting base on Mars. An initial power supply based on a combined solar and wind energy conversion and storage system would be the best option for leveraging the best available information for system sizing and storage capacity to guaranty that minimal energy requirements can be met. The system should be designed in such a way as to benefit from additional wind, if available. A mobile system that could be relocated to local areas with more wind would be ideal. As mentioned previously, areothermal energy extraction will require additional surface and space-based exploration to identify useful sources. This is considered to be a long term solution with the potential to allow large scale development on Mars. As such, geothermal exploration should be an important aspect of the first missions, but a non-nuclear power supply for these missions will necessarily depend strictly on solar and wind energy.

The extraction of solar energy on Mars has several advantages and few significant obstacles. More than any other resource, the availability, distribution, and seasonal variation of solar energy on Mars is well known (or well modeled). Even if more information was obtained to further characterize atmospheric properties and dust particle size/composition distributions, this would not affect, at this point, the ability of current models to characterize the worst-case scenario. If a solar power system is designed for robustness and reliability, then it must provide acceptable power production in the worst-case. This is possible, now, without requiring new atmospheric data. Another advantage is that the latest technology points to solar power becoming cost competitive with nuclear, due to higher specific powers, especially with thin film solar cells. Independent of development and material costs, the cost of mass delivered to the Martian surface per kW is perhaps the best measure of economical power. Solar and nuclear are becoming close competitors in this regard. Furthermore, solar energy is reliable, sustainable, and expandable, well beyond the lifetime of a nuclear power source. Models indicate that while solar energy availability decreases during dust storms, losses are limited to about a 60% drop in output, due to the conversion of direct light into scattered light. Doubling the capability for ideal conditions would assure required energy supplies are satisfied.

While the availability and distribution of wind energy on Mars is not well known at this time, it is possible to achieve reliable, continuous power, with reduced energy storage requirements using a system based on combined solar and wind resources. As discussed earlier, wind energy can be more significant on Mars than is commonly assumed. Large diurnal temperature variations and massive topographical relief produce reliable, consistent winds. Since power is proportional to the cube of velocity, high local winds can make up for the loss of density as compared to Earth. Moreover, wind energy may be complementary to solar, due to day/night cycles, dust storms, and the fact that slope winds and global circulation are increased during times when solar energy is decreased. Since solar power alone requires cyclic energy storage to provide continuous power, and energy storage involves a significant weight penalty, wind energy would help to reduce cyclic storage requirements. Using the models described in this paper for solar and wind resource variation, current research aims to develop trade study codes to determine the optimal combination of solar, wind, and energy storage elements as a function of landing site location and atmospheric parameters. Diurnal and seasonal variations in solar flux and wind near the surface, combined with mass estimates for power and storage systems, can be used to determine optimal specific power configurations with robust margins above given energy requirements. In considering energy storage devices, it is preferable to use liquid fuels with high energy content and useful products of reaction. A dual use concept that combines life support and surface transportation systems with energy storage will reduce overall mission cost and provide redundancy. Regenerative fuel cells (H_2-O_2) and/or methane/oxygen systems are likely candidates.

Aspects of the Martian climate may dictate a solar/wind power system design that differs from Earth-based systems. One of the major problems with solar power on Mars is the degradation of efficiency due to Martian dust. Either solar panels would require periodic maintenance or an automatic system must be devised to remove dust accumulation from panel surfaces. Another option, however, might be to remove the panels from the dustiest environment, which is in the saltation flow at ground level. Another driver for elevating the power system is that the highest wind energy on Mars is expected to be found at about 25m off the ground. This is difficult to accomplish, however, with conventional terrestrial wind mill designs. Higher winds may also help to keep solar panels clean. These considerations suggest a new approach for a combined solar/wind system that is airborne and tethered to a mobile base station. A similar concept, called an Aerobot, has already been examined by (Romohalli 1998), consisting of a balloon made from a substrate material for thin film solar cells. Such a system could use solar and wind power to process the Martian atmosphere physically or chemically to control buoyancy and/or store energy. A mobile airborne power platform would be especially desirable for exploration, rover support, and for relocating the system to ideal locations for wind energy, once those sites are locally determined.

CONCLUSIONS

In summary, this paper has discussed alternatives for locating and extracting native resources that are essential to survive on Mars without nuclear energy. A nuclear-free Mars mission is possible and achievable within the bounds of current technology. The latest solar cell technology has demonstrated capabilities that are comparable to nuclear systems, due to specific power performance and the cost of payload delivery to Mars. Models of wind are encouraging for the utilization of wind resources combined with solar. Implementation issues and the Martian environment suggest the possibility of a combined airborne solar/wind design for maximum efficiency. While more research is needed to characterize aerosols (dust particle composition and size distribution), this should have little impact on solar energy system designs aimed to guaranty robustness in the worst-case conditions. More research is needed, however, to identify and characterize wind resource distribution. The concept of a space-based Coherent Doppler Lidar system is one promising approach for detailed mapping of the global Martian winds. Significant space-based and surface-based research will be required to determine the feasibility of utilizing Areothermal energy resources. This has huge potential, however, for long term energy supply and for supporting large scale development on Mars. Precursor missions designed to conduct specific research aimed at identifying these resources can expedite the possibility of a nuclear free Mars mission and the development of a permanent human base.

REFERENCES

Acuna, M. H., *et al.*, "Magnetic Field and Plasma Observations at Mars: Initial Results of the Mars Global Surveyor Mission," *Science*, vol. 279, March 13, 1998, pp. 1676-1680.

Appelbaum, J. and Flood, D. 1989a. *Photovoltaic Power System Operation in the Mars Environment*, NASA TM-102075.

Appelbaum, J. and Flood, D. 1989b. Solar Radiation on Mars. *Solar Energy* 45:353-363, and NASA TM-102299.

Appelbaum, J. and Flood, D. 1990. *Solar Radiation on Mars - Update 1990*, NASA TM-103623.

Bowen, R., *Geothermal Resources*, John Wiley and Sons, New York, 1979.

Collins, P., "SPS-2000 and Its Internationalization," *Proceedings of the Fifth International Conference on Engineering, Construction, and Operations in Space*, Albuquerque, NM, June 1-6, 1996, pp. 269-279.

Colston, B. W., "Nuclear Power Supplies: Their Potential and the Practical Problems to Their Achievement in Space Missions," working group papers of the Manned Mars Missions Workshop, Huntsville AL, June 10-14, 1985, NASA M002.

Covault, C., "Mars 98 to seek water near Martian south pole," *Aviation Week and Space Technology*, 47-52, December 9, 1996.

Crisp, D., Micro Weather Stations for In-Situ Measurements in the Martian Planetary Boundary Layer, *LPI Technical Report*, 94-04, 1994.

Criswell, D. R., "World and Lunar Solar Power Systems Costs," *Proceedings of the Fifth International Conference on Engineering, Construction, and Operations in Space*, Albuquerque, NM, June 1-6, 1996, pp. 293-301.

Duke, M. B., "Mars Resources," working group papers of the Manned Mars Missions Workshop," Huntsville AL, June 10-14, 1985, NASA M002.

Esposito, P. B., W. B. Banerdt, G. F. Lindal, W. L. Sjorgren, M. A. Slade, B. G. Bills, D. E. Smith, and G. Balmino, Gravity and topography, in *Mars*, ed. by H. H. Kieffer, B. M. Jakosky, C. W. Snyder, and M. S. Matthews, University of Arizona, Tuscon, pp. 209-248, 1992.

Fogg, M. J., "Geothermal Power on Mars," *Journal of the British Interplanetary Society*, Vol. 49, No. 11, November 1996, pp. 403-422. Also published as AAS 97-383, in *From Imagination to Reality: Mars Exploration Studies of the Journal of the British Interplanetary Society* (Part II: Base Building, Colonization and Terraformation), ed. R. M. Zubrin, Vol. 92, AAS *Science and Technology Series*, 1997, pp. 187-227.

Geels, S., Miller, J. B., and Clark, B. C. 1989. "Feasibility of using solar power on Mars: Effects of Dust Storms on incident solar radiation." AAS 87-266, in *The Case for Mars III*, ed. C. R. Stoker, Vol. 75, AAS *Science and Technology Series* (San Diego: Univelt), pp. 505-516.

Greeley, R., "Rate of Wind Abrasion on Mars," *Journal of Geophysical Research*, Vol. 87, No. B12, pp. 10009-10024, Nov 1982.

Haberle, R. M. et al., "Atmospheric Effects on the Utility of Solar Power on Mars," in J. Lewis, M. S. Matthews, and M. L. Guerrieri, *Resources of Near-Earth Space*, pp. 845-885, University of Arizona Press, Tuscon, 1993.

Haslach, H. W., Jr., "Wind Energy: A Resource for a Human Mission to Mars," *Journal of the British Interplanetary Society*, Vol. 42, No. 4, April 1989, pp. 171-178. Also published as AAS 97-376, in *From Imagination to Reality: Mars Exploration Studies of the Journal of the British Interplanetary Society* (Part II: Base Building, Colonization and Terraformation), ed. R. M. Zubrin, Vol. 92, AAS *Science and Technology Series*, 1997, pp. 53-70.

Huguenin, R. L. and S. M. Clifford, "Remote sensing evidence for regolith water sources on Mars," *Journal of Geophysical Research*, 87, No. B12, 10,227-10,251, 1982.

Kavaya, M. J. and Emmitt, G. D., The Space Readiness Coherent Lidar Experiment (SPARCLE) Space Shuttle Mission, to appear in *Proceedings of the SPIE Laser Radar Technology and Applications III*, Orlando, FL., April, 1998.

Kieffer, H. H., T. Z. Martin, A. R. Peterfreund, and B. M. Jakosky, "Thermal and albedo mapping of Mars during the Viking primary mission," *Journal of Geophysical Research*, 82, No. 28, 4249-4291, 1977.

Landis, G. And Appelbaum, J., 1990. "Design Considerations for Mars photovoltaic power systems." In *Proceedings of 21st IEEE Photovoltaic Specialists Conference*. (New York: Inst. of Electrical and Electronic Engineers), pp. 1263-1270.

Lee, S. W., "Wind Streaks in Tharsis and Elysium: Implications For Sediment Transport by Slope Winds," *Journal of Geophysical Research*, Vol. 87, No. B12, pp. 10025-10041, Nov 1982.

Luchitta, B. K., "Young volcanic deposits in the Valles Marineris, Mars?," *Icarus*, 86, 476-509, 1990.

Magalhaes, J., and Gierasch, P., "A Model of Martian Slope Winds: Implication for Eolian Transport," *Journal of Geophysical Research*, Vol. 87, No. B12, pp. 9975-9984, Nov 1982.

Mankins, J. C., "The Space Solar Power Option," *Aerospace America*, pp. 30-36, May 1997.

Meyer, T. R. and McKay, C. P., "The Resources of Mars for Human Settlement," *Journal of the British Interplanetary Society*, Vol. 42, No. 4, April 1989, pp. 147-160. Also published as AAS 97-373, in *From Imagination to Reality: Mars Exploration Studies of the Journal of the British Interplanetary Society* (Part II: Base Building, Colonization and Terraformation), ed. R. M. Zubrin, Vol. 92, AAS *Science and Technology Series*, 1997, pp. 3-29.

Meyer, T. R. and McKay C. P., "Using the Resources of Mars for Human Settlement," AAS 95-489, also Chapter 19 of *Strategies for Mars: A Guide to Human Exploration*, ed. by Stoker, C. R. and Emmart. C., American Astronautical Society, Vol. 86, AAS *Science and Technology Series*, San Diego, CA, 1996, pp. 393-442.

Mouginis-Mark, P. J., Wilson, L., and Head III, J. W., "Explosive volcanism on Hecates Tholus, Mars: Investigation of Eruption Conditions," *Journal of Geophysical Research*, 87, No. B12, 9890-9904, 1982.

Mouginis-Mark, P. J., "Recent Water Release in the Tharsis Region of Mars," *Icarus*, 84, 362-373, 1990.

NASA, The Reference Mission of the NASA Mars Exploration Study Team, Technical Report EX13-98-036, NASA, June 1998.

Nelson, M., and Dempster, W. F., "Living in Space: Results from Biosphere 2's Initial Closure, and Early Testbed for Closed Ecological Systems on Mars," AAS 95-488, also Chapter 18 of *Strategies for Mars: A Guide to Human Exploration*, ed. by Stoker, C. R. Emmart, C., American Astronautical Society, Vol. 86, AAS *Science and Technology Series*, San Diego, CA, 1996, pp. 363-390.

Romohalli, Kumar N., Solar-Powered Aerobots with Power-Surge Capabilities, NASA Tech Brief Vol. 22, No. 4, Item #189, JPL New Technology Report NPO-20155, April 1998.

Schofield, J. T., Barnes, J. R., Crisp, D., Haberle, R. M., Larsen, S., Magalhaes, J. A., Murphy, J. R., Seiff, A., and Wilson, G., "The Mars Pathfinder Atmospheric Structure Investigation/Meteorology (ASI/MET) Experiment," *Science*, vol. 278, Dec. 5, 1997, pp. 1752-1758.

Scott, D. H., and K. L. Tanaka, "Ignimbrites of Amazonis Planitia of Mars," *Journal of Geophysical Research*, 87, No. B2, 1179-1190, 1982.

Smith, B. A., "Knowledge about Mars set to expand rapidly," *Aviation Week and Space Technology*, 53-57, December 9, 1996.

Sutton, J. L., Leovy, C. B., and Tillman, J. E., "Diurnal Variations of the Martian Surface Layer Meteorological Parameters During the First 45 Sols at Two Viking Lander Sites," *Journal of the Atmospheric Sciences*, Vol. 35, pp. 2346-2355, Dec 1978.

Tanaka, K. L., "The stratigraphy of Mars," *Journal of Geophysical Research*, 91, No. B13, E139-E158, 1986.

Thomas, P., "Present Wind Activity on Mars: Relation to Large Latitudinally Zoned Sediment Deposits," *Journal of Geophysical Research*, Vol. 87, No. B12, pp. 9999-10008, Nov 1982.

Voss, S. S. "Topaz II System Description," *Proceedings of the Fourth International Conference on Engineering, Construction, and Operations in Space*, Albuquerque, NM, February 23 - March 3, 1994, pp. 717-728.

Wright, P. M., S. H. Ward, H. P. Ross, and R. C. West, "State-of-the-art geophysical exploration for geothermal resources," *Geophysics*, 50, No. 12, 2666-2699, 1985.

Zubrin, R., and Wagner, R., *The Case for Mars*, Free Press, New York, 1996.

Zurek, R.W., Comparative Aspects of the Climate of Mars: An Introduction to the Current Atmosphere, In *Mars*, Kieffer, H. H. *et al.*, University of Arizona Press, 1992.

MAR 98-059

NEAR-TERM, LOW-COST SPACE FISSION SYSTEMS

Michael G. Houts,* David I. Poston,†
Marc V. Berte‡ and William J. Emrich, Jr.**

The Heatpipe Power System (HPS) is a potential, near-term, low-cost space fission power system. The Heatpipe Bimodal System (HBS) is a potential, near-term, low-cost space fission power and/or propulsion system. Both systems will be composed of independent modules, and all components use existing technology and operate within the existing database. The HPS and HBS have relatively few system integration issues; thus, the successful development of a module is a significant step toward verifying system feasibility and performance estimates. A prototypic HPS module was fabricated, and initial testing was completed in April 1997. All test objectives were accomplished, demonstrating the basic feasibility of the HPS. Fabrication of an HBS module is underway, and testing should begin in early 1999.

INTRODUCTION

Fission systems can enhance or enable numerous space missions of interest. Fission systems scale well to extremely high power levels (>10 MWe) and are not affected by solar proximity or orientation. They are well suited for high-power missions in Earth orbit (such as an electric propulsion tug for moving large payloads between orbits), missions on the lunar or Martian surface, and deep space missions requiring more power than is available from practical radioisotope systems. Fission systems can enable a power-rich environment on the surface of Mars, which will be very important to manned exploration.

The Heatpipe Power System (HPS) is a potential, near-term, low-cost space fission power system. The Heatpipe Bimodal System (HBS) is a potential, near-term, low-cost space fission power and/or propulsion system. The HPS and the HBS incorporate lessons learned from previous space fission power development programs to reduce both development cost and time. Both systems have the following 16 important features:

1. Safety. The HPS and HBS are designed to remain subcritical during all credible launch accidents without using in-core shutdown rods. This passive subcriticality results from the high radial reflector worth and the use of resonance absorbers in the core. The systems also passively remove decay heat and are virtually nonradioactive at launch (there is no plutonium in the system).

2. Reliability. The HPS has no single-point failures and is capable of delivering rated power, even if several modules and/or heatpipes fail. The HBS has very few single-point failures, which are limited to ex-core components (e.g., the propellant tank).

---

* Los Alamos National Laboratory, MS K551, Los Alamos, New Mexico 87545. E-mail: houts@lanl.gov.
† Los Alamos National Laboratory, MS K551, Los Alamos, New Mexico 87545.
‡ Massachusetts Institute of Technology, Nuclear Engineering Department, Cambridge, Massachusetts 02139. E-mail: mvberte@mit.edu.
** NASA Marshall Spaceflight Center, PS05, Huntsville, Alabama 35758. E-mail: bill.emrich@msfc.nasagov.

3. Long life. The low power density in the HPS and HBS cores and the modular design give the potential for long life. At 100 kWt, fuel burnup limits will not be reached for several decades.

4. Modularity. The HPS and HBS consist of independent modules, and most potential engineering issues can be resolved by testing modules with resistance heaters (used to simulate heat from fission).

5. Testability. Full HPS/HBS tests can be performed using resistance heaters, with very few operations required to replace the heaters with fuel and ready the system for launch. The HBS can be tested in the bimodal mode using resistance heaters. Flight qualification is accomplished with resistance-heated system tests and zero-power criticals. No ground nuclear power test is required. The HPS or HBS flight unit can be tested at full power prior to launch. This is not possible in systems where the fuel cannot be readily replaced by resistance heaters.

6. Versatility. The HPS and HBS can use a variety of fuel forms, structural materials, and power converters.

7. Scalability. The HPS design approach scales well to >1000 kWt. Very high power (>10 MWt) systems based on the HPS approach are possible, but are much more complex than lower power versions (large number of heatpipes) and would most likely suffer a system mass penalty compared with other options.

8. Simplicity. There are few system integration issues. There is assured shutdown without in-core shutdown rods, no hermetically sealed refractory metal vessel or flowing loops, no electromagnetic pumps, no coolant thaw systems, no gas separators, and no auxiliary coolant loop.

9. Fabricability. The HPS has no pumped coolant loops and does not require a pressure vessel with hermetic seals. There are no significant bonds between dissimilar metals, and thermal stresses are low. There are very few system integration issues, thus making the system easier to fabricate. The HBS may require a pressure vessel for the propulsion mode.

10. Storability. The HPS and HBS are designed so that the fuel can be stored and transported separately from the system until shortly before launch. This capability will reduce storage and transportation costs significantly.

11. Early Milestones. Several milestones early in the development of the HPS and HBS will prove the viability of the concepts. The most significant early milestones were the development and testing of an HPS module.

12. Near term. An HPS and/or HBS capable of enhancing or enabling missions of interest can be built with existing technology.

13. Bimodal. The basic approach can be used to provide either a power-only or bimodal system.

14. Dual use. Technology utilized by the HPS and HBS has military, commercial, and civilian uses in both aerospace and terrestrial applications.

15. Acceptable mass. The HPS and HBS have a high fuel fraction in the core, which reduces core, reflector, and shield mass for criticality-limited systems. The HPS has no pumped coolant loops and relatively simple system integration, further reducing mass.

16. Reduced program expense. The attributes of the HPS should allow for inexpensive (<$100 M) development. After development, the unit cost should be <$20 M. The attributes of the HBS (especially the ability to test bimodal operation with resistance heaters used to simulate heat from fission) will significantly reduce cost as compared with other bimodal concepts.

HPS DESCRIPTION

The HPS and HBS use similar (or identical) modules to create a core with the performance and lifetime required for a given mission. A wide variety of core layouts have been evaluated, using 12 to more than 100 modules. A schematic of the 12-module HPS is shown in Figure 1, and a schematic of a 4-fuel-pin HPS module is given in Figure 2. The fuel pins are bonded structurally and thermally to a central heatpipe, which transfers heat to an ex-core power conversion system. The heatpipe also provides structural support for the fuel pins. Modules are independent during normal operation. If a heatpipe fails, some thermal bonding between modules is desirable to reduce peak temperatures. Thermal radiation provides some module-to-module thermal bonding, which can be enhanced by (1) adding helium or lithium to the interstitial spaces, (2) brazing modules to adjacent modules, or (3) adding refractory metal wool to the interstitial spaces.

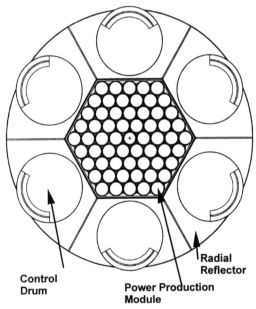

Figure 1 Schematic of HPS Showing Fuel Pins and Radial Reflector.

Two fuel types have been evaluated for use in the HPS: uranium nitride (UN) and uranium dioxide (UO_2). The use of UN results in the most compact core. However, UN fuel pins must be sealed hermetically, and the peak fuel temperature should be limited to ~1800 K (Matthews 1994). For conservatism, the peak UN fuel temperature is limited to 1600 K in all HPS designs, assuming a worst-case heatpipe failure. In the nominal case (all heatpipes working) peak UN fuel temperature is significantly less than 1600 K. UO_2 has a lower uranium loading than UN; however, the pins do not have to be sealed hermetically and can be operated at a higher temperature than UN pins. UO_2 fueled designs assume that a peak clad operating temperature of 1800 K is acceptable. Peak temperatures are calculated assuming a worst-case heatpipe failure has occurred. Fuel burnup limits (based on experimental results) are not reached for several decades in most designs (Makenas *et al.* 1994). During power operation, there will be some asymmetry in the fuel radial temperature profile because heat primarily is removed from one section of the fuel clad.

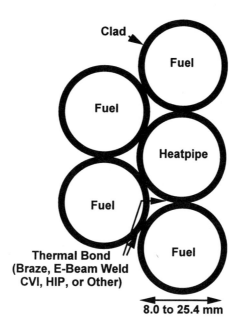

Figure 2   Schematic of HPS Power Module.

However, the temperature asymmetry will not be severe because of the low power density. Carbide and advanced fuels could also be used by the HPS.

The HPS primary side can provide heat to a power conversion subsystem at temperatures up to 1500 K. One HPS power conversion option (especially at relatively low power) is thermoelectric power conversion. Unicouple thermoelectric converters that are well suited for use with the HPS have been designed (Raag 1995). These converters have a hot-shoe temperature of 1275 K, and they reject waste heat at 775 K. This general type of thermoelectric converter has been used extensively by the space program and has demonstrated an operational lifetime of decades (Ranken *et al.* 1990). If desired, thermoelectric converters identical to those used by radioisotope thermoelectric generators (RTGs) could be coupled to the HPS. Close-space thermionic converters, alkali metal thermoelectric converters (AMTECs), and Stirling and Brayton power conversion are also options.

The specific mass of the HPS ranges from 40 kg/kWe at 20 kWe using AMTEC power conversion down to 10 kg/kWe (or less) at 250 kWe using Brayton cycle power conversion. It may be possible to further reduce specific mass for Mars surface applications by using in-situ materials for radiation shielding. Equipment needed for moving Martian soil into a shield configuration could be used in other aspects of the mission.

Low-temperature heatpipes are routinely used in computers, satellites, and other applications. In addition, there is considerable experience with heatpipes operating at or above the temperatures and heat fluxes required by the HPS and HBS. A molybdenum-TZM / sodium heatpipe operated for 53,000 hours at 1390 K without failing. A molybdenum / lithium heatpipe operated for 25,216 hours at 1700 K before failing. The purity of the liquid metal working fluid (especially lithium) is of primary importance to high-temperature heatpipe lifetime (Merrigan, 1997). Techniques have been developed for obtaining and maintaining a high purity working fluid, which will help ensure long heatpipe lifetime in future systems.

There is also considerable irradiation data on high-temperature liquid metal heatpipes. A total of 29 liquid metal heatpipes have been irradiated to significant fast neutron fluences, with no failures. A stainless steel / sodium heatpipe operated at 1100 K to a fast fluence of 2.2×10^{22} n/cm^2—more than an order of magnitude higher than that required by most potential near-term HPS or HBS missions (Merrigan, 1997).

Fuel can be removed easily from the HPS at any time, which will facilitate fabrication and handling greatly. The HPS is inherently subcritical during launch accidents and has no single-point failures. The HPS can undergo full system testing (using resistance heaters to simulate heat from fission) at existing facilities. Each of the HPS modules is independent, allowing most technical issues to be resolved with inexpensive module tests. Heatpipes have demonstrated >50,000-h lifetime at an operating temperature of 1400 K.

Mechanical bonding within the HPS modules is achieved by methods, such as a tack weld, an electron beam weld, chemical vapor infiltration (CVI), or hot isostatic pressing. For low-power cores (<100 kWt), radiation heat transfer will be adequate if finned (or small) heatpipes are used and if some reduction in power is acceptable following the loss of a heatpipe. If needed, thermal bonding can be accomplished by methods such as an electron beam weld, braze, helium bond, use of a refractory metal wool, or CVI. Structural support of the core is provided by the module heatpipes, which are anchored to a molybdenum or Nb/1Zr tie plate.

The pins are confined laterally on the opposite end of the core but are allowed to move freely in the longitudinal direction to allow for differential expansion. Neutron shielding is provided by lithium hydride; tungsten gamma shielding may or may not be required, depending on the thermal power level, payload separation, and allowable dose. For lunar and planetary applications, the shielding probably will consist of an optimal mix of material brought from earth and indigenous material. Because of its small size and the lack of activated coolant in its radiator, the HPS can be well shielded, with relatively little extra mass needed from earth. For manned missions, it may be desirable to shield the HPS so that no radiation-related exclusion zone is needed.

The HPS is designed to remain subcritical during all credible launch accidents. This has been accomplished by keeping the system radius small, keeping the reflector worth high, and strategically placing neutron absorbers in the core. The positive reactivity effect of core flooding or compaction is offset by (1) the negative reactivity worth of the control drums in the reflector or (2) the negative reactivity effect of losing the reflector and surrounding the reactor with wet sand or water. This effect eliminates the need for in-core safety rods. For deep-space or planetary surface missions where reentry after reactor startup is impossible, passive launch safety can be ensured by fueling the reactor in space or using retractable boron wires to provide shutdown. This allows the removal of resonance absorbers from the core and reduces system mass and volume. The HPS is virtually nonradioactive at launch (no plutonium in the system).

The HPS scales to 300 kWt with no significant increase in reactor mass. Above 300 kWt, the system is no longer criticality-limited, and the heatpipe-to-fuel ratio must be increased. The baseline HPS approach can be used in systems providing over 1000-kWt. At very high powers

(>>1000 kWt) it may be desirable to design modules that consist of cylindrical fuel pins surrounded by noncylindrical heatpipes. Although little data are available on noncylindrical heatpipes, such data can be obtained inexpensively by testing electrically heated modules.

The baseline HPS has refractory metal heatpipes and fuel cladding. If the HPS is to be used on a planetary surface (e.g., Mars), it may be desirable to eliminate all refractory metals from the system. A several-hundred-kilowatt (thermal) stainless-steel or superalloy HPS can be built with cylindrical heatpipes and cylindrical fuel pins, all operating within the existing database. Thermal power levels of a few megawatts can be achieved using cylindrical fuel and noncylindrical heatpipes, again without the use of refractory metals.

HPS Development Status

An HPS module has been fabricated and tested. The HPS module that was tested was designed to be 1 of 12 identical modules required by a 100 kWt HPS optimized to make maximum use of existing hardware and facilities. The module consisted of three electron discharge machined (EDM'd) molybdenum pieces brazed to a central molybdenum/lithium heatpipe. The EDM'd pieces simulated the molybdenum fuel pins that would be used in an actual system. The pin's outer diameter was 2.54 cm, which allowed existing resistance heaters to be used for testing (Izhvanov 1995). The heated length of the pins was 0.30 m. Fabrication cost for the first module, including the central heatpipe, was ~$75k. The use of existing resistance heaters reduced the cost of testing the module; in the future, different module sizes can be tested if additional money is available for new heaters.

Fabrication of the first HPS module was completed in January 1997, and initial tests were completed in February 1997. Initial testing of the HPS module was performed at the New Mexico Engineering Research Institute (NMERI) in Albuquerque, New Mexico, using resistance heaters and test equipment purchased from Russia during the TOPAZ International Program. After the NMERI tests, the module was removed from the test chamber and found to be in excellent condition. The initial tests demonstrated the following.

- High power heatpipe operation against gravity
- High temperature heatpipe operation against gravity
- Heatpipe operation with high, nonuniform radial heat fluxes
- Multiple restart capability
- Advanced refractory metal bonding techniques.

The initial tests were unable to demonstrate the very high operating power (8.5 kWt) that was originally set as a goal. Because the module was in excellent condition, diagnostics and additional tests were performed at Los Alamos National Laboratory (LANL).

The LANL tests were performed using radio frequency (RF) heating instead of resistance heating. The RF heated tests repeatedly demonstrated module operation at 1400 K and 4 kWt. Thermal power was limited to 4 kWt by the amount of heat that could be radiated from the heatpipe condenser at 1400 K. Diagnostics show that the module and heatpipe are functioning well. Resistance heaters capable of providing 8.5 kWt to the module are being procured, and high power testing should resume in late 1997. A gas-gap calorimeter will be used to enable the full 8.5 kWt to be removed from the condenser end of the heatpipe. A summary of all HPS tests to date is given in Table 1.

HBS DESCRIPTION

The HPS readily evolves to the HBS, which is capable of providing both power and thermal propulsion. A key attribute of the HBS is the ability to test bimodal operation using resistance heaters to simulate heat from fission. This attribute will allow flight qualification without a ground nuclear power test, saving both development time and money. Resistance-heated tests can be performed on the actual flight unit (nuclear power tests cannot be), further helping to ensure system reliability.

Table 1
SUMMARY OF HPS MODULE TESTS PERFORMED THROUGH AUGUST 1998

Parameter	Value
Peak Operating Power (transported to condenser-end)	4.0 kWt
Peak Heatpipe Operating Temperature (during module test)	>1400 K
Peak Heatpipe Operating Temperature (during module fabrication)	>1500 K
Number of Module Startups (frozen to >1300 K and/or >2.5 kWt)	9

A schematic of a five-pin HBS module is shown in Figure 3. Hydrogen propellant flows through the interstitial spaces and out through a nozzle. Thrust levels of up to 400 N at exhaust velocities >7500 m/s can be achieved. A vacuum gap isolates the heatpipe from the hydrogen flow, allowing electric power to be generated during the propulsion mode. The vacuum gap also reduces heatpipe cooling of the propellant at the hot end of the core. Detailed analysis of HBS performance is presented by Poston (1996).

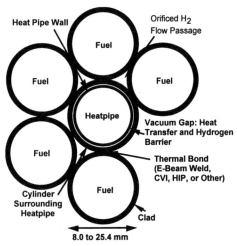

Figure 3 Five-Pin HBS Module.

The baseline HBS uses UO_2 fuel clad with polycrystalline tungsten. Programs in the United States (US) (such as the Thermionic Fuel Element Verification Program) have shown that tungsten-clad UO_2 has excellent dimensional stability and burnup capability at high temperatures. A second-generation HBS could take advantage of advanced US or Russian fuel. For example, using single-crystal, tungsten-alloy-clad, uranium-tantalum carbide fuel could increase the HBS specific impulse and decrease the HBS mass. This fuel/clad combination showed good performance during recent tests in a hydrogen environment (Bremser and Moeller 1996). Other advanced fuel could also be tested and used in HBS modules. If high thrust is required, cylindrical cermet fuel could be used in the HBS without changing the basic configuration and balance of plant. The primary drawback of using cermet fuel is that it eliminates the ability to test bimodal operation using resistance heaters to simulate heat from fission.

HBS DEVELOPMENT STATUS

Thermal hydraulic, neutronic, and general design calculations have been completed for the HBS. The fuel clad / heatpipe interface has been designed, and required heatpipe operating conditions shown to be within the existing database. Calculations have also shown that it will be possible to keep hydrogen from affecting heatpipe operation. An HBS module has been designed, and fabrication will completed in late 1998. The module is testable and closely simulates a module that could be used in an actual flight system. Testing of the HBS module should begin in early 1999. Goals of the HBS module test will include demonstrating HBS module fabricability, demonstrating that a radiatively coupled heatpipe can operate with large axial differences in radial heat flux, and demonstrating that the HBS configuration allows propellant to be heated to temperatures close to that of the maximum module temperature.

FUTURE WORK

If the HBS module test is successful, both power-only and bimodal module performance will have been demonstrated. At that stage, the next logical step in HPS/HBS development is to fabricate and test a quarter of an HPS core. Testing the quarter core will allow system-level issues to be investigated, including system startup, operation with a failed heatpipe, and operation under other off-normal conditions. Upon completion of the quarter-core test, a full core, including fuel, a reflector, and a control system, should be fabricated. Zero-power critical experiments then would be performed to verify nuclear-related safety and operational calculations. If all of these steps are successful, flight system fabrication then could begin.

ACKNOWLEDGMENTS

Funding for this research was provided by NASA's Marshall Space Flight Center. This paper was prepared with the help of numerous individuals within the space power community who contributed comments and suggestions related to the HPS and HBS concepts. The designs have benefitted greatly from this input.

REFERENCES

Bremser, A. H. and H. H. Moeller (1996). "High Temperature Fuel/Emitter System for Advanced TFE's," Babcock & Wilcox (RDD:96:43336:01), Lynchburg, VA: Lynchburg Research Center.

Izhvanov, O. L. (1995). Personal Communication, NM Engineering Research Institute, Albuquerque, New Mexico, March 1995.

Makenas, B. J., D. M. Paxton, S. Vaidyanathan, and C. W. Hoth (1994). "SP-100 Fuel Pin Performance: Results from Irradiation Testing," in *Proceedings of the 11th Symposium on Space Nuclear Power and Propulsion* (DOE CONF-940101), M. S. El-Genk, ed., Albuquerque, NM: American Institute of Physics (AIP CP 301), 1: 403-412.

Matthews, R. B., R. E. Baars, H. T. Blair, D. P. Butt, R. E. Mason, W. A. Stark, E. K. Storms, and T. C. Wallace (1994) "Fuels for Space Nuclear Power and Propulsion: 1983-1993," in *A Critical Review of Space Nuclear Power and Propulsion 1984-1993*, M. S. El-Genk, ed., American Institute of Physics.

Merrigan, M. (1997) Personal Communication, Los Alamos National Laboratory, Los Alamos, NM, May 1997.

Poston, D. I. and M. G. Houts (1996) "Nuclear and Thermal Analysis of the Heatpipe Power and Bimodal Systems," in *Proceedings of Space Technology & Applications International Forum (STAIF-96)*, DOE CONF-960109, M. S. El-Genk ed., American Institute of Physics, AIP CP 361, Albuquerque, NM, 3:1083-1093.

Raag, V. (1995) Personal Communication, Thermotrex, Waltham, MA, August 1995.

Ranken, W. A., P. J. Drivas, and V. Raag (1990) "Low Risk Low Power Heat Pipe/Thermoelectric Space Power Supply," in *Proceedings of the 7th Symposium on Space Nuclear Power and Propulsion*, M. S. El-Genk, ed., University of New Mexico's ISNPS, Albuquerque, NM, 1:488-496.

Chapter 13
ACCESSING MARTIAN RESOURCES

NASA Ames researcher, Carol Stoker, and MIT Technology Review reporter, James Oberg, seem to be enjoying the conference.

ARTESIAN BASINS ON MARS: IMPLICATIONS FOR SETTLEMENT, LIFE-SEARCH AND TERRAFORMING

Martyn J. Fogg*

Hydrological models that take into account global topography variations predict the existence of artesian basins on Mars where pressurized groundwater may exist at comparatively shallow depths. It is possible that such basins are extensive and could involve Hellas and much of the northern plains. The implication of this is that the *liquid* water resource on Mars might be easier to probe and exploit than commonly assumed. Such a ready supply could finally answer the question of life on Mars and enhances the realism of schemes for colonization and terraforming.

1. INTRODUCTION

All the water on Mars of which we currently have direct evidence is in the form of vapor or ice. The liquid phase—so crucial for life in all its range—is either non-existent, or is predicted to exist as groundwater at great depth. Thus, proposals for exploring Mars, and scenarios of settlement and terraforming, rarely invoke the utilization of indigenous *liquid* water. The reason for this reticence is not difficult to grasp when one considers that the Martian hydrological system is commonly represented as a diagram similar to Figure 1. This shows a crustal cross section of Mars from north to south poles with zones of stability of surface ice, ground ice and ground water, predicted from thermodynamic, geologic and gravitational criteria [1]. A glance at Figure 1 gives a strong impression that liquid water on Mars will only be obtainable by first drilling down several kilometers, through cryosphere and dry rock to a deep water table, and then pumping it this same distance to the surface. Looked at in this way, it is not surprising that groundwater has appeared too inaccessible to be relevant to any foreseeable human activities on Mars.

However, some of the difficulties inherent in this problem may be perceptual rather than real. This is because simple crustal models, such as illustrated in Figure 1, do not include variations in topography. Since the size of topography variations on Mars are of the same order as those hydrological zones predicted at depth, it can be seen that the quantity and proximity of liquid water beneath the surface might vary considerably depending on geographic location. As water tables have a tendency to relax to a surface of constant geopotential, it may be that drilling beneath the Martian highlands will encounter no liquid water at all before reaching impermeable bedrock. On the other hand, low-lying plains and basins may actually be under artesian pressure from the groundwater beneath. In this case, a substantial flow of water might be obtained, with little or no pumping, by drilling through the cryosphere to a confined aquifer directly beneath.

A ready supply of water on Mars, that does not require melting or condensation, would be of great scientific and economic significance. In this paper, the concept of artesian basins on

---

* Probability Research Group, c/o 44, Hogarth Court, Fountain Drive, London SE19 1UY.
E-mail: mfogg@cix.compulink.co.uk.

Mars is discussed, along with implications for settlement, the search for indigenous life, and terraforming.

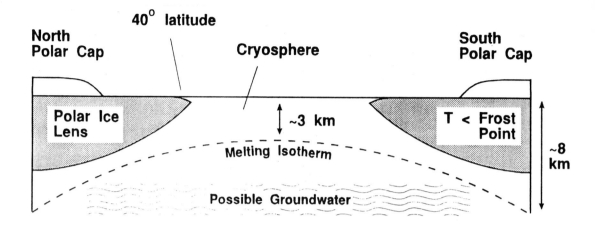

Figure 1 Theoretical cross-section of the Martian regolith from north to south poles, showing zones of long-term stability for ground-ice, polar ice and groundwater. The cryosphere, also referred to in some texts as 'permafrost', is the upper layer of the Martian crust that remains permanently below freezing: its upper bound coincides with the surface and its lower bound is determined at depth by the geothermal gradient. Once formed, the cryosphere is a barrier to liquid water and behaves as a 'cap rock' present everywhere on Mars. The tropical cryosphere may be depleted of ground-ice by evaporation to the atmosphere, due to the local surface temperature rising above the frost point.

2. MARTIAN HYDROLOGY

Liquid water has not always been absent from the Martian surface. Extensive valley networks visible in the ancient cratered terrain in the south (thought to date mainly from the Noachian Epoch, in excess of 3.5 Gyr ago) hint at a time when Mars was warmer than at present [2]. Their primary mode of erosion may have been pluvial runoff or groundwater sapping; but whichever of these was dominant, it seems clear that at that time the Martian cryosphere was incomplete. During the Noachian therefore, water tables in certain areas intersected with the surface feeding streams and rivers that drained wide areas. The hydrology of Mars at this time may have been more Earth-like, with water being resupplied to the highlands either by precipitation or by hydrothermal re-circulation through the regolith.

At times during the succeeding Hesperian Epoch (roughly 3.5 - 1.8 Gyr ago) liquid water was again to be found on the Martian surface, but the geomorphological evidence for its presence is very different. Catastrophic flooding appears to have replaced the more gradual process of valley network formation [2]. Enormous outflow channels are to be seen emerging from areas of chaotic terrain associated with Mars's plains/uplands boundary. Their features are best explained by the sudden draining of perched, confined, aquifers in the southern uplands. Mars may have been cooler at this time than during the earlier Noachian Epoch and it is certain that the planet's geothermal heat flow would have declined. A global cryosphere would have formed, thinner than estimated at present, that could have confined groundwater in lowland aquifers under considerable artesian pressure. When disturbed, perhaps by impacts or magmatism, the permafrost cap may have failed, permitting brief but powerful inundations of the surface [3]. That the northern plains of Mars were artesian basins back in the Hesperian Epoch seems certain as this is where the great floods ponded and came to rest. It has even been suggested that a boreal ocean may have formed briefly, and episodically, due to this process [4]. Recent data returned by Mars Global Surveyor have revealed the northern plains to be extraordinarily flat [5], perhaps providing circumstantial evidence in favor of the ancient ocean hypothesis.

The incidence and extent of outburst flooding appears to have declined with the ageing of Mars, but not to have completely ceased. Such features are known from the Amazonian Epoch (1.8 Gyr ago - present) and even the youngest surfaces on Mars (Upper Amazonian: roughly 0.25 Gyr ago - present) are not totally untouched by what appear to be smaller-scale, eruptive, water-cut features [6]. Given that Mars is a water-rich planet, where did the vast groundwater discharges of the past disappear to? What can be said about the present abundance and distribution of groundwater on Mars?

2.1 The Clifford Model

The most comprehensive model of Martian hydrology, in which these questions can best be framed for discussion, has been developed over a number of years by Steven Clifford of the Lunar and Planetary Institute, Texas [7]. It is briefly outlined in this Section.

It has long been appreciated that ground ice at low latitudes on Mars (< 30 - 40°) is thermally unstable due to mean temperatures being above the frost point [8]. Unless isolated from the atmosphere by a diffusive barrier, such ice will gradually evaporate, and become redeposited in the colder polar regions. A net transport of water from the tropics to the poles would occur, causing the progressive desiccation of the tropical regolith to depths of ~ 100 - 1000 m. Clifford however has proposed that this process is just the atmospheric leg of a Martian global hydrological cycle (see Figure. 2) whereby low latitude ground-ice is replaced from below by a net pole → equator flow of groundwater. The hydraulic head which drives this return leg results from a process Clifford termed "polar basal melting" [9]. Essentially, the presence of an ice cap insulates the crust against the loss of geothermal heat. Local isotherms are raised such that, when the caps grow thick enough, they start to melt at their base, raising the local water table to form a "groundwater mound", hence causing a pressure that drives groundwater towards temperate and tropical latitudes. In these zones, where the cryosphere has become unsaturated due to the evaporation of ground-ice, this ice can be replenished via upwards vapor transport from the water table below. In this way, most of the lost low latitude ground ice might be replenished and a ground-ice → vapor → polar ice → groundwater → ground-vapor → ground-ice cycle is established.

Clifford's model is covered in considerable theoretical and observational detail in his original paper [7]. Here, it suffices to make the important point that the model is based on just two assumptions: *"(1) that the physical properties of the Martian crust, including porosity, permeability, and crustal thermal conductivity, are no different than those which characterize the Earth and Moon, and (2) that Mars possesses an inventory of water that exceeds the pore volume of the cryosphere by as little as a few percent. Given these conditions, basic physics dictates that the processes of surface deposition, basal melting, groundwater flow, and the thermal transport of H_2O, will thermodynamically and hydraulically link the atmospheric, surface, and subsurface reservoirs of water on Mars into a single self-compensating system."*

The outflow channels are particularly suggestive of the validity of these assumptions. They provide strong evidence for the past existence of aquifers and groundwater on Mars and the relative youth of a minority of these features suggests that such systems might persist to the present day. The model also explains the apparent disappearance of the water that carved the channels. Assuming maximum possible flooding, "sea level" on Mars would have approximated to the topographic contour where the hydraulic head of the global aquifer intersected the surface. Once groundwater discharge ceased, the cryosphere would have re-sealed itself and the frozen lakes and seas pooled at the surface would have eventually evaporated. An increase in the rate of polar basal melting caused by their substance condensing at the poles would have returned this water to the global aquifer from whence it came.

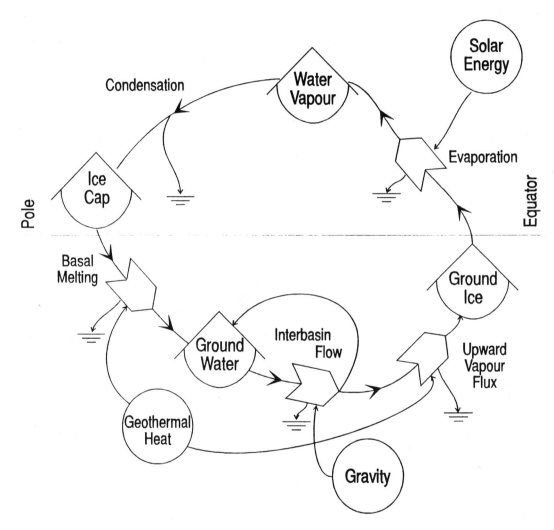

Figure 2 A representation of the Clifford Model of Martian hydrology, detailed in the text.

It is thus the presence of a ubiquitous cryosphere on Mars which makes its hydrological cycle so different from that on the Earth. The "normal" state on Mars is to have a dry surface, ice at the poles and the shallow subsurface, and groundwater confined at depth. In some low-lying areas, groundwater is confined under artesian pressure: i.e. the hydraulic head is above the base of the cryosphere. When the permafrost cap breaks and outburst flooding is initiated, these areas become temporary "ocean basins" containing large volumes of free-standing water. Essentially, when the cryosphere is complete the surface is dry; when it is incomplete then the surface can be wet. Outburst flooding seems to have occurred several times on Mars but has waned in scale and frequency since the Hesperian Epoch. This is much as might be expected given that the decline in Martian geothermal heat flow should result in a progressive increase in the thickness of the cryosphere.

Whilst the Clifford model is geologically reasonable, and can readily be described in a qualitative fashion, a quantitative approach inevitably requires estimates of its various physical parameters. Clifford did this in his original paper [7], bracketing his numerical description of the Martian regolith between two extremes and then proceeding with a discussion based on a nominal intermediate. The parameters of this "best guess" model are listed in Table 1.

Table 1
PARAMETERS OF CLIFFORD'S NOMINAL REGOLITH MODEL

Geothermal Heat Flow	0.03 W m^{-2}
Regolith Thermal Conductivity	2.0 W m^{-1} K^{-1}
Melting Isotherm of NaCl brine	252 K
Cryosphere Depth at Equator	2.27 km
Cryosphere Depth at Poles	6.53 km
Self-Compaction Depth (Bedrock)	~ 11 km
Cryosphere Pore Volume	936 m global ocean equivalent
Total Regolith Pore Volume	1.4 km global ocean equivalent

Of particular interest here are his derivations of a 1.4 km (global ocean depth equivalent) pore volume for the regolith as a whole and 936 m depth equivalent volume for the cryosphere. The crucial point is that the cryosphere is a thermodynamic sink which must be saturated with ice before there is any groundwater left over. In fact we might define a term called the "groundwater excess" to be equal to the global water inventory minus the cryosphere pore volume of 936 m. Now we know that copious groundwater was available on Mars ~ 2 Gyr ago when outburst flooding was ongoing, but we also expect that the planet's geothermal heat flow was greater and the cryosphere perhaps half its present thickness and pore capacity. One could thus certainly conjecture that the increase in cryospheric volume since then might have locked up all Mars's groundwater excess. However, it is just as possible that groundwater remains on Mars, especially if its original water inventory per unit mass was similar to that of the Earth. In this case, Mars would have started off its evolution with the equivalent of a ~ 1200 m deep global ocean [10]: allowing for some modest losses due to mineralization and atmospheric escape etc., it would still be possible for Mars to possess a groundwater excess as great as 250 m. Clifford showed that even a small groundwater excess allows for the existence of groundwater under large areas of the Martian surface [7]. In Table 2, his estimates for the extent of these areas and the depth of the hydraulic head below the planetary mean radius are listed for groundwater excesses of 10, 100 and 250 m. These data suggest that, unless the excess is small, groundwater will be present beneath most of the Martian surface.

Table 2
CLIFFORD MODEL PREDICTIONS OF GROUNDWATER ABUNDANCE ON MARS

Groundwater Excess (Global Ocean equiv.)	Depth of Water Table Below Mean Radius	Global Areal Coverage of Subsurface Groundwater
10 m	-10.1 km	34%
100 m	- 5.5 km	91%
250 m	-2.5 km	98%

3. WATER PROSPECTING ON MARS

A general rule of thumb emerges from Clifford's model as to what phases of water might be found by drilling through the Martian regolith. In order of increasing depth, for three generalized elevations, we find:

1. "Highlands": ground-ice → dry regolith → basement
2. "Intermediate areas": ground-ice → dry regolith → groundwater → basement
3. "Lowlands": ground-ice → pressurized groundwater → basement.

Elevation type 3 offers particularly interesting possibilities if the groundwater excess is substantial. Since the basement is more or less a constant depth below the surface, the water table is raised relative to the surface beneath a lowland site. If the hydraulic head is locally above the base of the cryosphere, then the groundwater is confined under pressure and we have what is effectively an artesian basin. Over the lowest sites the hydraulic head may actually be *above* the surface and little or no initial pumping would be required once a well is sunk into the aquifer.

In addition to local topography, the latitude of the site is also important. Even beneath the surface of an artesian basin, the minimum depth of drilling will always be the thickness of the cryosphere. Since the cryosphere thickness increases from equator to pole, it follows that the optimum locales for obtaining liquid water on Mars will be beneath low-lying areas in the Martian tropics (see Table 3). Such a resource would be especially useful beneath such areas as shallow ground-ice is expected to be depleted at low latitudes.

These points can be more fully appreciated by studying the cross section of the Martian regolith illustrated in Figure 3. The section is based on the Clifford model and elevations derived from Viking data [11], and runs from north to south along longitude 45°W and has been chosen as it illustrates many key features of Martian geomorphology and inferred stratigraphy and hydrology. The dichotomy between southern uplands and northern plains is shown along with the Argyre basin in the south, the Valles Marineris complex and the Chryse basin at the foot of the plains-uplands boundary. Water tables for the three different values of groundwater excess listed in Table 2 are also included along with four different well sites (for the case of groundwater excess = 250 m) that are to be the subject of discussion.

The most striking aspect of Figure 3 is that, although groundwater is predicted to occur beneath wide areas of Mars, its accessibility varies significantly on location. Even with a groundwater excess of 250 m, extraction of this water from beneath most of the southern uplands would entail drilling to and pumping from depths of > 5.5 km. Topographic depressions improve this situation somewhat. Drilling beneath the Argyre basin (A) would hit pressurized water at ~ 4 km depth which would require pumping for the last ~ 3.5 km of its way to the surface. Capri Chasma (B) is even better as it is nearer the equator: the cryosphere here is only ~ 2.5 km thick and the hydrostatic head just ~ 1.5 km below the ground. The best water prospecting sites along 45°W are however in the north. The floor of the Vastitas Borealis at 50°N is at an elevation of ~ -3km: once the ~ 4km thick cryosphere is drilled through here, water would be expected to rise to the surface without any pumping, resulting in a flowing well. Chryse, at 25°N, is perhaps the optimum site. Here, a well would only need to be ~ 2.5 km deep and little pumping would be required initially.

Table 3
AREAS BETWEEN ± 30° LATITUDE AT ELEVATION <-2 km

Area	Contours Present	Coordinates†	Geothermal Potential [17]?
Amazonis Planitia	-2, -3	c. 20°N, 160°W	✓
Chryse Planitia	-2	c. 25°N, 40°W	
Xanthe Terra Complex			
Eos Chasma	-2, 3	c. 12°S, 40°W	
Hydrates Chaos	-2	c. 0°, 34°W	
Tiu Vallis	-2	c. 10°N, 32°W	
Simud Vallis	-2	c. 7°N, 38°W	
Ares Vallis	-2	c. 15°N, 30°W	
Aram Chaos	-2	c. 2°N, 21°W	
Iani Chaos	-2	c. 3°S, 18°W	
McLaughlin Crater	-2	c. 22°N, 22°W	
Bequerel Crater	-2	c. 22°N, 8°W	
Isidis Planitia	-2	c. 15°N, 270°W	
Elysium Planitia	-2	c. 25°N, 250°W	
Cerberus plains	-2, 3	c. 5°N, 190°W	✓

† *Coordinates for purposes of location only. Some areas are extensive.*

When the groundwater excess is 100 m, groundwater is still abundant, but is deeper and artesian systems are less common. In this case, water could be obtained from depths of 9 km beneath most of the southern uplands to an optimum of ~ 3.5 km beneath Chryse. The Vastitas Borealis remains an artesian basin and wells must be sunk to ~ 4 km and pumped initially from a depth of ~ 3 km. Even if the groundwater excess is as little as 10 m, water can still be obtained from beneath the northern plains. A minimum extraction depth of ~ 7 km seems indicated in this case.

Studies of other model cross sections of the Martian regolith at other longitudes would show a similar picture to that in Figure 3. Groundwater might be obtained most efficiently on Mars at the foot of the plains-uplands boundary and northwards beneath the low-lying northern plains. The one anomaly in this generalization is Hellas Planitia, due south of Syrtis Major and halfway to the South Pole. This giant impact basin has the lowest elevation anywhere on Mars and a cross section of the deeper western half of the basin, along longitude 300°W, is shown in Figure 4. Here, we see that the floor of Hellas drops as low as 5 km below the zero datum, creating an enormous potential artesian basin. Towards the northern half of the basin (wells B and C), well depths of 3 km would be required to strike groundwater and artesian pressure would obviate much initial need for pumping. If groundwater excess is as little as 10 m, then Hellas is the best place on Mars to prospect for groundwater, a well of ~ 5.5 km depth being needed before reaching the water table.

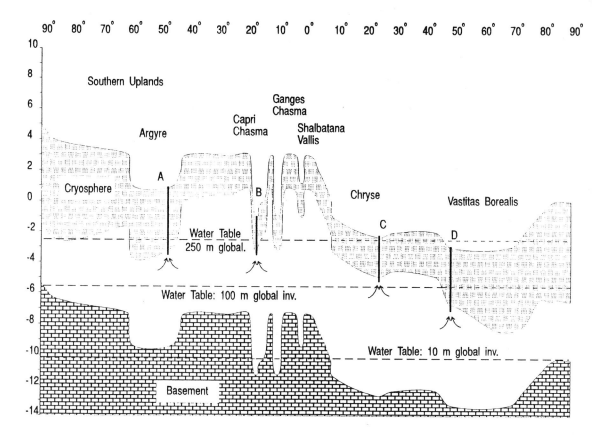

Figure 3 A south to north cross-section of the Martian regolith along longitude 45°W, showing vertical relief in km, cryosphere and basement (both assumed to be impermeable to liquid water), and interleaved potential aquifer rock. Water tables for three different groundwater excesses are marked along with four possible well sites, appropriate for the case of groundwater excess = 250 m.

The model presented in Figures 3 and 4 is still a very simple one. In reality, the suitability of any particular site for the extraction of water will depend on other local factors, such as stratigraphy, the lithology of water-bearing rocks and variations in geothermal heat flow. It is however an advance on the even simpler arrangement shown in Figure 1 which, although based on the same physics, ignores topography, predicts groundwater at depths of > 4-6 km, and fails to predict artesian systems. The model outlined here specifically predicts certain areas on Mars where groundwater is not only closer to the surface, but is pressurized as well, hence facilitating extraction. The implications of this are perhaps crucial to the realism of the goal of colonizing a planet which is currently a global desert.

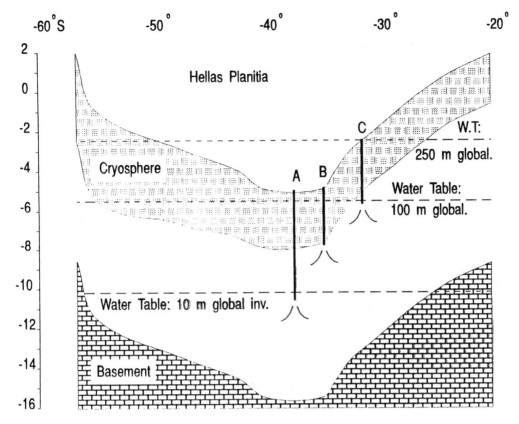

Figure 4 Cross-section of Hellas Planitia along longitude 300°W. Conventions as in Figure 3.

4. THE UTILITY OF ARTESIAN WELLS ON MARS

4.1 Human Settlements

There will be a number of difficulties in obtaining an actual flow of water through any well sunk on Mars, but none of these are likely to be insuperable—especially given the considerable experience we have performing this feat on the Earth. The principal problem will likely be to prevent water freezing in the borehole on its way through the cryosphere to a surface at -60°C. However, regolith is thought to have a low thermal conductivity so, once a flow is established, such that a surrounding collar of rock is warmed above freezing, all that would be required to keep the well free of ice would be to maintain the flow. As with any well, drawdown of the surrounding hydrostatic head would eventually limit its output, but this represents a minor deterrent compared with those that afflict other proposed methods of obtaining water on Mars.

Initial manned missions to Mars may transport their water supply from Earth, but any settlement must obtain its supply from native sources. These have been proposed as:

1. Manufacture of water from imported hydrogen [12];
2. Condensation of water from the Martian atmosphere [13-14];

3. Mining ice from the regolith [12,15];

4. Quarrying ice from the polar caps [16].

Method (1) can be ruled out as any realistic long-term solution as it runs counter to the principle of maximum self-sufficiency which is essential to the economics of any extra-terrestrial colony. Method (2) has the advantages of producing a continuous supply of water once its machinery (imported from Earth) is set running, but huge quantities of the dry Martian air must be processed. It produces a trickle, rather than a torrent. Method (3) involves digging up ice, or ice-rock mixtures, at ~ -60°C and transporting them to the colony where they are melted to produce the valuable liquid. Large quantities are potentially available but only at the cost of an energy and labor-intensive process. In contrast, the drilling of a well is expensive and labor-intensive at first, but produces a relatively inexpensive and reliable source of water for some time afterwards. To melt ice at -60°C requires 5.8×10^5 J kg^{-1}; to raise liquid water from a depth of 2.5 km requires just 9.8×10^3 J kg^{-1}, about 2% of the energy to melt surface ice. Various inefficiencies are of course neglected in this preliminary analysis, but the point is made that if colonists want water for there homes, available at the turn of a spigot, then pressurized groundwater is likely to be by far the best option.

Since water is at such a premium on Mars, the best place of all to start building for the future might be over an artesian basin where groundwater is associated with a geothermal hotspot. There, not only will the water table be expected at shallower depths due to local thinning of the cryosphere, but geothermal heat might also be extracted from the well's output in addition to its other uses [17].

4.2 The Search for Life

The Viking missions did not rule out the possibility of life on Mars but did seem to be indicative of a sterile surface. The leading hypothesis therefore for where life on Mars might survive today places it at depth within the regolith, beneath the permafrost, within water-bearing rocks [18]. As well as water being present in these deep oases, there would be carbon dioxide and, in areas associated with magmatic activity, a supply of reduced volcanic gases such as H_2, CO and H_2S. These would permit chemosynthetic autotrophic bacteria to establish themselves and to be the primary producers of an ecosystem that is independent of any need for photosynthesis.

Excitement as to the potential of Mars to host subsurface life has increased recently following a suggestion that the SNC meteorite ALH 84001 (thought to originate from Mars) shows traces of fossil microbes [19]. This interpretation is open to doubt, but the carbonate minerals found within the rock are indicative of an aqueous environment where life might have thrived irrespective of whether it actually did. Investigations of the subsurface on Earth have revealed a surprising diversity of habitats that support microbial ecosystems. Such communities have been found in deep ocean sediments [20], volcanic vents on the ocean floor, in oil reservoirs [21], limestone caves [22], and plutonic rocks such as granite [23]. Even though the biomass density of these systems is very low, the enormous volume of such potential habitats has lead some to speculate that the total mass of living matter in this "deep hot biosphere" may rival or exceed that of life at the surface [24]. In a fascinating development, anaerobic chemosynthetic microbes have been discovered living in 3-5 km thick basalt aquifers, formed from lavas 6-17 million years ago. One would expect such formations to be sterile even to chemotrophs as any supply of volcanic gases is long gone: yet samples of crushed basalt and water prepared in the laboratory supported microbial growth [25]. The explanation for this turns out to be that the simple act of weathering of basalt minerals by anaerobic water produces hydrogen which can be used by bacteria such as methanogens to provide energy for their metabolism:

$$CO_2 + 4 H_2 \rightarrow CH_4 + 2 H_2O + energy. \tag{1}$$

This suggests that microbial life can survive underground anywhere there is water, CO_2, and rock that is out of chemical equilibrium with these fluids. Volcanic gases enrich the environment with nutrients but are not absolutely necessary for life to persist.

On Mars we strongly suspect that aquifers exist, at certain depths they are above freezing and contain water and CO_2. Basalt is thought to be common on Mars and there may still be extant volcanism. These are conditions for life. Life that is fit to survive aeons of winter above ground. But at the surface this life will not survive and most traces of it will be degraded. Careful exploration of the surface may discover its evidence; but to be certain that life persists will require drill sampling of the deep Martian water table.

The static model in Figure 1 gives no hint of any preferable location where to look. The Clifford model however (Figure 2) has significant implications. Since he proposes a global circulation of groundwater through the Martian regolith, life, or some signature of life, must either be found everywhere or nowhere. That is life either exists or it doesn't. A core sample from a wet aquifer anywhere on Mars constitutes a definitive test for the existence of life on Mars. The search for the answer to this age-old question will doubtless be the primary motivating factor behind any future Mars deep-drilling project.

4.3 Terraforming

Providing Mars with bodies of surface water is a frustrating problem for proponents of terraforming Mars [26]. Groundwater is not normally an issue in these scenarios, attention being usually focussed on the much greater reserves of water thought to exist in the polar caps and cryosphere. A difference of opinion has emerged between proponents of passive or active methods of releasing this water.

The former rely on warming the surface of Mars to a point where hardy anaerobic microorganisms can survive, typically those found in cold deserts on the Earth. This process of *ecopoiesis* is envisaged as being feasible by releasing CO_2 from crustal and polar reservoirs and relying on climate feedback to warm the surface and drive further volatile release [27]. Optimistic scenarios, after 200 years of this runaway, have Mars in a new "warm and wet" climatic regime with a ~ 2 bar CO_2 atmosphere and a mean surface temperature above freezing [28]. Warm this may be, but unfortunately, not wet. The crust of Mars is frozen to great depths, and rock is such a poor conductor of heat, that enormous lengths of time are predicted for any thermal wave from the surface to be conducted through the cryosphere. Mars thaws with excruciating slowness. According to one model, after 200 years, a 1 meter global equivalent of water is predicted to have been thawed from the regolith, after 1000 years, 2 m [26]. Just how dry this will leave Mars can be appreciated by comparing these depths with the ~ 2700 m global equivalent depth of the terrestrial hydrosphere. Such water will be mostly trapped in the soil, little will pool at the surface or be available to the cycles so crucial to a vigorous biosphere. To passively thaw the Martian cryosphere to its depths might take a million years.

Other arguments may affect this timescale, but even if things are improved by a factor of ten, it makes little difference in terms of a human life-span. Proponents of active terraforming therefore have examined technological solutions to more rapidly liberate the needed water. However, since the task of manufacturing a substantial hydrosphere is so immense, enormous macroprojects are proposed in response: including melting the base of the cryosphere [29], or the polar caps [30] with linear nuclear explosives, devolatilizing the regolith with huge space-based mirrors [31,32], and impacting Mars with comets [32]. Whilst some such solution is undoubtedly required for the rapid terraforming of Mars, they have understandably not proved popular with the many who question the taste of an aggressively engineered universe.

If there appears to be no acceptable way of rapidly flooding Mars, then the Clifford model predicts an option not available to the "flat planet" model of Figure 1—the draining of confined groundwater onto the surface. It is a compromise between the two positions outlined above. The reserves of water in the permafrost and polar caps are left to passively thaw; instead technology is used to mobilize groundwater to a gravitationally more stable position. Any technology that exploits an instability within a system can always drive change more efficiently. The Clifford model holds out the possibility of swiftly creating lakes and small seas on a terraformed Mars with little more than well-sited boreholes and equipment for their placement and maintenance.

Crucial to the feasibility of such a "hydrospheric recovery" technique is that the groundwater excess should be high, preferably > 200 m. If this is the case, then a number of areas in the northern lowlands, as well as the Hellas basin, have their surfaces below the hydraulic head of the global aquifer (see the examples in Figures 3 and 4). Thus, any clear connection between any underlying, charged, aquifer and the surface will naturally result in that aquifer draining to the surface. Any attempt to flood thousands of square kilometers in this way will require a great many boreholes. If powerful terraforming methods capable of massive fracturing of the cryosphere are approved (such as mentioned above) then these will certainly facilitate water recovery, but artesian groundwater alone will only be sufficient to dot the northern plains with great lakes, rather than provide the quantity required to fill the boreal ocean envisaged in other scenarios. It is hard to predict the stability of such a sparse hydrosphere, but its much greater volume alone recommends it over any purely passive plan. In addition, little energy must be put in to melting this water and once it can percolate down into the cryosphere from *above* it might substantially increase the thawing of ground ice. Thus, by disrupting the integrity of the cryosphere beneath artesian basins, it seems possible that an element of positive feedback might be introduced into any scheme for creating a hydrosphere on Mars. Planetary engineers might therefore be emulating on a smaller scale the causative mechanism of outburst flooding, suspected to have a played a major part in abrupt, "warm and wet," climatic episodes in the Hesperian Epoch of Martian history.

Far too little is known about Mars to assess its true potential for terraforming. Deep drill sampling of the regolith's waters may thus show the way forward.

5. CONCLUSIONS

Artesian aquifers may be widespread on Mars. They may be charged with groundwater and lie closer to the surface than often assumed. These aquifers would represent a great scientific and economic resource, especially if associated with a geothermal hotspot. They could answer the question of indigenous life, provide water and heat for habitation and industry, and perhaps provide the key that terraformers need to establish future life on Mars.

It is suggested that the wherewithal to commence exploring Mars for such resources be included in future manned missions.

ACKNOWLEDGEMENTS

Julian Hiscox and Richard Taylor are thanked for their encouragement.

REFERENCES

1. F. P. Fanale, "Martian volatiles: Their degassing history and geochemical fate," *Icarus*, 28, 179-202 (1976).
2. V. R. Baker et al., "Channels and Valley Networks," in H. H. Keiffer et al. (Eds.) *Mars*, pp. 493-522, University of Arizona Press (1992).
3. M. H. Carr, "Formation of Martian flood features by release of water from confined aquifers," *J. Geophys. Res.*, 84, 2995-3007 (1979).
4. V. R. Baker et al., "Ancient oceans, ice sheets, and the hydrological cycle on Mars," *Nature*, 352, 589-594 (1991).
5. M. C. Malin et al., "Early views of the Martian surface from the Mars Orbiter Camera of Mars Global Surveyor," *Science*, 279, 1681-1685 (1998).
6. P. J. Mouginis Mark, "Recent water release in the Tharsis region of Mars," *Icarus*, 84, 362-373 (1990).
7. S. M. Clifford, "A model for the hydrologic and climatic behaviour of water on Mars," *J. Geophys. Res.*, 98, 10973-11016 (1979).
8. C. B. Farmer and P. E. Doms, "Global and seasonal variation of water vapour on Mars and the implications for permafrost," *J. Geophys. Res.*, 84, 2881-2888 (1979).
9. S. M. Clifford, "Polar basal melting on Mars," *J. Geophys. Res.*, 92, 9135-9152 (1987).
10. C. P. McKay and C. Stoker, "The early environment and its evolution on Mars: Implications for life," *Rev. Geophys.*, 27, 189-214 (1989).
11. *Topographic maps of the polar, western, and eastern regions of Mars*, 1:15,000,000, Map I-2160, US Geological Survey (1991).
12. R. Zubrin with R. Wagner, *The Case for Mars. The plan to settle the Red Planet and why we must*. The Free Press, New York (1996).
13. T. R. Meyer and C. P. McKay, "The atmosphere of Mars-Resources for the exploration and settlement of Mars," in P. J. Boston (Ed.), *The Case for Mars*, AAS *Science and Technology Series*, 57, 209-232 (1984).
14. J. E. Finn, K. R. Sridhar and C. P. McKay, "Utilization of Martian atmosphere constituents by temperature swing absorption," *JBIS*, 49, 423-430 (1996). Also published as AAS 97-360, in *From Imagination to Reality: Mars Exploration Studies of the Journal of the British Interplanetary Society* (Part I: Precursors and Early Piloted Exploration Missions), ed. R. M. Zubrin, Vol. 91, AAS *Science and Technology Series*, 1997, pp. 127-142.
15. L. Phillips, "Utilizing the permafrost on Mars," in C. P. McKay (ed.), *The Case for Mars II*, AAS *Science and Technology Series*, 62, 567-603 (1985).
16. D. Jones et al., "The retrieval, storage, and recycling of water for a manned base on Mars," in C. P. McKay (ed.), *The Case for Mars II*, AAS *Science and Technology Series*, 62, 537-556 (1985).
17. M. J. Fogg, "The utility of geothermal energy on Mars," *JBIS*, 49, 403-422 (1996). Also published as AAS 97-383, in *From Imagination to Reality: Mars Exploration Studies of the Journal of the British Interplanetary Society* (Part II: Base Building, Colonization and Terraformation), ed. R. M. Zubrin, Vol. 92, AAS *Science and Technology Series*, 1997, pp. 187-227.
18. P. J. Boston, M. V. Ivanov and C. P. McKay, "On the possibility of chemosynthetic ecosystems in subsurface habitats on Mars," *Icarus*, 95, 300-308 (1992).
19. D. S. McKay et al., "Search for Past Life on Mars: Possible Relic Biogenic Activity in Martian Meteorite iALH84001," *Science*, 273, 924-930 (1996).
20. R. J. Parkes et al., "Deep bacterial biosphere in Pacific Ocean sediments," *Nature*, 377, 410-413 (1994).
21. S. L'Haridon et al., "Hot subterranean biosphere in a continental oil reservoir," *Nature*, 377, 223-224 (1995).
22. S. M. Sarbu, B. K. Kinkle, L. Vlasceanu and T. C. Kane, "Microbiological characterization of a sulfide-rich groundwater ecosystem," *Geomicrobiol. J.*, 12, 175-182 (1994).
23. U. Szewzyk, R. Szewzyk and T-A. Stenstrom, "Thermophilic, anaerobic bacteria isolated from a deep borehole in granite in Sweden," *Proc. Natl. Acad. Sci. USA*, 91, 1810-1813 (1994).
24. T. Gold, "The deep, hot biosphere," *Proc. Natl. Acad. Sci. USA*, 89, 6045-6049 (1992).
25. T. O. Stevens and J. P. McKinley, "Lithoautotrophic microbial ecosystems in deep basalt aquifers," *Science*, 270, 450-454 (1995).

26. M. J. Fogg, *Terraforming: Engineering Planetary Environments*, SAE International, Warrendale, PA (1995).
27. R. H. Haynes, "Ecce Ecopoiesis: Playing God on Mars," in D. MacNiven (ed.), *Moral Expertise*, pp. 161-183, Routledge, London and New York (1990).
28. C. P. McKay, O. B. Toon and J. F. Kasting, "Making Mars habitable," *Nature*, 352, 489-496 (1991).
29. M. J. Fogg, "A synergic approach to terraforming Mars," *JBIS*, 45, 315-329 (1992).
30. C. Jack, Correspondence, *JBIS*, 45, 330 (1992).
31. P. Birch, "Terraforming Mars quickly," *JBIS*, 45, 331-340 (1992).
32. R. M. Zubrin and C. P. McKay, "Technological requirements for terraforming Mars," JBIS, 50, 83-92 (1997). Also published as AAS 97-390, in *From Imagination to Reality: Mars Exploration Studies of the Journal of the British Interplanetary Society* (Part II: Base Building, Colonization and Terraformation), ed. R. M. Zubrin, Vol. 92, AAS *Science and Technology Series*, 1997, pp. 309-326.

MAR 98-061

DRILLING OPERATIONS TO SUPPORT HUMAN MARS MISSIONS

Brian M. Frankie, P.E.,* Frank E. Tarzian,* Scott Lowther* and Trevor Wende†

Water, both as a source of hydrogen and for life support, will be one of the most valuable commodities for Martian operations. Large quantities of water are available as ice at the Martian poles, but access to these sources will be restricted by the necessarily complex transport, mining, and solids handling infrastructure that is required. Drilling for subsurface liquid aquifers may provide a cost effective alternative large scale supply of water, will enable bases to access geothermal power, and will also be of considerable scientific interest. Although there is little data on the depth of Martian aquifers, it is expected that large amounts of water can be found at depths between 1 and 5 km. Terrestrial drilling operations reaching this depth are massive, power intensive industrial efforts, but some of the latest technological advances hold promise to reduce the equipment and power requirements to a level that would be feasible for a Martian drilling operation. This paper outlines the technical design of a proposed low mass Martian drilling mission capable of reaching depths of more than 1 km.

INTRODUCTION

Water is one of the more useful compounds on space missions, and has numerous uses. Among other applications, it is required for life support for astronauts, can be electrolyzed to provide hydrogen/oxygen bipropellant, provides a compact and easily handled material for hydrogen storage, and is required for most construction activities. On Mars, the surface environment is more arid than the driest deserts on Earth, and water collection will be one of the primary difficulties. Water supplies for early missions to Mars will undoubtedly be based on importation of water or synthesis of water from imported hydrogen and Martian atmospheric carbon dioxide. These missions will also rely on near complete closure of the crew life support system to recycle water for repeated use.

As Mars efforts expand, the expense of imported hydrogen will become prohibitive, and water will be acquired directly from the Martian atmosphere. Data returned by the Viking landers shows atmospheric water content of approximately 300 ppm by volume.[1] This is extremely dry, but water can still be extracted by using extreme desiccants as regenerable adsorbents, or by condensing the vapor on a cryogenically cooled plate. Water will also be extracted from the soil, which may in some places be as much as 1% water by weight. In terms of amount of water that can be recovered, these techniques are feasible for support of relatively large bases. However, they are extremely power intensive, requiring more than 100,000 kJ energy input per kilogram of water recovered.[2] Thus, the early bases will probably not be limited in terms of actual physical amount of water that can be recovered as much as by the amount of power available for water recovery operations.

---

* Pioneer Astronautics, 445 Union Blvd., Suite 125, Lakewood, Colorado 80228.
† University of California, Berkeley, California 94720.

For these reasons, it is expected that at a relatively early period in the lifetime of the Mars bases, the crews will start to look for large sources of water that will require minimal power input, yet be able to support a growing population nearly indefinitely. When faced with these restrictions, there are only two possible sources of water on Mars. The first is the ice at the polar caps. There are vast reserves of water located at the poles, and they can be recovered relatively efficiently simply by mining the ice and putting in the required melt energy. However, by definition, these reserves are highly localized. Either the bases will have to be situated near the poles, with all the attendant disadvantages, or a large mining, solids handling, and transportation infrastructure will need to be built.

The alternative to polar ice is to find subsurface water, either ice locked in permafrost reserves, or warm liquid water in deep aquifers. Drilling to the liquid water aquifers would provide Martian bases with an essentially limitless water supply and, once the borehole is completed, would require little maintenance and energy input. However, completion of a successful well is not a trivial task. On Earth, deep drilling operations are massive and power hungry heavy industrial efforts, and it would seem that this technique would have few advantages over simply scooping up polar ice and transporting it to the required locations. However, this paper will show techniques and equipment adapted from terrestrial practices that promise to make drilling to the Martian water table a manageable task, while the benefits of a well to a base or colony will make this the preferred method of large scale water acquisition.

MARTIAN WATER

Evidence that large amounts of water have existed on Mars in the past is well documented, and it is expected that huge quantities remain in the subsurface cryosphere and hydrosphere. Estimates for total inventories of water range from a low of under 10 meters of water evenly spread over the planetary surface to a high of several kilometers[3] (for comparison, the Earth water inventory is about 3 kilometers spread evenly over the Earth's surface). The lowest estimates are based on the enrichment of deuterium in hydrogen, and assume constant hydrogen loss rates and atmospheric enrichment factors.[4] Adjusting any of the assumptions in the model can dramatically increase the estimated inventory of water. Estimates based on the geological evidence, particularly from the amount of water needed to cut the deep erosion features seen on the surface near Chryse and in Elysium and Hellas, require a minimum of 400 meters of water spread evenly over the surface.[5] Based on all the available evidence, and the accuracy of the various estimates, it is generally assumed in the scientific community that the current Martian water inventory is at least several hundred meters, but probably less than one kilometer. However, the combined known reserves in the polar caps, the regolith, and the atmosphere cannot account for more than about 40 meters. The remaining 85 - 95% of the Martian water inventory is assumed to be in the subsurface cryosphere and in the underlying groundwater system.

The Martian cryosphere (also known as the permafrost zone) is the portion of the megaregolith that is permanently frozen. Based on the current Martian temperatures, the cryosphere can start very close to the surface, and is expected to penetrate to several kilometers. The depth at which the cryosphere starts is determined by how fast underground ice sublimes, and how quickly the water vapor can diffuse through the soil to the atmosphere. While much of the data necessary for a precise calculation has not yet been collected, and models are fairly ideal, most calculations have shown that near the equator the cryosphere starts about 100 to 200 meters below the surface, while more than 30 degrees latitude away from the equator the cryosphere is stable all the way to the surface.[6] Likewise, we do not have data required for the calculation of the bottom of the cryosphere, but ideal models have shown that it is expected to be on order of 2.3 km at the equator and up to 6.5 km deep near the poles.[7] Underlying the cryosphere is the expected region of liquid water, the hydrosphere.

The cryosphere will certainly contain plenty of water, and it is fair to ask the question: Since the cryosphere is relatively close to the surface, why can't this be used for a water source, instead of drilling all the way to the hydrosphere? In fact, it is likely that early bases, particularly those located more than 30 degrees from the equator, where the cryosphere approaches within a couple meters of the surface, will collect water by mining and melting permafrost. However, there are two basic drawbacks to this approach. First, it is difficult to get the permafrost from the ground. An operation equivalent to a terrestrial strip mine would be required. Since permafrost at Martian temperatures is as hard as basalt and as sharp as glass, the mining equipment would have to be very rugged, and the entire operation would be labor intensive and dangerous for the personnel involved. Second, the permafrost would not replenish itself. Once an area had been mined, the facility would have to be packed up and moved off to a new site. Both of these requirements mean that for any large scale water recovery effort, a permafrost mine will involve continuous, hard labor for many people.

TO THE MARTIAN HYDROSPHERE

So, again we are confronted with the question: What is the most economical way to access the Martian hydrosphere? First, we need to determine how deep we must go to reach the hydrosphere. Based on the preliminary calculations referred to above, the cryosphere in equatorial regions starts at 200 meters and extends to 2.3 km beneath the surface. To access the hydrosphere here, we would have a drilling profile consisting of 200 meters of dry rock (probably layered impact ejecta, volcanic flows, and sedimentary deposits), 2.1 kilometers of permafrost bound rock, followed by water saturated rock to the desired final well depth. However, this is the expected average profile of the megaregolith in these regions, and does not account for local variations. Mars has a long history of volcanic, impact, and possibly tectonic activity, all of which may cause extreme local variations in the thickness of the cryosphere. There are many processes that can cause thin local regions in the cryosphere. For example, there may be local regions where there are higher concentrations of salts that depress the freezing point of water, or low thermal conductivity rock, which increases the temperature at depth. Either of these processes would decrease the depth of the cryosphere. Two of the most likely processes that would decrease the depth of the cryosphere would be seismic pumping and hydrothermal circulation.

Seismic pumping occurs when a geologic event or an impact sends shock waves through the surface material, which can fracture the cryosphere or cap rocks and/or compress deep water reservoirs. For example, the Alaska earthquake of 1964 caused dozens of shallow aquifers to break through the (admittedly thin, relative to Mars) permafrost and discharge to the surface. Mars has likely had thousands, if not millions, of events as large or larger than the Alaska earthquake. It is certainly possible, even likely, that some of these events have forced water from the hydrosphere into reservoirs above the nominal depth of the cryosphere.

Hydrothermal activity is a semi-stable circulation pattern that develops whenever a thermal anomaly occurs in an area where groundwater is present. Hot water and steam rise from depth to the top of the hydrothermal anomaly and cool and condense. The cool water moves away from the hot anomaly in a radial pattern, and sinks back to the bottom of the plume. On Mars, this circulation is likely to have been frequently induced by volcanic activity.[8] The proximity of the hydrothermal circulation to the surface of Mars is a function of the time since the last active episode of volcanism. It has been suggested that more than 10 percent of the total surface area of Mars has been volcanically active within the last 500 million years, and as much as 1 percent within the last 50 million years.[9] Land that has been active within the past 500 million years will have temperatures above 253 Kelvin (the expected freezing point of the briny Martian subsurface water) at a depth of about 1.5 kilometers, while land that has been active within the past 50 million years will have 253 K temperatures below about 500 meters. It is even expected that a small percentage of the land area, that active within the last half million years or so, will reach liquid water temperatures within 250 meters of the surface. Hydrothermal sites are expected to be the most important and likely places to search for near surface water.

Thus, we see that there is almost certainly liquid water within reach of current technologies. A properly sited drilling rig capable of reaching one kilometer depths will not only be able to penetrate the cryosphere, but will be able to reach depths at which generation of geothermal energy is viable. Thus, once the effort is made to complete the bore, not only will liquid water be available to the base without net power input, but it will provide a generous power supply. In addition, the suspected hydrothermal activity is of particular biological interest as potential sites for possible Martian life. So the total benefits of drilling into a Martian hydrothermal vent will be threefold: an effectively infinite liquid water supply, plentiful power, and outstanding scientific value for both exobiologists and planetary scientists.

WHERE TO DRILL

To reap the benefits of a drilling operation, the rig will have to be sited properly. The costs of a dry hole on Mars will be exorbitant relative to that on Earth, so it will be imperative to characterize the subsurface fully before beginning the actual drilling operation. Fortunately, it is an easy process to tap terrestrial technologies, with little modification, for this purpose. In fact, terrestrial technologies will have a much easier time on Mars, as there are not expected to be any large deposits of oil, natural gas, or other fluids with densities significantly different from water. The primary technologies that will be used to pinpoint potential drilling sites will be visual imaging, spectroscopy, ground penetrating radar, and seismic mapping.

Orbital spectroscopy and visual imaging will be used to determine the mineralogy and contours of the surface.[10] Hydrothermal alteration zones will show tell-tale signs of unique mineralogy, topographical irregularities, and possibly ground warmth (detected by thermal emission sensors). Visible and near-IR reflectance spectroscopy will be useful tools for determining the mineralogical composition of the surface. The mid-IR band, from 2.08 to 2.35 μm, will reveal areas rich in clay type minerals that have resulted from hydrothermal alteration. Gamma ray spectroscopy will be able to map chemical variations across the surface. Intensive spectroscopic measurements will be concentrated on areas that seem likely candidates for hydrothermal activity based on high resolution visible and IR photo imagery, such as that currently being gathered by Mars Global Surveyor. The photo imagery will be particularly useful for detection of geological lineaments indicating underlying fracture zones.

Spectroscopy is the tool of geoscientists, and experiments will be placed on various orbiters whether or not there is ever any interest in searching for hydrothermal sites for use by a base. Regardless of the initial goal of the spectroscopy experiments, once the data has been collected and analyzed, sites of hydrothermal activity should be apparent. At this point, satellites specially made for detection of subsurface water with ground penetrating radar (GPR) will need to be launched. If we can expect to find water at about 1 km, and radar can penetrate about 10 wavelengths into the dry Martian soil, then long wavelength radar, on the order of 3 MHz, will need to be used. With two receiving antennas in orbit, a multiple kilowatt transmitter should be able to generate a decent two dimensional map of desired areas (squares approximately 50 km per side) within a few months.

If the GPR is able to verify liquid water within about a kilometer of the surface, the next step would be several landers equipped with extremely sensitive seismometers. These instruments should be emplaced as deeply as possible (a minimum of two meters) in the regolith in an array around the desired area. A minimum of four landers will be required to get a reasonable three dimensional picture of the subsurface. Once the instruments are in place, a number of seismic events will have to be induced. Terrestrially, this is accomplished with small chemical explosives emplaced up to several hundred meters underground. On Mars, this is not likely to be cost effective. Larger charges can be fired from remote locations on the Martian surface by a mortar type apparatus, or missiles can be fired from orbit. In addition, the landers can have auxiliary geophysical equipment, including thermal conductivity and electrical resistance sensors. The data transmission rate or storage capacity of the landers will have to be quite high relative to the spectroscopic or radar satellites, so these missions will be considerably more expensive.

For the exploration program outlined above, a total of four launches will be required, not including the high quality orbital imaging that will be completed long before missions to search specifically for subsurface liquid water are launched. The spectroscopy experiments at all required wavelengths will be the equivalent of one complete discovery class mission with a Delta II launcher. The GPR satellites (one transmitter and two receivers) can be made with one Titan equivalent launch vehicle. The seismology experiments will likely require about two launches - a Delta class for a high data rate orbital relay, and a Titan class for the four soft landers. The total costs of these missions, figuring $250 million for the Delta class payloads and about $1 billion for the Titan class payloads, will be on order of $2.5 billion. Compared with the cost of the drilling mission outlined below, the exploration phase is roughly an order of magnitude cheaper. Thus, a thorough mapping of the drilling site will be warranted.

DRILLING MISSION DESIGN AND EQUIPMENT

Once liquid water is located with precision, a drilling mission may be considered. For the purposes of this paper, the authors have baselined a drilling rig capable of reaching 1000 meters depth, yet one that can be launched with the available capacity of a single landing based on the NASA design reference mission (DRM) architecture. Proven off the shelf terrestrial drilling equipment and technologies are used to the greatest possible extent.

Transportation Vehicles

The drilling mission outlined in this paper is designed with the express purpose of being able to launch and land on Mars using no more than the vehicles proposed for the NASA DRM architecture. The benefits of scaling a drilling mission to fit on vehicles that will be required anyway for the baseline human Mars missions are obvious.

The NASA DRM currently proposes using two launches of the Shuttle derived Magnum launch vehicle per vehicle launched to Mars. Each Magnum launch can put about 70 tonnes of payload into LEO. One of the launches will contain the TMI stage, which will have three 15,000 lbf thrust NTR engines and about 50 tonnes of hydrogen fuel. The other launch will contain the Mars payload module, which will mate with the TMI stage and perform the burn to Mars. The Mars DRM payload module for crew transfer has two 14 meter tall decks. The bottom deck is about 40 tonnes of fully fueled landing stage to perform the Mars terminal descent. The top deck is about 30 tonnes of habitat payload.

The Mars payload module will consist of a braking and landing stage similar to that proposed by the DRM for crew transfer, with internal modifications. The same aeroshell module can be used as in the DRM module, but the position of the landing rockets will have to be modified. The drill rig module will have a large and massive wellhead, 18" in diameter, sitting precisely in the middle of the landing module floor. The wellhead must be on the bottom of the lander so that it can be placed firmly into the ground. This required placement causes havoc with the baseline DRM landing stage. The four RL-10 class engines will have to be placed in a rectangle around the wellhead and the plumbing will have to be modified appropriately. It is proposed to place the propellant tanks on the top deck. Although this is comparatively far away from the engines, it allows the drilling rig equipment to be placed in the open space around the wellhead on the lower deck, which will expedite construction and start up of the drilling rig after arrival on Mars.

The total landed payload is approximately 30 tonnes. The drill rig landing stage will double as the support structure for the drill rig and pressure enclosure for the work crew. Current NASA plans are to provide landing modules with wheels for mobility. This may be a significant advantage but is probably not necessary, strictly speaking, since a drilling module will only land after there is enough human presence on the planet to provide terminal landing guidance.

Drilling Fluid

The primary difficulty in drilling on Mars is the lack of readily available drilling fluid. On earth, water based muds are commonly used for cleaning and cooling the bit, breaking up rock under the bit, removing cuttings from the bore, and internal support of the bore. Terrestrial mud settles in a pond, allowing removal of most of the cuttings. Obviously, settling ponds are impossible on Mars, and water based drilling fluids are impractical at best, at least until a large source of water is available. Water could be imported, but this approach is extremely mass intensive (a 4 inch diameter, 1 kilometer bore is about 8 cubic meters volume; about 3 times this much mud would be required for the drill rig) and any significant lost circulation in the bore would bring operations to a screeching halt until more water could be shipped in.

Table 1
THERMODYNAMIC PROPERTIES OF LIQUID CARBON DIOXIDE[11]

Temperature (K)	Vapor Pres (Bar abs)	Density (kg/m3)	Enthalpy (kJ/kg)	Entropy (kJ/kg,K)	Heat capacity (kJ/kg,K)	Notes
216.6	5.180	1178.7	386.3	2.656	1.707	Triple point
220.0	5.996	1166.3	392.6	2.684	1.761	
230.0	8.935	1129.2	411.1	2.763	1.879	
240.0	12.830	1089.6	430.2	2.842	1.933	
250.0	17.860	1046.7	450.3	2.923	1.992	
260.0	24.190	1000.0	471.6	3.005	2.125	
270.0	32.030	947.0	494.4	3.089	2.410	
280.0	41.600	885.0	519.2	3.176	2.887	
290.0	53.150	805.8	547.6	3.271	3.724	
304.2	73.830	466.2	636.6	3.558	0.000	Critical point

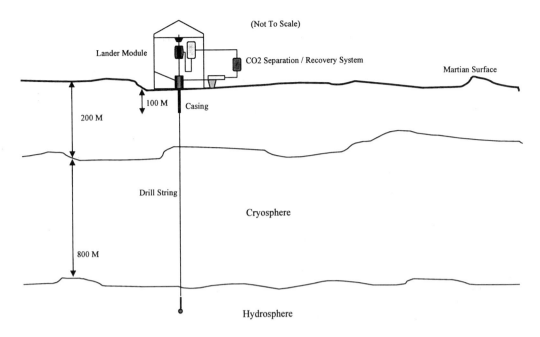

Figure 1 Assumed drilling profile for near-equatorial region with hydrothermal anomaly.

To solve the drilling fluid problem, the authors suggest using liquid carbon dioxide gathered *in situ* in place of a water based mud. As the Martian atmosphere is 95% CO_2, using liquid carbon dioxide means that no drilling fluids need to be transported, and any lost circulation can be replenished *in situ*. The energy cost of acquiring carbon dioxide on Mars impose the requirement of a closed circulation system, but this will be necessary no matter what drilling fluid is used, and the benefits of making up the initial inventory of the system with readily available fluid are enormous. Thermodynamic properties of carbon dioxide are shown in Table 1.

The properties of liquid CO_2 turn out to be precisely what is required for slimhole drilling on Mars, although borehole pressure must be maintained at a fairly high level. A complete drilling rig has been designed using terrestrial slimhole techniques with a non-rotating string, hydraulic motor, and carbon dioxide drilling fluid. Similar mechanical systems have been thoroughly tested in terrestrial laboratories, and have been used for terrestrial drilling for years. The modifications of the systems to work with carbon dioxide are relatively minor, with the higher operating pressure being the most troubling. An overview of the entire drilling rig and baseline drilling profile, assuming a near-equatorial drilling location, is shown in Figure 1. A preliminary equipment list is shown in Table 2.

Table 2
MASS ESTIMATES FOR PROPOSED DRILLING SYSTEM

System	Item	Specifications	Qty	Mass each (kg)	Total mass (kg)	Total Mars weight (N)
Drilling	Drill string	3"OD x 2.5" ID x 4 m L, Ti/6Al/4V, flush fit connections	260	27.5	7150	26654
	Drill bit	Diamond bit, high carat fine tooth, 4"	4	20	80	298
	Spudding bit	Quad spade bit	1	30	30	112
	Drill motor	4' length turbine, 3" OD, 900 psi DP	2	50	100	373
	BOP	10000 PSI	1	400	400	1491
	Cutting charges	Suitable for cutting drill string	24	0.625	15	56
	Perf gun	4 charges per perf gun	2	35	70	261
	Reamer	4" W carbide	1	80	80	298
	Fish hook	And/or overshot for fishing out material	2	20	40	149
	Toroidal charges	For fishing application	12	1.25	15	56
	Electrical cable	12 gauge, 1200 m	1	110	110	410
	Blasting caps	Standard caps for charge ignition	50	0.05	2.5	9
	Packer	4" OD x 3" ID	2	50	100	373
	Wellhead	12" ID x 18" OD, Ti/6Al/4V, 3 m high	1	1220	1220	4548
	Butterfly valve	6", wellhead isolation	1	70	70	261
	Ball valve	6", bore pressure isolation	1	85	85	317
Casing	Casing	5"OD, 4.5" ID, Al tubing, 4 m lengths	25	26.2	655	2442
	Casing bit	Roller cone bit, 5.5" diam.	1	52	52	194
Rigging	Cable	5/16" braided steel, 1200 m	1	272	272	1014
	Crown block	2 sheaves	1	100	100	373
	Crown cleat	Tie off for crown block cable	1	15	15	56
	Travelling block	2 sheaves	1	88	88	328
	Drawworks	Low torque, 50 kW	1	150	150	559
	Hook	Travelling block hook for attachment to CO2 feed	1	20	20	75
	Swivel	Swivel for connection to CO2 feed	1	20	20	75
	Side beams	S4x9.5 Ibeams, Ti/6Al/4V for derrick support	4	112.5	450	1678
	Bolting hardware	For tying beams	1	25	25	93
	Roof beams	S4x9.5 Ibeams, Ti/6Al/4V 19.1 ft long	6	46.7	280.2	1045
	Roof beams	S4x9.5 Ibeams, Ti/6Al/4V 3 ft long	2	7.35	14.7	55
	EPDS	Electric power distribution system	1	200	200	746
	HADS	Hydraulic fluid (CO2) actuation distribution system	1	300	300	1118
	Catwalks/ladders	Three levels	1	200	200	746
Mud	Globe valve	3", CO2 tank/ heat exch. isolation, CO2 flow main shutoff	4	50	200	746
	Butterfly valve	3", pump outlet flow control	1	60	60	224
	PRV	3" pressure relief from wellhead/separator	1	50	50	186
	4" OD pipe	CO2 flow system, stainless steel, per meter	54	16.1	869.4	3241
	4" OD flexline	CO2 flow from pump to bore, 10 m length + spare	20	16.1	322	1200
	4" flanges	Extra for piping repairs	8	9	72	268
	Cuttings removal	Underflow/ screen cuttings removal from CO2	1	500	500	1864
	Fines removal	Magnetic separator from filtered CO2	1	70	70	261
	Heat exchanger	Heat rejection to atmosphere; duty approx 100 kW	1	500	500	1864
	Mud storage tank	CO2 storage tank, 800 psi design	1	2300	2300	8574
	Mud pump	Modification for descent stage RL-10 LOX pump for CO2	2	75	150	559
	Electric motor	For mud pump - 50 hp, 3525 rpm	1	170	170	634
	Sorption pump	160 kg/day makeup for CO2 inventory	1	1270	1270	4734
	Compressor	CO2 boost from SP discharge to CO2 system makeup	1	180	180	671
Tools	Welding machine	TIG and stick welding, 50A, 240 V	1	150	150	559
	Separation unit	Membrane unit; 80% Ar, 20% N2 from CO2 resid	1	150	150	559
	Power tongs		2	200	400	1491
	Hand tools	Pipe wrenches, torque wrenches, sledgehammers, etc.	1	180	180	671
	Production pump	3" ID water pump to produce non-flowing well	1	21	21	78
			Total 1 =		20024	36343

Borehole Design and Drilling

Numerous parameters come into play in determining the desired borehole dimensions. However, the primary constraint is the mass of the drill string, which is the heaviest single component of the entire drilling system. For the Mars drilling mission, the string material must be strong yet lightweight, able to support the entire weight of the string in tension without excessive stretching, or to transmit compression without buckling. The string must also possess adequate corrosion resistance, particularly with CO_2 drilling fluid, as any water encountered during the drilling will cause immediate and severe acid corrosion to non-resistant materials. Faced with these constraints, Ti / 6 Al / 4 V alloy (density = 4450 kg/m^3) was selected as the drill string material. A 3" outer diameter string with 1/4" wall thickness is the practical minimum to allow sufficient downhole fluid flow. For 15 cm at the end of each tube length, the wall thickness is 3/4" to allow a tapered threading for flush fit connections between string segments; flush fit connections are required for the linear feedthrough on the wellhead. Four meters was determined to be about the optimum for each string segment. This means that each tube masses about 27.5 kg and weighs a little more than 100 Newtons on Mars. Thus, these tubes can be maneuvered relatively easily inside the 14 meter tall drilling module, and can be manhandled, if necessary, by a single worker. A total of 260 drill string tube segments were included in the baseline design.

For a 3" OD, 2.5" ID drill string, a 4" borehole will provide annular flow area approximately equal to the string internal flow area. Thus, a 4" diameter borehole was selected for the base design. This leaves only 1/2" around the string, so bits must be chosen with care to ensure fine cuttings traveling uphole. A high carat fine tooth diamond bit, such as the American Coldset Super Trigg double concentration diamond bit, would be appropriate for this duty. A bit of this nature can be expected to last approximately 500 meters when cutting basalt, so four bits were included on the equipment list.

Casing was also included for the first 100 meters of the hole. The lower part of the hole, in the cryosphere and in solid rock layers, is expected to be nearly immune from severe wear and lost circulation, and casing was eliminated for this portion of the hole to reduce launch mass and labor requirements. However, the first portion of the bore, which penetrates compacted unconsolidated materials, may be prone to caving. Casing the first 100 meters was baselined as a reasonable estimate of the amount of casing required. Welded aluminum casing, 5" OD, 1/4" wall thickness, 25 four meter lengths, was selected to case this portion of the bore, and a single 5.5" diameter roller cone bit was included for drilling the cased region. The casing will be free hanging, welded to a neck in the wellhead.

Using a non-rotating string lightens the drill rig considerably by eliminating the rotary table, kelly, and supporting equipment, and by reducing the wall thickness of the string. However, power to the bit must be supplied by a downhole motor. A hydraulic motor has significant advantages over an electric motor, and has been used extensively in terrestrial applications. The baseline system is a 3" OD motor (with counter torque stabilizers to the bore diameter, of course) with 1" shaft. If the carbon dioxide provides a 62 bar pressure drop and has a density of 1000 kg/m^3, then the motor should provide about 20 - 25 kW to the bit. For a 4" bit, this translates to 55 - 70 kgf-m (400 - 500 ft-lbs) torque at 300 rpm. Two motors were provided in the equipment list. Figure 2 shows a terrestrial hydraulic motor with flush fit connections and diamond bit.

Mud System Design, Energy Balance

Once borehole dimensions have been established, the mud system can be designed for the necessary duty. The complete design of the mud system is shown in Figure 3. The two parameters of greatest interest are the uphole mud velocity and the mud temperature profile. Since liquid CO_2 at temperatures of interest has a relatively low viscosity (1.14 x 10^{-4} Pa s at 260 K), a high velocity was used to ensure adequate cuttings entrainment and removal. With the given annular flow area of 31.7 cm², a flow of 3.4 liters/sec of CO_2 with a 960 kg/m³ density (temperature = 268 K) gives a velocity of 0.97 m/s (about 190 ft/min). This flow rate is considered excellent by terrestrial standards. The corresponding mass flow rate is 3.3 kg/s. Assuming the downhole mud flow is at 260 K, the downhole flow rate is 1.04 m/s, or 3.3 liters per second at a 1000 kg/m³ density. This provides a downhole Reynolds number of 580,000 and an uphole Reynolds number of 414,000, so flow in both directions is nicely turbulent. From the Reynolds number and flow path geometry, assuming bore surface roughness of 1 mm (fanning friction factor = 0.025), we calculate the total frictional pressure drop for a 1000 meter borehole of about 16.5 bar.

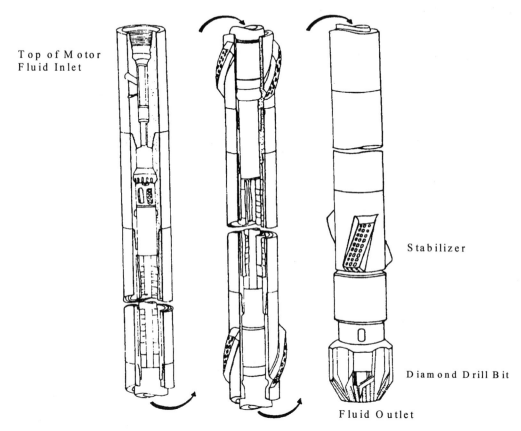

Figure 2 **Hydraulic downhole motor with diamond bit.**

The drill bit will have bottom cleaning jets, which clean the bit as well as assist in breaking up rock. The assumed pressure drop across these jets is 27.5 bar. The total pressure drop for the CO_2 liquid in the circulation loop, including the 62 bar in the downhole motor, is thus 106 bar.

Therefore, the mud pump has to move 3.3 kg/s of 1000 kg/m³ density fluid through a 106 bar pressure rise. Assuming the fluid is essentially incompressible and the pump achieves 70% efficiency, this means that 53 kW (71 hp) of pumping power is required. Maintaining the pump inlet at substantially below the bubble point reduces the required NPSH for the pump, and eliminates the possibility of impeller damage from cavitation. A 55 bar suction pressure maintains 31 degrees of subcooling at the pump inlet and was chosen as the baseline for the system.

The minimum system pressure is maintained with a high pressure inert (Ar, N_2) atmosphere in the CO_2 storage tank. The pump discharge is 161 bar.

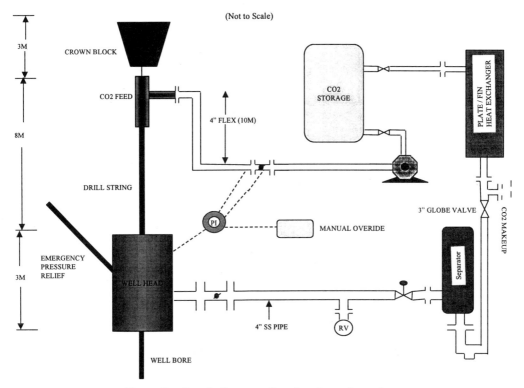

Figure 3 Rough diagram of mud system schematic.

The drill rig module, based on the NASA DRM, uses four RL-10 engines to soft land on the Martian surface. Each RL-10 engine has a common shaft with the liquid hydrogen (or methane) pump, the liquid oxygen pump, and the gas expander. The liquid oxygen pump performs approximately 90 kW pumping duty on a high density cryogenic liquid while the engine is in use. Unfortunately, the discharge pressure is only about 110 bar. However, if the pump casing thickness can be increased to take the maximum pressure of the mud system, the pump internals should be adequate to handle the CO_2 flow. For the baseline mass estimate shown in Table 2, two of the four RL-10 LOX pumps (primary and spare) were modified to handle the CO_2 flow. The drive shaft was modified to be able to disconnect from the gas generator and connect to a commercial 100 hp (75 kW) electric motor during use on the drill rig.

The temperature profile of the mud flow is a function of total power input and mud flow rate. Power input will come from the mud pump (53 kW) and heat leak. Since the inlet temperature of CO_2 into the system is 260 K, the drilling fluid should always be above the temperature of the bore, based upon the estimate that liquid water will occur at temperatures above 253 K (i.e., we should hit water before the bore reaches a temperature of 260 K). Thus, in reality, we would expect heat to be leaking *from* the drilling fluid into the bore, resulting in cooling for the mud as it flows downhole. However, to be conservative, the authors assumed no net heat leak from the bore into the fluid, and thus the total heat input into the drilling fluid is the 53 kW from the mud pump, most of which will show up as frictional heating of the bit on the rock. This power input into 3.3 kg/s of 260 K liquid CO_2 will result in a temperature rise of 7.5 degrees K (heat capacity = 2.12 kJ/kg,K). Thus, the CO_2 will flow through the pump and downhole at 260 K, will heat up to 268 K while cooling the bit, will return to the surface, and be cooled back to 260 K in the external air cooled heat exchanger.

Since the carbon dioxide flows in a closed loop, solid separation from the uphole flow must take place in an in-line separator. A diagram of the proposed solid separation apparatus is shown in Figure 4. In this apparatus, the mud first flows first into a chamber with an overflow baffle. The residence time for the fluid behind the overflow baffle will be about two minutes (400 liters). Assuming the fluid behind the baffle is more or less quiescent, and the average solid particle density is 3500 kg/m^3, then this time is sufficient to settle any particle larger than about 0.1 cm, based on an ideal Stokes' Law calculation. Particles that settle to the bottom of the overflow chamber will accumulate in collection canisters. These canisters can be sealed from the solid separation chamber with a gate valve, which will allow them to be emptied without shutting down and depressurizing the drill rig.

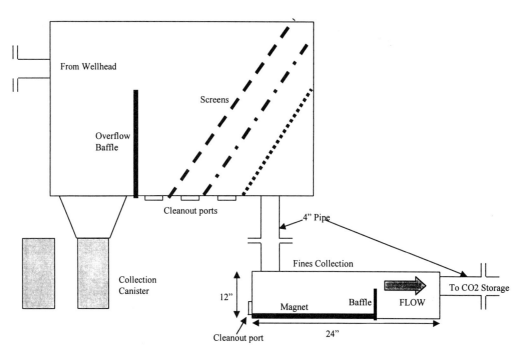

Figure 4 Rough diagram of in-line solids separator.

The fluid that pours over the baffle will go through a series of fine screens, starting with about a 20 mesh, and going to close to a 100 mesh. The screens will be near vertical with a slight bias. Particles collected from the screens will fall into collection canisters similar to those used in the overflow chamber.

Immediately below the primary separator there is a small magnetic separator. It is unknown exactly how magnetic Martian material will be; the sample return mission should help settle these questions. If there is significant magnetism, then a magnetic separator may be very useful in removing the finest material from the liquid. A cleanout port is provided at one end of the magnetic separator to allow an operator to reach in with a raking tool and scrape out any collected material. Before the magnetic separator is cleaned out, the separator system must be isolated and depressurized.

After the separation system, a heat exchanger is required to remove the enthalpy gained during the drilling fluid circulation (53 kW total). Figure 3 indicates a plate fin heat exchanger, but this may not be the optimal design for the thin Martian air. It is possible that a finned air cooler may be more appropriate for the required duty. Design work is required to optimize this heat exchanger.

Following the heat exchanger, the cool liquid CO_2 flows back to the storage cylinder, ready for another circuit through the bore. The carbon dioxide storage cylinder has a total of 24 cubic meters volume, which is three times the total volume of a 1000 meter deep bore. This volume will provide margin for the system in case of lost circulation. Liquid CO_2 is collected for the system by a large sorption pump with 1000 kg of sorbent material. This should allow collection of about 160 kg of high pressure carbon dioxide per night. During the day, the sorbent bed is heated and the relatively high pressure carbon dioxide is pumped into the mud system with a small, electrically powered positive displacement compressor. During the sorption phase, this same compressor operates as an induced draft fan to prevent the buildup of a diffusive barrier by the inert gases (primarily Ar and N_2) in the Martian atmosphere. These gases, pumped to the high pressure side of the compressor, can be collected for use as pressurant for the mud system and welding shield gas.

Wellhead, Feedthrough, CO_2 Feed

The wellhead is the most critical piece of equipment on the drill rig. A wellhead detail is shown in Figure 5. For the proposed drill rig, the wellhead will have four primary functions. The first function is to handle the return flow of cuttings-laden drill fluid. The second is to provide a feedthrough for the string as the bit penetrates the rock. The third is to seal the bore from the working environment, yet allow access to the drill string. The final function is to prevent overpressure accidents in the event of a kick.

As designed, the wellhead is 3 meters tall, 12" ID, with 3" thick walls made out of Ti/ 6 Al/ 4 V alloy. The wellhead is significantly overdesigned for any expected service conditions, but the authors felt it would be prudent on a first of a kind mission to include this margin. The top of the wellhead is a removable flanged piece that can be lifted to allow access to the drill string. In the center of the top is a 3" diameter linear feedthrough with labyrinth seals. The drill string can push through the feedthrough, but only a minimal amount of CO_2 will leak through the seals into the working environment. Support chocks will hold the string in place while tubing is added.

Approximately 1.5 meters below the wellhead top is the 6" bore isolation ball valve. The string can be pulled up through the linear feedthrough with the bit and downhole motor still in the wellhead but above the ball valve. Then the bore isolation ball valve can be closed, allowing pressure to be retained in the bore while the top portion of the wellhead is depressurized (depressurization valve not shown in Figure 5) and the flanged top is removed. Immediately below the ball valve is the drilling fluid outlet to the separator with a 6" butterfly isolation valve. Below the fluid outlet is a crush type blowout preventer for emergency use in the event of a kick. At the very bottom of the wellhead are supports for the flared end of the top casing segment.

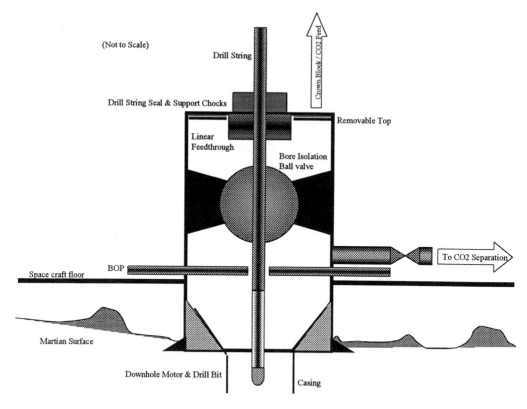

Figure 5 Proposed wellhead detail.

The base of the wellhead is fixed to the ground with anchors. In addition, the wellhead is an integral part of the spacecraft, which is fixed to the ground with anchors in the landing legs. Any force in the wellhead will be transmitted both to the anchors at the base of the wellhead, as well as through the spacecraft frame to the anchors in the landing legs.

The CO_2 feed is located at the top of the drill string, as shown in Figure 3. It has a swivel connection between the 4" flex line feed from the drilling fluid pump and the flush fit connection to the drill string. Although not shown in Figure 3, it also has a vapor feed connection at the top to allow a feed of high pressure vapor to force liquid carbon dioxide fluid out of the drill string pipe.

Support Structure for Drill Rig

The landing module will provide the basic support required for the drilling apparatus, but this has been supplemented by some additional supporting beams. The beams are made of titanium, which provides excellent strength for very low mass. The maximum tensile mass on the center of the drill rig structure is estimated to be about ten tonnes, which is a load of about 37 kN. Four vertical titanium I-beams, type S4 x 9.5, are used to support this load in addition to the landing module structure. Similar beams are used for horizontal support of the crown block and other items of tackle. A total of 305 linear feet of the beams have been provided. Catwalks and ladders made of normal steel grating have been provided to allow crew members access to all levels of the drill rig structure.

Typical equipment used for drilling has been provided for working the rig. This includes the crown and traveling blocks, and a small drawworks. A small selection of drilling tools has been provided, including a 4" reamer, fishing tools including an overshot, packers for bore completion, and a selection of useful charges. A wide selection of hand tools and power tools, including two complete pairs of power tongs has been provided for the crew. Mass considerations make it impossible to include all the tools typically provided for terrestrial drilling rigs, so the selection of the tools attempts to provide the most useful all around items.

Numerous subsystems are required for various miscellaneous tasks on the rig. The valves in the mud system require an actuation method. Electric motors are likely to be too heavy for actuation of the valves, so the mud system was provided with an inert gas distribution and actuation system. High pressure normal Mars air can be used for this system, or a small amount of the vapor in the mud system can be tapped for this purpose.

A small (50 amp, 240 V) welding machine capable of stick and TIG welding has been provided for repair jobs. The residual gas pumped from the sorption pump during the nighttime carbon dioxide sorption phase is about 60% nitrogen and 40% argon, which is not good enough for welding shield gas. However, this gas can be sent through a single pass membrane separator, which will produce about 80% argon in the retentate. This gas will be accumulated and used for TIG welding shielding.

Total power requirements of the drill rig are expected to be a maximum of about 500 kW. It will be utterly impractical to power this drilling operation using solar power, particularly if work proceeds on a 24 hour schedule. A 500 kW nuclear reactor is the baseline power system for the drill rig.

DRILLING MISSION OPERATIONS

The design and equipment of the drill rig mission outlined in the previous section are all well and good, but how will the equipment be operated once it reaches Mars? This section briefly describes the working environment, operations, and procedures enabled by the lander previously described.

Working Environment

The drill rig is located in an enclosed module similar to the habitat modules in NASA's design reference mission. This is a pressure enclosure and will allow the module to be shipped out pressurized, and to operate in a positive pressure relative to Mars ambient. This has numerous advantages.

During the ten month interplanetary shipping phase, about 1 psi absolute of nitrogen will be sufficient to prevent the loss of lubricants to vacuum. This will also prevent a thin layer of lubricant grease from forming on every exposed surface in the module, and will allow relatively easy maintenance of the equipment by the crew after arriving on Mars.

During operation on Mars, the internal pressure can be boosted to 2 - 5 psi absolute, and will be a mixture of carbon dioxide, nitrogen, and argon made up from the Martian atmosphere. Again, this prevents loss of lubricants, and also allows better thermal convection and sound conduction from working equipment. In addition, it will help alleviate the crew fatigue caused by an overly stiff pressure suit (assuming about 4 psi pure oxygen in the pressure suits). The drill rig module will be protected by a pair of relief devices, including a relief valve set to slightly above desired cabin ambient, and a large volumetric flow burst disk set to 20% below the module bursting pressure.

Note that the crew will have to wear life support suits. As desirable as a shirt sleeve working environment may be, it is impossible to provide one cost effectively. Leakage from the bore, wellhead, and mud system will be high enough to raise the concentration of carbon dioxide in the small module volume above toxic levels in an extremely short time (order of minutes). There is no practical way to maintain a breathable atmosphere in the drilling module. In addition, the pressure suits are an important safety consideration. In the event of an accident (a sudden kick, a blowout, stuck bit, etc.), numerous things can happen which would make it advisable for the crew to leave the rig rapidly. Pressure suits allow the crew to evacuate quickly without immediately dying on Mars. The major disadvantages of the pressure suits are the unwieldiness for the working crew and the requirement for a pressure lock and external rover/living quarters.

The temperature of the drill rig module during operations will be determined by the temperature of the carbon dioxide in the storage cylinder. It will be desirable to maintain rig temperature at or below this temperature to avoid any heat leak into the fluid or the need for refrigeration units. A temperature of 255 - 260 K will be cold for the crew, but not too cold to work in. Insulation in the pressure suits and/or parkas will be required.

Set Up Operations

After the drilling module has arrived on Mars, the crew will travel out to the rig in the mobile living quarters. If the lander is designed to be mobile, it will be precisely maneuvered to the desired location; if the rig is not designed with surface mobility, considerable care will have to be provided for a precision landing.

After rendezvous and positioning of the drilling module, the crew will remove equipment that will be external to the module. This equipment includes the CO_2 collection system (sorption pump and other equipment), the heat exchanger, the mud system particulate separators, and associated piping and valving. The engine bells will be removed from the underside of the lander. The ground under the lander will be leveled and the lander will be lowered until it rests firmly on the ground. Both the wellhead and the landing legs will be anchored to the ground. Design and setting of these anchors is critical, as a 10,000 psi kick, for example, will transmit 63 tons of upward force to the wellhead through a 4" bore. The wellhead must be able to safely distribute this potential load to the anchors.

When the lander is settled and anchored, the CO_2 acquisition system, and drilling fluid system will be constructed and piped. The acquisition system will be started as soon as it and the CO_2 storage vessel are completed, which will allow accumulation of drilling fluid while other set up work is proceeding. Since the acquisition system is designed to accumulate about 160 kg of carbon dioxide per day, 25 days of production will be about four cubic meters of liquid CO_2, which should be enough to start drilling. While the acquisition system is accumulating carbon dioxide for the specified period, the remaining parts of the drilling fluid system and the auxiliary systems (electrical, hydraulic) can be completed and checked out. The estimated time from arrival of the crew at the rendezvous to ready for start up (RFSU) is 60 days.

Drilling Operations

At RFSU, a quad fishtail spudding bit will be attached to the motor and first segment of drill string. The wellhead isolation ball valve will be opened and the well spudded using a very low fluid circulation. After spudding, the bit will be changed over to the 5.5" roller cone casing bit, which will drill until it is consumed. The bore will be completely depressurized while the 5" casing is set, with each section welded to the next. After casing is set, the 4" diamond bit will be attached and drilling will proceed, changing bits as required, until final depth is reached.

During normal drilling operations, adding string will use the following procedure:

1. Trip mud pump and close pump outlet valve. This will stop drilling fluid circulation.

2. Use the high pressure inert gas feed at the top of the CO_2 feed to force liquid CO_2 through the check valves in the drill motor into the bore.

3. When the entire string, from pump outlet valve to motor check valves, is emptied of liquid CO_2, depressurize the string by venting to atmosphere.

4. Pull up and chock string in linear feedthrough. Detach CO_2 feed from top string segment and lift to the level of the crown block.

5. Attach a new string segment and pull up to specified torque with power tongs. Reattach the CO_2 feed.

6. Repressurize the string with high pressure inert gas. Open mud pump outlet valve, restart pump, and ramp to desired flow.

7. Bleed excess pressure, if any, caused by inert gas repressurization to maintain desired bore pressure.

When the bit or motor needs to be changed during normal drilling operations, the following procedure will be used:

1. Follow the procedure above to empty and depressurize the string, and remove the CO_2 feed.

2. Pull the string up, detaching each section as the joint passes the feedthrough.

3. When only one drill string segment is attached to the drill motor, the motor will be pulled up to the feedthrough and the bore isolation valve will be closed.

4. High pressure inert gas will force liquid CO_2 from the wellhead above the isolation valve into the line to the separator. After the liquid is gone, the top portion of the wellhead will be depressurized.

5. The top of the wellhead will be removed and lifted and the bit or motor changed.

6. When desired changes are complete, the process will be reversed to repressurize the bore and start drilling again.

Normal operations should be quite straightforward, with few technical problems. A crew of two will be sufficient to run the rig, although it may be desirable to have a crew of three on duty for safety reasons. However, drilling is a very complex operation and problems and non-standard operation are inevitable. The mass constraints on the system prevent sending a large number of specialized tools to solve whatever problems may arise, so a selection of good all-around tools has been included in the rig equipment list. These tools should be sufficient for the vast majority of common problems that may be encountered.

Stuck bits and stuck pipe are commonly encountered in drilling operations. To some extent, this problem should be partially alleviated by using 3" titanium string with 1/4" walls, a non-rotating string, and a straight bore. The strength of the string relative to the desired drilling depth (string length) and drilling forces should eliminate buckling. Nevertheless, it would not be wise not to include equipment and procedures for dealing with these eventualities. The first step, as in any terrestrial rig, would be to try to work the obstruction free using the tackle and drill fluid. Vary the pump pressure and flow, reverse the motor, pulse pressure, and apply different axial forces with the crown block. However, if the unit is well and truly stuck, numerous charges have been included in the cargo for eliminating the problem. For a stuck bit, cutting charges would sever the string above the motor, and the good string would be withdrawn. The motor and bit would then be fished with hook and/or overshot as in terrestrial applications. As a last resort, a series of toroidal charges can be dropped to destroy both motor and bit so that the pieces can be fished. For stuck pipe, the string can be cut above and below the obstruction. The good string can be withdrawn, the obstruction destroyed with toroidal charges, and the bottom segment (with motor and bit) fished. The hole will be reamed out and drilling can restart.

Lost circulation (loss of drilling fluid to the surrounding formation) is another problem that may arise. In one respect, this factor alone is what drove the authors to consider liquid CO_2 for drilling fluid, since lost circulation with any Earth imported fluid is a potential show stopper. If lost circulation remains within a manageable range, it can simply be replaced with daily make up from the acquisition unit. However, if losses exceed the capacity of the acquisition unit, bore pressure will have to be reduced. Reducing bore pressure means that the system has to be run at a lower temperature to maintain the subcooled margin of the fluid (alternatively, margin can be trimmed, with the potential consequence of damaging the mud pump), but this can be managed without undue difficulty. The lower temperature may reduce the maximum duty of the heat exchanger, which may slow circulation and reduce drill penetration rate. The exact operating pressure and temperature will be that to achieve the maximum penetration rate, and will have to be determined daily by the crews on site.

Potentially the most serious problem to be encountered by the operating crew will be kick, which is when the pressure of the bore exceeds the mud system pressure plus the static head of the liquid at the bottom of the bore. Kick has numerous adverse effects, the severity of which is mainly dependent on the rate (and, to a lesser extent, the type) of fluid inflow from the formation to the bore. A severe kick threatens a blowout, which can damage or destroy the bore and equipment, and endanger the lives of the crew. The lower gravity on Mars reduces the overburden on fluid reservoirs, which should lessen the severity of any kick.

The first line of defense against kick will have to be in the training of the crew. Consequences of kick are lowered by recognizing the phenomenon quickly and, conversely, heightened by not recognizing it in time. The crew of the Mars drilling mission will have to be one of the best trained and most experienced groups of roughnecks ever to work a rig. Once a kick is recognized, excess pressure can be bled from the system by removing bore fluid. Then the mud pump pressure and/or density (via particulate injection or a lower working temperature) will have to be increased at the pump discharge. If none of the countermeasures work, or if the kick is too rapid and threatens a blowout, the wellhead has been provided with a crush type blowout preventer as a last resort. Mass constraints preclude sending equipment that will allow the well to be restarted after the BOP crushes the string, so, in this event, operations will have to be suspended until additional recovery equipment can be flown in.

Completion and Production Operations

The drill rig has been designed as a one-shot piece of equipment. After drilling operations, the drill string can also be used as production tubing, which eliminates a considerable mass from the mission requirements. After reaching the final depth, the string can be depressurized and withdrawn to remove the motor and bit assembly. Alternatively, a cutting charge can be sent down to blow loose the motor and bit. In any event, the string desired for production tubing will have to be withdrawn to set the packers, two of which have been included on the equipment list. Once the packers are set, the well can be produced, if freeflowing, through the drill string. If stimulation is required, two perf guns and numerous charges have been included. In addition, the mud pump can be used with a heater for hot water/steam or supercritical CO_2 injection. If pressure is not sufficient for a freeflowing well, a 3" production pump can be attached to the bottom string section. Water will tend to freeze at Mars ambient temperature, so heaters may also be required along the length of the bore to prevent ice formation. Alternatively, the well can be continuously produced, venting water when it is not required for other purposes.

Failure Modes and Countermeasures

There are too many possible failure modes of drilling rig equipment and operations for a complete analysis in this paper. However, some of the more common and expected failures are briefly described along with steps to prevent them or mitigate the consequences.

Pump impeller wear, as well as erosion in piping systems, can be caused by erosive solids in the drilling fluid. Eliminating this problem may require finer mesh screens in the solids separator or, perhaps, different separation techniques. In case of a pump failure, a spare pump has been included in the mass estimates.

A similar problem will be corrosion. Water and carbon dioxide make carbonic acid, which can very quickly corrode even high grade alloys. It is likely that during drilling through the cryosphere, some ice will get into the drilling fluid. This ice must be maintained as a solid by careful monitoring of the temperature. If problems still arise, the temperature of the drilling fluid may have to be reduced further, or the piping system may have to be glass lined. Extensive experiments to determine the effect of a carbon dioxide drilling fluid on various alloys will need to be performed during the development phase of the drill rig.

Thread galling of the flush fit connections between piping sections can be caused by overtightening or by downhole stresses. The drilling module is actually tall enough that two drill string sections can be removed from the bore, which may solve the problem. However, if this method fails, the string will have to be cut and the bad connection removed.

Operator error is probably the most likely cause of system failure. So many things can go wrong, from overtightening connections to overspeeding the pump, that it seems inevitable that there will be errors along the way. This will have to be solved with training, not only in how to operate the system, but in all imaginable failure modes, and in the design tolerances and capability of each tool and piece of equipment in the system. This will allow the operators to assess a failure accurately, and quickly develop solutions using available resources. The required exceptional level of training cannot be overstressed.

CONCLUSIONS

Water is an essential commodity on Mars, particularly if we wish a large and expanding human presence on the Red Planet. The majority of potential sources of water on Mars either require a significant amount of power or a large crewed effort to collect appreciable quantities. The one exception to this is the hydrosphere, which requires a relatively large up front effort, but provides nearly unlimited water after a bore is completed.

It seems likely that water will be found within a kilometer or so of the Martian surface. Terrestrial drilling techniques can be adapted to find and reach water at this depth using very lightweight drilling rigs, capable of being launched with no more than the vehicles proposed for the NASA design reference mission. The key to making a lightweight Martian drilling rig, as with many Mars efforts, is to use the materials available *in situ* to reduce the material that must be brought from Earth. Since the most massive portion of a drilling system is the drilling fluid, using liquid carbon dioxide from the Martian atmosphere will reduce the imported mass considerably. In addition, strong lightweight alloys can reduce the mass of components, and eliminating unnecessary tools and equipment will reduce the total mass of the proposed drill rig to 10 - 33% of an equivalent terrestrial rig, not including the mass savings from the drilling fluid. In summary, the mass of a Martian water drilling mission can be reduced to a level that can be launched with near future systems and relatively soon in a series of crewed Mars missions.

NUMBERED REFERENCES

1. Ref. from Viking mission.
2. Coons, S. C., J. D. Williams, and A. P. Bruckner, "Feasibility Study of Water Vapor Adsorption on Mars for In Situ Resource Utilization," AIAA 97-2765, Proceedings of The 33rd AIAA/ASME/SAE/ASEE Joint Propulsion Conference, Seattle, WA, July 1997.
3. Carr, Michael H., *Water on Mars*. Oxford University Press: New York, NY, 1996, p. 167.
4. Yung, Y. L., *et al.*, "HDO in the Martian atmosphere: Implications for the abundance of crustal water," *Icarus*, 76 (1988), pp. 146-159.
5. Carr, Michael H., "Mars: A water rich planet?," *Icarus*, 56 (1986), pp. 187-216.
6. Fanale, *et al.*, "Global distribution and migration of sub-surface ice on Mars," *Icarus*, 67 (1986), pp. 1 - 18.
7. Clifford, S. M., "A model for the hydrologic and climatic behavior of water on Mars," *Jour. Geophys. Res.*, 98 (1993), pp. 10,973 - 11,016.
8. Squyres, S. W., D. E. Wilhelms, and A. C. Moosman, "Large scale volcano ice interactions on Mars," *Icarus*, 70 (1987), pp. 385 - 408.
9. Fogg, M. J., "The Utility of Geothermal Energy on Mars," *Jour. Brit. Interplanetary Soc.*, 49, no. 11, (1996), pp. 403-422. Also published as AAS 97-383, in *From Imagination to Reality: Mars Exploration Studies of the Journal of the British Interplanetary Society* (Part II: Base Building, Colonization and Terraformation), ed. R. M. Zubrin, Vol. 92, AAS *Science and Technology Series*, 1997, pp. 187-227.
10. Legg, C. A., *Remote Sensing and Geographic Information Systems*, Ellis Horwood Ltd., West Sussex, UK, 1992.
11. Green, D. W., ed., *Perry's Chemical Engineers' Handbook*, 6th Edition, McGraw-Hill Book Company, New York, 1984.

OTHER REFERENCES

Chugh, C. P., K. Steele, and V. M. Sharma. *Design Criteria for Drill Rigs: Equipment and Drilling Techniques*. A. A. Balkema, Brookfield, VT, 1996.

Chugh, C. P. *High Technology in Drilling and Exploration*. A. A. Balkema, Rotterdam, 1992.

Moore, W. W., ed. *Fundamentals of Rotary Drilling*. Energy Publications: Dallas, TX, 1981.

NASA Mars Exploration Study Team. *Reference Mission Version 3.0. Addendum to the Human Exploration of Mars: The Reference Mission of the NASA Mars Exploration Study Team.* Report EX13-98-036, Exploration Office, Advanced Development Office, Johnson Space Center, Houston, TX. June, 1998.

Tsytovich, N. A., *The Mechanics of Frozen Ground.* Scripta Book Company, Washington, D.C., 1975.

EXTRACTION OF ATMOSPHERIC WATER ON MARS IN SUPPORT OF THE MARS REFERENCE MISSION

M. R. Grover,* M. O. Hilstad,† L. M. Elias,† K. G. Carpenter,†
M. A. Schneider,† C. S. Hoffman,* S. Adan-Plaza† and A.P. Bruckner‡

The University of Washington has designed an in situ resource utilization system to provide water to the life support system in the laboratory module of the NASA Mars Reference Mission, a piloted mission to Mars. This system, the Water Vapor Adsorption Reactor (WAVAR), extracts water vapor from the Martian atmosphere by adsorption in a bed of type 3A zeolite molecular sieve. Using ambient winds and fan power to move atmosphere, the WAVAR adsorbs the water vapor until the zeolite 3A bed is nearly saturated and then heats the bed within a sealed chamber by microwave radiation to drive off water for collection. The water vapor flows to a condenser where it freezes and is later liquefied for use in the life support system. In the NASA Reference Mission, water, methane, and oxygen are produced for life support and propulsion via the Sabatier/Electrolysis process from seed hydrogen brought from Earth and Martian atmospheric carbon dioxide. In order for the WAVAR system to be compatible with the NASA Reference Mission, its mass must be less than that of the seed hydrogen and cryogenic tanks apportioned for life support in the Sabatier/Electrolysis process. The WAVAR system is designed for atmospheric conditions observed by the Viking missions, which measured an average global atmospheric water vapor concentration of $\sim 2 \times 10^{-6}$ kg/m^3. WAVAR performance is analyzed taking into consideration hourly and daily fluctuations in Martian ambient temperature and wind speed and the corresponding effects on zeolite performance.

INTRODUCTION

As part of a NASA program called Human Exploration and Development of Space, University Partners, or HEDS-UP, a team of University of Washington Aeronautics and Astronautics students conducted an eight month study of a method of obtaining indigenous water on Mars in support of NASA's Mars Reference Mission. This report presents the results of the study along with new analysis carried out on the utility of ambient winds in aiding the process.

NASA's current plan to send humans to Mars rest on the mission architecture of the Mars Reference Mission [1]. With concepts derived from Zubrin *et al's* Mars Direct mission architecture [2], the Reference Mission utilizes a strategy known as *in situ* resource utilization, or ISRU, which is defined as the use of indigenous resources at the site of an interplanetary mission for the production of life support consumables and/or rocket propellant [3]. In the Reference Mission, an ISRU process called the Sabatier reaction produces water from seed hydrogen

---

* Graduate Student, Department of Aeronautics and Astronautics, University of Washington, Box 352400, Seattle, Washington 98195-2400.

† Undergraduate Student, Department of Aeronautics and Astronautics, University of Washington, Box 352400, Seattle, Washington 98195-2400.

‡ Professor and Chair, Department of Aeronautics and Astronautics, University of Washington, Box 352400, Seattle, Washington 98195-2400.

brought from Earth and carbon dioxide from the Martian atmosphere [2]. This water is partially used for life support and the remainder is used for the production of rocket propellants.

Water needs on Mars in the Reference Mission require the production of 23,200 kg of water for life support from 2,600 kg of seed hydrogen imported from Earth [4]. This cache of water is intended to supply the water needs of three missions and is produced entirely by an original ISRU plant landed with the first cargo flight two years prior to the arrival of the first crew. While simple in principle, the importation of seed hydrogen to Mars is extremely challenging due to the need to cryogenically store liquid hydrogen for extended periods of time. A cryogenic hydrogen system having a boil-off rate of 0.5% per day requires leaving Earth with 7,008 kg of liquid hydrogen in order to reach Mars with 2,578 kg after a 200-day journey. This does not include boil-off that occurs on Mars. To make boil-off amounts tolerable, a presently unobtainable evaporation rate on the order of 0.1% per day needs to be attained. With such a rate, delivering 2,600 kg of liquid hydrogen to Mars requires leaving Earth with 3,200 kg. NASA's current plan for liquid hydrogen storage rests on super-thermal cryogenic tank research that will maintain liquid hydrogen with no boil-off using active refrigeration [4], however, the mass and power required for this alternative may ultimately prove to be prohibitive.

Initially the Mars Reference Mission is completely dependent on seed hydrogen for water; however, as pointed out by its architects, a source of indigenous water is needed for the long term success of human Mars exploration. The purpose of this study is to examine how an ISRU concept called the Water Vapor Adsorption Reactor, or WAVAR, might be incorporated into the Reference Mission to meet this indigenous water need.

WAVAR is a process conceived and developed at the University of Washington's Department of Aeronautics and Astronautics under the guidance of A.P. Bruckner [5]. It obtains indigenous water by extraction from the Martian atmosphere. The atmosphere of Mars is the most highly characterized and global water source on the planet [6-8]. Both seasonal and daily cycles have been observed and the amount of water vapor has been found to vary strongly with latitude. The column abundance of water vapor was determined as a function of latitude for a period of nearly 1½ Mars years (~1000 days) by the Viking Orbiters [8]. The amount ranged from less than 1 pr μm (precipitable micrometers) at high southern latitudes in midwinter to 100 pr μm at high northern latitudes in midsummer. The seasonal variation of local humidity at the two Viking Lander sites was found to be in the range of ~1.8×10^{-7}–2×10^{-6} kg/m^3 at VL-1 and ~4×10^{-10}–3×10^{-6} kg/m^3 at VL-2 [9]. More recently Pathfinder measured a column abundance of ~10 pr μm [10]. These numbers appear to indicate an extremely dry atmosphere compared to Earth's, but on the average, the atmosphere of Mars is holding as much water as it can on a daily basis, i.e., 100% relative humidity at night throughout the lowest several kilometers, at most seasons and latitudes [11]. The global average of atmospheric water is 0.03% by volume [6], corresponding to saturation at about 200 K, i.e., a concentration of ~2×10^{-6} kg/m^3. At the north polar regions during summer the concentration may exceed 10^{-5} kg/m^3. For this study the humidity data of Ryan *et al.* at VL-1 and VL-2 were used [9]. In addition, hypothetical sites near the north pole and elsewhere showing enhanced humidities were also used, as described later.

Key to the WAVAR concept is the use of a molecular sieve adsorbent called zeolite, a strongly hydrophilic crystalline alumino-silicate commonly used in industrial dehumidifiers. As illustrated in Figure 1, the WAVAR process is conceptually very simple. Martian atmosphere is drawn into the system through a dust filter by the fan. The filtered gas passes through the adsorbent bed, where the water vapor is removed from the flow. Once the bed has reached saturation, the water vapor is desorbed from the bed, condensed, and piped to storage. The design has only seven components: a filter, an adsorption bed, a fan, a desorption unit, a bed rotating mechanism, a condenser, and an active-control system.

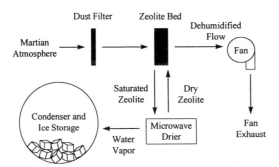

Fig. 1 The WAVAR process.

The WAVAR fan has to move a low humidity (~0.03% by volume), low temperature (~210 K), low pressure (~5 torr) gas, deal with frequent off-design operational periods, and work continuously and reliably for long periods of time (500-600 sols typical surface stay for low-energy Mars transfers). Because the flow will already be rigorously filtered to minimize fouling of the adsorption bed by Martian dust, abrasive wear on the fan can be kept to a minimum. The motor used for the WAVAR fan must operate over a range of loadings because of the variable nature of the ambient density [12].

Adsorption is a process which removes a species (the adsorbate) from a fluid as the fluid passes through a bed (the adsorbent). The adsorbent in WAVAR is zeolite 3A, a material which adsorbs water vapor but allows the other atmospheric gases (primarily carbon dioxide) to pass through. Section 2 provides further details about zeolites and the adsorption process. In the current WAVAR design, the pelletized adsorbent is packed into a bed placed in a radial flow configuration. This design is discussed in detail in Section 3.

Desorption of the bed is achieved by thermal swing desorption, which involves heating the bed until the thermal energy of the adsorbed molecules is greater than the adsorbent/adsorbate bond strength [13]. Thermal swing desorption is well suited for strongly adsorbed species such as water and can be accomplished either through resistive heating or with microwaves. The use of microwaves for the regeneration of zeolites has been demonstrated by Roussy, et al. [14], and Whittington, et al. [15]. The major advantage of using microwave energy over conventional conductive heating is that it provides rapid uniform heating for reduced desorption time and can be tailored to specifically heat water molecules.

The use of WAVAR on Mars has been the topic of past studies at the University of Washington, with most attention focused on its use in robotic sample return missions [5,16-19]. However, WAVAR is a process that is easily scaleable and has been included in one previous human Mars mission study [20]. In the present study, as a starting point for the incorporation of WAVAR into the Reference Mission, the water requirement needed to replace regenerative life support losses is set as a top-level design requirement. For a crew of eight, estimated losses amount to 6.5 kg per sol [21] over a typical surface stay duration of approximately 600 sols. Design of the physical configuration of a WAVAR system to meet this requirement is subject to several constraints. Among these constraints are system mass and footprint limitations, the adsorption capacity of zeolite 3A, the water needed to make up for life support regenerative losses, power limitations, minimization of moving parts, ease of integration into the NASA Reference Mission, and the overall simplicity and maintainability of the system and components.

The WAVAR configuration proposed by Williams, *et al.* [5] was used as a starting point for the design. Redesign and optimization of the WAVAR is focused around four goals. First, the WAVAR must collect 3.3 kg of water per sol to make up for the water lost through life support regenerative processes. The WAVAR arrives at Mars with the laboratory habitat module, and begins operation immediately. The system then collects water for the next two years before astronaut arrival, as well as during the 500-600 sol human surface mission. This total of almost 4 Earth years of operation time reduces the daily water collection requirement by a factor of two as compared to a 500-600 sol operation during the human surface mission only. The mass flow rate of water vapor through the zeolite bed must be high enough to ensure an average net gain of 3.3 kg of water per sol during its operation time, enough to supply the astronauts with the water needed during the nominal surface mission. Second, the power drain of the system must be kept to a minimum. Power requirements are dominated by the need to transport large volumes of air through the filter and the zeolite bed (up to 1×10^9 m^3/kg-H$_2$O during the driest seasons), and the power required to desorb water from the zeolite. In order to minimize the pressure drops at the filter and bed and the corresponding fan power needs, flow velocities are kept low and the zeolite bed and dust filter are kept as thin as possible. Third, the WAVAR must be sized to fit on top of the current Reference Mission laboratory module to facilitate integration with the Mission and to simplify collection of the water for use in the life support system. Fourth, the mass of the WAVAR system must be less than that of the seed hydrogen it replaces in the current NASA Reference Mission. Table 1 summarizes the major quantitative design restrictions.

Table 1. Summary of design constraints.

Characteristic	Restriction
Net water gain	$\geq$ 3.3 kg/sol
Average power drain	$\leq$ 16 kW
Footprint	$\leq$ 7.5 m diameter
System mass	$\leq$ 1200 kg

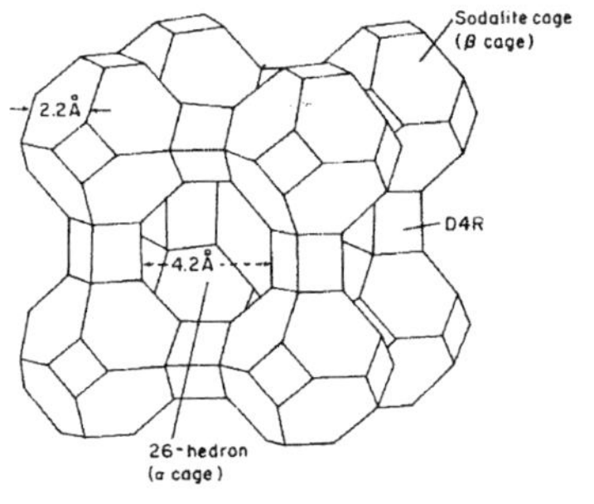

Fig. 2 The molecular structure of zeolite 4A [21]. The SiO$_4$/AlO$_4$ structure of the cage is the same for zeolite 3A and zeolite 4A.

ZEOLITE

Zeolites are crystalline alumino-silicates with a three-dimensional interconnecting network structure of silica and alumina tetrahedra that contain many micropores (Figure 2). Since zeolites have a crystalline structure, the pore openings are uniform and therefore permit adsorption discrimination based on the size and configuration of molecules in a system. This is a property unique to zeolites, and forms the basis for the name "molecular sieve." The chemical composition for the naturally occurring sodium zeolite is Na$_{12}$[(AlO$_2$)$_{12}$(SiO$_2$)$_{12}$]·27 H$_2$O, where 27 is the number of water molecules adsorbed per unit cell of fully saturated zeolite [22]. The tetrahedra are formed by oxygen atoms surrounding a silicon or aluminum atom.

Each oxygen has two negative charges and each silicon has four positive charges. The trivalency of aluminum causes the alumina tetrahedron to be negatively charged, requiring an additional cation to balance the system. Thus, cations such as potassium, calcium, lithium or sodium are the exchangeable ions of the zeolite [22].

Type A zeolites have two types of void spaces where adsorbed molecules are stored: the outer cages, called β-cages, and, the inner cages, called α-cages (Figure 2) [5]. The size selectivity takes place at these spots [22]. By controlling the ratios of cation exchange and the cation used, it is possible to synthesize zeolite crystalline structures with varying pore size, and thus regulate adsorption selectivity. In both the α-cages and β-cages the water molecules are held by van der Waals forces [16].

For the WAVAR, a zeolite must be chosen that adsorbs water molecules but not other species in the Martian atmosphere. The major constituent of the Martian atmosphere is CO_2 (95.3% by volume) and is the primary species to be excluded. As can be seen in Figure 3, the only zeolite that can exclude CO_2 is the K type (type 3A), which is a zeolite with most of the naturally occurring smaller sodium cations replaced by larger potassium cations. This reduces its average pore size to 3 Å, which excludes the 3.3 Å size of CO_2 but accepts the 2.65 Å size of water [22]. Therefore, zeolite 3A was chosen to be the adsorbent for the WAVAR unit.

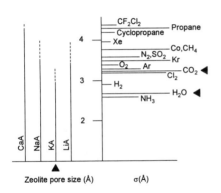

Fig. 3 Chart showing a correlation between effective pore size of various zeolites in equilibrium adsorption over temperatures of 77 K to 420 K (range indicated by ---), with the kinetic diameters of various molecules as determined from the L-J potential relation [22].

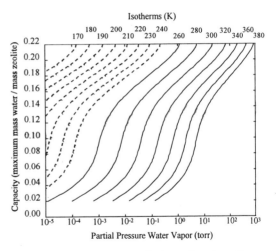

Fig. 4 Isotherms for capacity as a function of water partial pressure. The curves are from W.R. Grace Davison Molecular Sieves [24]. Dashed curves represent logarithmic extrapolations.

Zeolite Capacity

An important parameter of zeolite 3A is its capacity for water. Capacity is defined as the mass of water adsorbed per unit mass of dry zeolite. As can be seen in Figure 4, the capacity of zeolite 3A varies strongly with both the ambient vapor pressure of water and the temperature. These data were obtained from a chart published by W.R. Grace Davison Molecular Sieves [24], having isotherms down to 253 K. The isotherms down to 170 K, represented by dashed lines, were obtained by logarithmically extrapolating the available data. These low temperature isotherms will need to be experimentally confirmed.

During a typical Martian day the temperature varies significantly and thus so does the water capacity of zeolite. Figure 5 shows the diurnal temperature variation on Sol 1 at the VL-1 site and the corresponding variation in the water adsorption capacity of zeolite 3A. As can be seen, the diurnal capacity fluctuation is large, which poses a problem for continuous running of the WAVAR unit. If it continued adsorbing through one of the low points in capacity (maximum ambient temperature), the zeolite would desorb down to what the maximum capacity was during that time. The condition in which the zeolite is loaded beyond its capacity due to a temperature rise is termed super capacity. During super capacity periods, the zeolite bed must be thermally isolated from the Martian ambient temperature so the zeolite does not heat up and the water prematurely desorb. This scheme is illustrated in Figure 6, where the instantaneous water loading fraction is plotted over a period of four sols (Sols 4-8) at the VL-1 site. The two curves respectively show the capacity of the zeolite with its diurnal fluctuations, and the actual cumulative loading fraction with the bed insulated and inactive during the high temperature periods (short horizontal line sections). This problem increases the complexity of the WAVAR but it is unavoidable if the total adsorption time is more than one sol, which for all locations on Mars is the case.

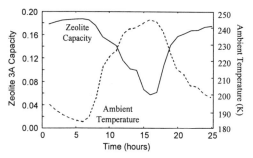

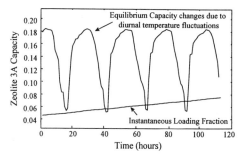

Fig. 5   Typical diurnal temperature variation (Sol 1 at VL-1) and corresponding zeolite equilibrium capacity.

Fig. 6   Simulation results showing times when water capacity of zeolite drops below the current loading fraction, necessitating a method for thermal isolation of the zeolite bed.

Figure 7 shows the capacity of zeolite 3A as a function of temperature for different partial pressures of water vapor. The dependence of capacity on temperature and partial pressure is key to the design of the desorption process and will be examined in depth in future studies to determine optimum conditions for the process.

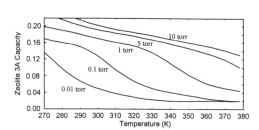

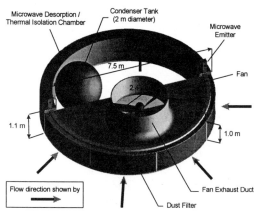

Fig. 7   Zeolite capacity vs. temperature at different partial pressures of water vapor.

Fig. 8   WAVAR geometry and dimensions.

SYSTEM DESIGN

WAVAR Geometry

The WAVAR is designed to minimize fan power requirements by providing a large area of zeolite through which the atmosphere can flow. A WAVAR design that operates efficiently and integrates cleanly with the NASA Reference Mission is shown in Figure 8. The WAVAR uses a single fan to draw Martian air radially through a curved filter and bed of packed zeolite pellets, both shown in Figure 9. The zeolite bed is a 180° arc, 10.8 m long, 0.93 m high, and 0.04 m thick, for a total bed flow area of 10.0 m² and mass of 240 kg. The annular structure that supports the four zeolite sections rests on rollers that are isolated from the dusty Martian environment. A stepping DC motor drives the rotation of the zeolite bed through a rack and pinion gear system, and a backup motor is available for emergency use.

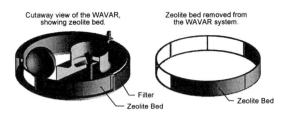

Fig. 9 Zeolite bed location and shape.

The airtight desorption chamber shown in Figure 8 is insulated from the temperature fluctuations of the ambient Martian environment, and is used for two purposes. The first use is for thermal isolation of the zeolite bed during the daily super-capacity hold cycle described above. When the bed reaches a super-capacity state due to an increase in ambient temperature, the fan stops and the bed rotates into the desorption chamber, located 180° about the WAVAR's central vertical axis. When the ambient temperature has dropped to beneath the super-capacity temperature, the bed rotates back 180° and the fan engages to continue the adsorption process.

The second use of the desorption chamber is for removal of adsorbed water from the zeolite bed. When a water loading fraction of 0.15 is reached, the zeolite bed rotates into the desorption chamber. During the desorption cycle, microwave emitters are used to heat and desorb the water from the zeolite bed. Initially the released water vapor freezes onto the walls of the desorption chamber, but further heating of the bed warms the walls of the chamber radiatively and sublimates this frost.

A variable-aperture valve links the desorption chamber with the 2 m diameter spherical condenser tank shown in Figure 8. A metal grid covering the valve opening prevents microwave radiation from entering the condenser. After the heating process begins, the valve opens to allow released water vapor to exit the chamber. The condenser is made of aluminum and remains exposed to the low temperature of the ambient Martian atmosphere. When the desorbed water vapor pressure reaches the saturation value, vapor begins to freeze on the cold condenser walls. This freezing maintains a pressure drop from the desorption chamber to the condenser, driving the vapor into the condenser. The rate of vapor transfer from the desorption chamber to the condenser is regulated by the variable-aperture valve to match the freezing rate so as to maintain this pressure difference between the desorption chamber and the condenser. When as much water as possible has been desorbed from the zeolite, the valve between the desorption chamber and condenser closes. The zeolite bed rotates back into the airflow to be cooled and then to continue the adsorption process.

Adsorption, with intermittent hold and desorption cycles, continues for six months. Every six months or when necessary, the condenser is heated resistively to increase the vapor pressure and produce liquid water. A valve at the bottom of the condenser then opens that leads to a heated, pressurized liquid water storage tank within the laboratory module. The condensation and liquid water storage process is diagrammed in Figure 10. Prior to astronaut arrival, the liquid collection cycles are performed remotely. Liquification of the contents of the condenser results in a loss of 4.2 m³ of habitat atmosphere as the atmosphere bubbles up through the valve into the condenser. This loss of atmosphere is not considered to be a problem because liquification need be performed only once every six months, and habitat atmosphere can be replenished relatively easily.

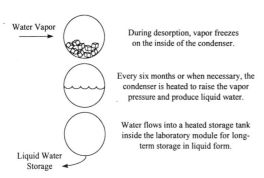

Fig. 10 Condensation and liquid water storage process.

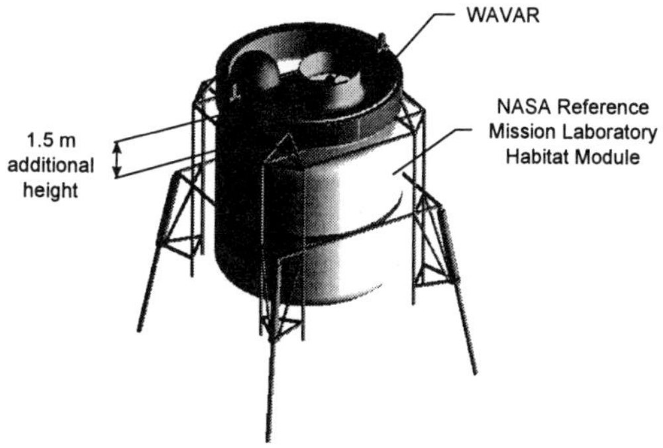

Fig. 11 WAVAR integration with the laboratory module of the NASA Reference Mission.

The WAVAR is designed to fit on top of the Reference Mission laboratory module with minor changes to the current configuration. After integration with the existing structural supports on top of the module, the WAVAR increases the height of the module by about 0.5 m at the edges and 1.5 m at the exhaust duct, as shown in Figure 11.

Desorption Process

To remove the adsorbed water from the zeolite bed, enough energy must be provided to break the bonds holding the water molecules in the bed. Thermal swing desorption is used due to its ease of implementation. The two types of heating processes considered were microwave and resistive wire heating. Heating by resistive wire is power and mass intensive due to the low thermal conductivity of zeolite. Microwave power was chosen for its controllability, specificity with water, and the relatively low mass necessary for its implementation. When heating the zeolite, there are two main considerations. The correct amount of power must be provided and the zeolite cannot be raised above the damage threshold temperature, ~600 K [25].

During desorption, the zeolite bed rotates into the insulated desorption chamber, where it is heated to 400 K. The desorption chamber is a microwave cavity resonator, and is sealed against the ambient environment for containment of desorbed water vapor. Heating reduces the water loading fraction in the bed to 1.5% (Figure 7) [26]. Attempting to desorb to a lower percentage would take more power than is justified. The microwaves also heat the walls of the desorption chamber to prevent the liberated water vapor from condensing on the walls. The aluminum honeycomb walls absorb less than 1% of the total microwave power, and this power is input within a skin depth thickness of the walls. The desorbed water vapor enters a condenser and is later stored in the habitat for the astronauts' use, as discussed above.

Power Requirements

To desorb the water, the zeolite bed is heated to an average temperature of 400 K. This temperature provides enough energy to break the adsorption bonds, while preventing serious degradation due to thermal cycling over a four-years operational lifespan. The heating process begins with a bed at thermal equilibrium with Mars ambient conditions, i.e., an average temperature of 210 K. Initially, the microwave must provide enough energy to raise the temperature of the water and break the adsorption bonds. The zeolite bed is also heated to 400 K during the process. The heat of desorption of water is assumed to be equal to the heat of adsorption, found experimentally to be 4.19 MJ/kg [26]. The specific heat of water vapor was extrapolated to low temperatures from low-pressure data [27]. Table 2 lists the parameters used to compute the desorption power. The power required for desorption of the water over a four-hour period is:

$$Energy = C_{p\,zeolite} \cdot m_{zeolite} \cdot \Delta T + m_{H_2O}(\Delta H + C_{pH_2O} \cdot \Delta T) = 317\ MJ$$

$$Power = \frac{Energy}{t_d} = 22 kW$$

Table 2. Desorption calculations constants.

$C_{p\,zeolite}$	Zeolite specific heat [20]	3.375 kJ/kg·K
$C_{p\,H2O}$	Water vapor specific heat	1.854 kJ/kg·K
ΔH	Water heat of desorption	4.19 MJ/kg
$m_{zeolite}$	Total zeolite mass	240 kg
m_{H2O}	Total water mass	36 kg
$T_{desorption}$	Maximum temperature	400 K
$T_{ambient}$	Mars avg. temperature	210 K
ΔT	$T_{desorption} - T_{ambient}$	190 K
t_d	Desorption cycle time	4 hours

Microwave Heating of Zeolite

Microwaves are electromagnetic (EM) waves operating in the gigahertz (GHz) frequency range. Water has a maximum absorption at 2.45 GHz; therefore the microwave operates at this frequency, similar to microwave ovens found in most kitchens. When heating a dielectric material with microwaves, certain considerations are needed in the design of an efficient, low mass system. The microwaves must penetrate throughout the volume of the bed, the power absorbed by the zeolite and water should be a substantial amount of the input power, the radiation impinging on the surface should be uniform, and the system that delivers the radiation should be as loss-free as possible.

As with most materials, zeolite is a dielectric [22]. EM wave propagation through dielectric materials can be represented by an oscillating electric field function that has an exponentially decaying amplitude. The complex dielectric constant, $\varepsilon' - j\varepsilon''$ [28] has a real and an imaginary permittivity term. The real term is the oscillation and the imaginary term is the exponential decay. The exponential decay represents the loss of power due to absorption by the dielectric. The distance from the input face to the location where the electric field is reduced by a factor of e^{-1} is called the skin depth, δ. The fraction of input the bed depth, L. power absorbed depends on this parameter as well as

$$\frac{P_{abs}}{P_{in}} = 1 - e^{-\frac{L}{\delta}}$$

Due to the lack of data on the electrical properties of zeolite 3A, data available for zeolite A was used. In general, ε' and ε" depend on the frequency, temperature, and water content in the zeolite [22]. The value of ε" represents the absorption by zeolite and water. At the frequency of interest, the permittivities become dependent on only one variable and become linear. At a water loading fraction of 0.15 and a temperature of 210 K, 40% of the input power is absorbed by a zeolite bed of 4 cm thickness.

Because only 40% of the input energy is absorbed during each pass, the desorption chamber is used as a microwave cavity to create a resonating field so that all the input power is absorbed by the water and the zeolite. Due to the shape and dimensions of the cavity, a specific field distribution resonates at 2.45 GHz. This is quantified by the mode numbers in the radial, azimuthal and axial directions. A waveguide transmits microwave power from the emitter and guides it to the desorption chamber, and an isolator prevents radiation from transmitting back and causing damage to the emitter.

Microwave Geometry

A magnetron microwave generator was chosen for its compactness and high power conversion efficiency of around 80%. Two magnetrons are used for redundancy and longevity. Each emitter is capable of supplying enough power for desorption by itself, but both are used simultaneously at half power to reduce wear from thermal cycling and overheating. The magnetrons are thermally isolated from the environment by an aluminum honeycomb shroud. The mass of a 25 kW magnetron is 20 kg, for a total of 40 kg for both. The magnetrons require a total input power from the main power grid of 27.8 kW, with each operating at half capability. In order to irradiate the large surface area of the zeolite in the desorption chamber, a combination of a waveguide and a cavity resonator (the desorption chamber) is used.

The magnetron emits a cylindrical wave from a cylindrical antenna [28]. A rectangular aluminum waveguide directs the transmission wave through an isolator. The waveguide is sized so that the microwave frequency is twice the geometric cutoff frequency of the waveguide. The waveguide cannot propagate EM waves at a frequency below the cutoff frequency. This frequency is obtained by solving Maxwell's equations for closed volumes, and depends only on geometry. The isolator prevents reflected waves from the resonating (desorption) chamber from transmitting back to the emitter by redirecting them into the terminator. The power is then transmitted through a small inlet waveguide and into the resonator. This input excites the resonant frequency of the desorption chamber, which is roughly 2.45 GHz, and the zeolite bed is bathed in radiation for four hours. Due to the geometry of the chamber, the actual resonant frequency will vary slightly from the optimum value. The resonant frequency also excites high mode numbers of 10-20 in the radial and axial directions.

Materials and Mass

The WAVAR system must have a mass less than that of the seed hydrogen and associated storage systems required to make up for water losses in the life support system of the NASA Reference Mission. Regenerative losses amount to 6.5 kg H_2O per sol [21] while the crew is on the surface of Mars. For a surface stay of 600 sols this amounts to a total of 3900 kg of replacement water, requiring the use of 433 kg of seed H_2 for water production. Assuming a boil-off rate of 0.5% per day in transit to Mars, a launch from Earth of 1200 kg of seed hydrogen would be needed. Taking into account additional boil-off after arrival on Mars, this figure would increase. The aim of this design was to limit the mass of the WAVAR to 1200 kg.

The WAVAR system (Figure 8) has a support structure made of two tubular aluminum circles spaced apart by tubular aluminum cross members, with the bottom plate of the adsorption chamber made of graphite-epoxy facesheets with Nomex honeycomb core [29]. The top of the adsorption chamber converges to the fan duct, which is made of the same honeycomb sandwich structure. The fan itself is made of graphite epoxy. The air filter, a Filtrete Type G from 3M [30], surrounds half the periphery of the WAVAR. All exterior components are flush mounted to the structural supports to prevent inflow of dust to the system. Inside the filter is the curved zeolite bed. An aluminum rack reinforced with steel facing under the zeolite bed is used with a stepping motor for rotation of the bed into the microwave desorption chamber.

The desorption chamber is completely encased and insulated from the rest of the WAVAR system and the Martian atmosphere. The walls of the chamber are composed of 0.25 mm thick sheet aluminum (inside), 3 cm thick Aerogel-Based Superinsulation and graphite-epoxy facesheets with a Nomex honeycomb core. These layers are illustrated in Figure 12. Aerogel-Based Superinsulation has a very low thermal conductivity, less than 0.1 mW/m-K, and a density of 12 to 35 kg/m^3 [31]. A density of 20 kg/m^3 was assumed for the mass estimate. Desorbed water vapor freezes in an aluminum condenser tank. The mass breakdown of the WAVAR system is summarized in Table 3.

Table 3. Summary of WAVAR system mass.

Component	Mass (kg)
Structural supports	60
Adsorption chamber panels and duct	85
Dust filter	10
Zeolite bed	240
Bed support structure and rack	120
Fan	30
Fan motor (10 kW)	30
Desorption chamber	150
Bed rotation motors (2)	10
Microwave emitters (2)	40
Condenser tank	100
Active control system	10
TOTAL	**885**

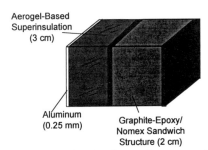

Fig. 12 Desorption chamber wall components.

PERFORMANCE

Fan Modeling

Achieving a high fan efficiency under WAVAR operational conditions is critical for the minimization of WAVAR power requirements. The design of the fan is driven by the need to efficiently transport large volumes of low density Martian atmosphere at the velocities required by adsorption design constraints. A fan was designed using momentum-blade element theory, with modifications intended to take into account the rotation induced in the flow by fan rotation [18].

The 2.4 m diameter WAVAR fan consists of 3 rectangular blades of 0.3 m chord length, as shown in Figure 13. Blade angle of twist varies from 68.6° at the root to 23.5° at the tip.

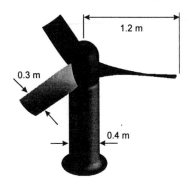

Fig. 13 WAVAR fan as modeled for simulations.

At the low Reynolds numbers characteristic of fan operation on Mars (on the order of 2×10^4-10^5), a circular arc airfoil with 5% camber provides a high lift to drag ratio [32], so this airfoil was chosen for the fan [18]. An analysis of the WAVAR fan using momentum-blade element theory and curve fit data for C_L and L/D leads to a fan efficiency of ~75% [18].

Pressure Drop Modeling

In order to calculate the power needed to drive the fan, it is necessary to determine the pressure drop of the flow across the filter and the zeolite bed:

$$\Delta P = \Delta P_{filter} + \Delta P_{bed}.$$

The pressure drop across the filter is proportional to the flow velocity and is dependent on the type of filter medium [30]. For the pressure drop calculations, a Filtrete Type G filter from 3M was chosen [30]. Filtrete is an electrostatically enhanced non-woven fiber and is available in numerous grades, each having a different filtration efficiency and associated pressure drop. For WAVAR applications on Mars, a Filtrete G-200 will provide at least 95% efficiency [17,30]. Based on Filtrete G-200 data, Coons, *et al.*, determined a linear pressure drop correlation across this filter to be [17,30]:

$$\Delta P_{filter} = 127.46 \rho V$$

This relation gives pressure drop in Pascals provided fluid density ρ and fluid velocity V are in SI units. Filtrete has been reported to have a longer life and greater temperature stability than similar media and should be acceptable for the ambient conditions that the WAVAR will encounter on Mars [30].

The pressure drop across the zeolite bed is calculated using the Ergun pressure drop model [33]. The Ergun model expresses the pressure drop across the bed as:

$$\Delta p = \frac{fL\rho(\varepsilon V)^2}{D_p}$$

where f is the friction factor, L is the bed depth, ρ is the average freestream density, ε is the void fraction, V is average flow velocity, and D_p is the pellet diameter. The friction factor is defined as:

$$f = \left[\frac{1-\varepsilon}{\varepsilon^3}\right]\left[\frac{150(1-\varepsilon)}{Re_p} + 1.75\right]$$

and the Reynolds number based on the average zeolite pellet diameter, D_p, is defined as:

$$Re_p = \frac{\rho V D_p}{\mu}$$

where μ is the viscosity of the Martian atmosphere.

The current WAVAR design incorporates a zeolite bed 4 cm deep with a void fraction of 0.4, and a 3 mm average pellet diameter. An average atmospheric density of $\rho = 0.017$ kg/m^3 was assumed, based on an average temperature $T_{avg} = 210$ K and an average pressure $P_{avg} = 5$ torr. The viscosity was curve fit as a function of temperature and found to be 1.08×10^{-5} N·s/m^2 at 210 K [34,35]. These parameters are sufficient to calculate both the filter and bed pressure drops as functions of flow velocity. The resulting total pressure drop as a function of Reynolds number at the zeolite bed, and the power required for fan operation are plotted in Figure 14.

The goal of the WAVAR is to produce an average of 3.3 kg of water per day, enough to replace the losses due to inefficiency in the life support system of the NASA Reference Mission. To meet this goal, the volume flow rate through the WAVAR must be sufficient to provide this average of 3.3 kg of water per day. Hence, the average atmospheric vapor concentration dictates the necessary volume flow rate through the WAVAR. However, the water vapor concentration depends strongly on temperature and varies with time of day, season, and latitude [9]. Thus it was necessary to find the average water vapor concentration at each of several locations on Mars and to calculate the average pressure drop at each of those locations, based on the average vapor concentration and corresponding volume flow rate. Once the pressure drop values were determined for certain flow rates, they were used to calculate the fan power required to pull the Martian atmosphere through the WAVAR's filter and zeolite bed. The results of these power calculations are discussed below.

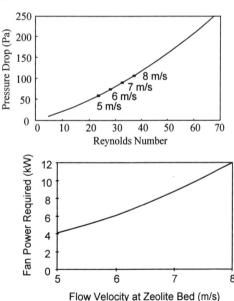

Fig. 14 Pressure drop as a function of Reynolds number at zeolite bed and power required for fan operation as a function of flow velocity at the bed. Pressure drop for the 4 cm deep bed was computed using a linear model for the filter pressure drop [17,30] and the Ergun model for the bed pressure drop [33]. Pressure drops at the zeolite bed corresponding to 5, 6, 7, and 8 m/s flow velocities are indicated.

Fan Power

Fan power requirements are determined from the fan efficiency and the pressure drop across the filter and zeolite bed. The fan efficiency and pressure drop are calculated as described above. Fan power is determined from:

$$Power = \frac{\Delta P \cdot Q}{\eta_m \eta_f}$$

where Q is the volumetric flow rate, η_m is the motor efficiency, and η_f is the fan efficiency. Values used in fan power calculations are summarized in Table 4. Results of fan power calculations are shown in Figure 14.

Table 4. Constants for performance calculations.

ρ_{amb}	Mars ambient density	1.7×10^{-2} kg/m³
--	Zeolite bed void fraction	0.4
--	Zeolite pellet diameter	3 mm
η_m	Motor efficiency (assumed)	0.95
η_f	Fan efficiency (calculated)	0.76

Surface Wind Power

Driven by solar power, the surface winds on Mars show a regular diurnal cycle both in speed and direction, as shown in Figures 15 and 16 respectively. The data shown in the figures were recorded during the first five sols of the Viking 1 Lander mission and were measured 1.6 m above the Martian surface [36]. The cycle shows a steady velocity increase from mid-morning until mid-day when the velocity reaches a maximum typically on the order of 8-10 m/s. The velocity then steadily decreases through the afternoon and evening hours until the cycle begins again at nearly static conditions in the early morning hours. The wind direction follows a similar cycle, sweeping through approximately 180° while lagging the velocity cycle, by a few hours. Although afternoon winds reach significant speed, the low density of the Martian atmosphere gives the flow relatively little power. A 10 m/s wind contains a kinetic power density of only 8.5 W/m². Given its low power density relative to the power requirements of the mission, ambient wind is considered to have a negligible impact on system performance.

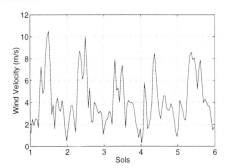

Fig. 15 Surface wind velocity variation during the first five sols of the Viking 1 Lander.

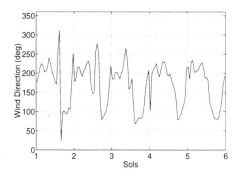

Fig. 16 Surface wind direction variation during the first five sols of the Viking 1 Lander.

Simulation Performance

The primary quantities that characterize the effectiveness of the WAVAR are the total mass of water collected, the power and energy required for operation, and the mass of the WAVAR. In order to characterize the WAVAR's performance under a wide range of Martian atmospheric conditions, simulations were performed that take into account seasonal and diurnal fluctuations in the temperature and vapor concentration, the characteristics of zeolite, and the limitations set on system power by the design constraints.

The energy required by the WAVAR for extraction of a given mass of water depends upon the amount of time that the fan operates and the number of desorption cycles that occur. An initial comparison of WAVAR performance under different atmospheric conditions was carried out with a constant flow velocity through the zeolite bed of 7 m/s, requiring a constant fan power of 8.6 kW. Additional simulations were carried out under varying atmospheric conditions with flow velocities of 5 m/s, 6 m/s and 8 m/s, each with a corresponding fan power. The energy required to desorb the water from the zeolite was assumed to remain the same for desorption cycles in all simulations since the loading fraction at which desorption begins is always the same.

Adsorption of water is dependent upon the instantaneous atmospheric water concentration, the instantaneous zeolite loading fraction, and the instantaneous zeolite capacity, as determined by the zeolite bed temperature. In the simulations it was assumed that all of the water vapor that passes through the zeolite bed is adsorbed.

To simulate variations in the atmospheric water concentration, data from the Viking Landers and Viking Orbiters were used. Since the water vapor concentration data on the surface were inferred [9] and are uncertain, simulations were run using different concentration fluctuation models to obtain an envelope of performance. The only locations for which both temperature and concentration data are available are the two Viking Lander sites. The water vapor concentrations inferred by Ryan, *et al.* are in good correlation with the MAWD measurements at VL-1 but not at VL-2 [9]. It is possible that the correlation disparities at VL-2 are due to a non-uniform vertical distribution of water vapor [9]. Regardless, both of these sites were used in the simulations.

Since the average atmospheric water concentration measured by VL-1 was below the global average of $\sim 2.0 \times 10^{-6}$ kg/m^3 [6], two additional concentration profiles were assumed in order to obtain an envelope of performance for the WAVAR. First, the vapor concentration for VL-1 was scaled up so that the average concentration was equivalent to the global average. This data set is termed New Houston. Second, the vertical column abundance at the northern polar region is about 10 times that at lower latitudes during the summer [9]. However, during the winter the water column is much lower at the north pole than at lower latitudes, so it was assumed that there is no appreciable water at any time other than summer. The simulation was run using a vapor concentration of eight times that measured by VL-1. The simulation was run for only 145 sols, in an attempt to simulate the high concentration during polar summer and extremely low concentration during other seasons.

The seasonal variations in water vapor concentrations from Viking 1 and 2, New Houston, and the northern polar region are shown in Figure 17. Extremely low vapor concentrations were also measured during the winter by VL-2. An analysis of WAVAR performance over a full Martian year showed that during the period from Sol 146 through Sol 500 of VL-2, the WAVAR would collect less than 15 kg of water, while still requiring a constant 8.6 kW for operation (with 7 m/s flow rate through the zeolite bed). Therefore, the simulation for VL-2 conditions was only run through Sol 145.

There are no diurnal water vapor concentration fluctuation estimates, but the daily maximum and minimum concentrations are known. The daily maximum is the concentration that Ryan, *et al.* [37] reported and the minimum is determined by using the 100% humidity restriction from the Clausius-Clapeyron equation [38] at the coldest part of the night. As described by

Ryan, *et al.* [9], when the temperature reaches the frost point, an inflection point occurs on a plot of temperature vs. time due to the energy released by water as it freezes. The atmosphere at the time of the inflection point may be assumed to be saturated. From the initial saturation time until when the temperature begins to rise the next morning, water vapor is being forced to precipitate out and is reported to be in the form of fog [9]. As the temperature begins to rise in the morning the fog and/or frost on the regolith returns to the atmosphere as vapor.

The assumption that the atmospheric water content follows the 100% humidity level below the frost point is probably pessimistic because as the temperature drops the water vapor precipitates out as fog. This fog would also be adsorbed by the WAVAR. In the simulations, the atmospheric water concentration is assumed to remain more or less constant over diurnal cycles, and is influenced mainly by seasonal effects. A sample of the data assumed for atmospheric water concentration is shown in Figure 18.

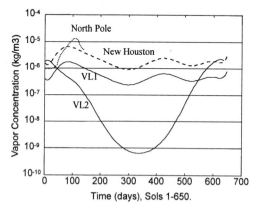

Fig. 17 Seasonal variation of vapor concentration as used in the simulations. Solid lines are actual data, dashed lines are estimated.

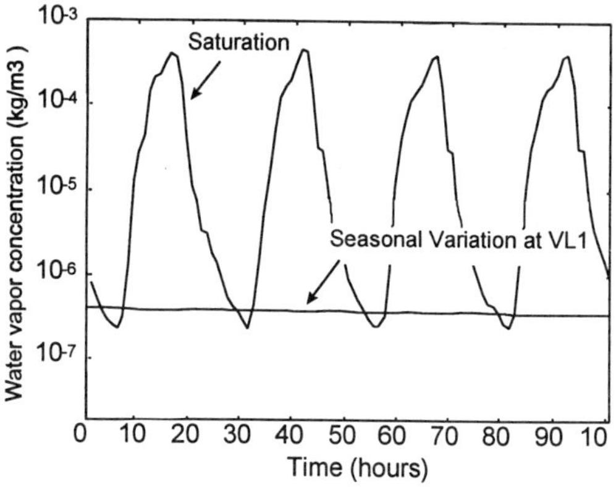

Fig. 18 Five sols of atmospheric water content data used in the VL-1 site simulation.

In the simulations, the desorption cycle was assumed to begin when the zeolite bed reached a loading fraction of 15%. Depending on temperature fluctuations, a loading fraction of over 18% could be achieved but the amount of time spent each day in thermal isolation during daytime capacity dips (daytime temperature peaks) increases at higher loading fractions, reducing the time available for adsorption and thus the mass of water adsorbed each day. A sample plot of the instantaneous amount of water adsorbed in the zeolite bed over one 300-hour period is shown in Figure 19. The nearly vertical line occurs during a desorption cycle, when the loading fraction of the zeolite drops from 0.15 to 0.015 in only four hours.

The specific energy required for water collection is calculated by dividing the total energy needed by the total mass of water extracted. The total energy is computed by multiplying the time spent adsorbing by the power draw of the fan and adding the number of desorption cycles times the energy needed to desorb the water from the zeolite bed.

Table 5 summarizes the WAVAR simulation results for the four site models under the above-mentioned atmospheric conditions. The simulation at the VL-1 site is for only 333 sols, about ½ of a Martian year, because accurate temperature data are not available for a full year [38]. There is a period from Sol 117 through Sol 133 of VL-1 for which no temperature data are available, so the simulation skips these sols. New Houston concentration and temperature are based on VL-1 data, so the simulation at New Houston also runs for only 333 sols. Since the data used for the VL-1 and New Houston simulations correspond to the first half of the year, during which the vapor concentration is higher than the yearly average (see Figure 17), the results may not be scaled linearly to obtain yearly results. North Pole concentrations and temperatures are based on Viking MAWD column abundance data. In using the MAWD data to obtain concentration, it is assumed the micron depth of precipitable water vapor in the atmosphere is uniformly distributed through a 10 km column, and that the concentration varies proportionally to measured seasonal column abundance data. North Pole temperature data are obtained from concentration levels using the Clasuis-Clapeyron relationship while assuming that above the Martian arctic circle diurnal temperature variation can be ignored because of low variation in solar radiation incidence angle during a summer sol. The simulations for the VL-2 site and the North Pole are for the summer only because the water concentration is too low during the rest of the year for efficient WAVAR operation. The results from VL-2 and the North Pole represent total yearly returns, because the WAVAR would not be in operation during seasons with extremely low vapor concentration.

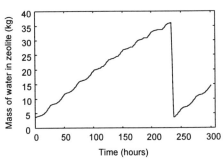

Fig. 19 Sample instantaneous water loading curve.

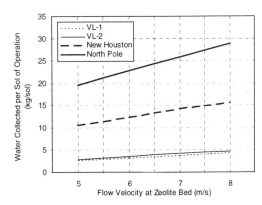

Fig. 20 Average mass of water collected by WAVAR per sol of operation.

Table 5. Simulation results with 7 m/s flow rate through zeolite bed.

Site	Number of sols simulated	Total mass of water collected (kg)	Mass per sol of operation (kg/sol)	Average power during operation (kW)	Specific energy (kW-hr/kg)
VL-1	333 *	1264	3.8	8.4	55.2
VL-2	145 †	616	4.3	8.7	52.6
New Houston	333 *	4730	14.2	9.8	17.3
North Pole	145 *†	3744	25.8	10.9	10.5

* Accurate temperature data is not available for Sols 117-133 or 351-640 of VL-1, so these sols were not included in the simulations based on VL-1 data.

† Simulations were run only during the summer, but correspond to a full year of operation since at these sites because the WAVAR only operates during seasons with high vapor concentration.

Overall performance at each site as a function of flow velocity at the zeolite bed is summarized in Figs. 20-22. Figure 20 shows the mass of water collected by the WAVAR per sol of operation. Figure 21 shows the average power required for WAVAR operation, and Figure 22 plots the specific energy required. As is apparent from Table 5 and Figures 20-22, WAVAR performance is highly dependent on atmospheric water content.

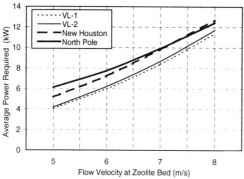

Fig. 21 Average power required for WAVAR operation.

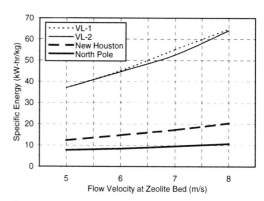

Fig. 22 Specific energy required for collection of water by the WAVAR.

The results of the simulations show varied performance with significant specific energy amounts for the water collected at VL-1 and VL2, and more reasonable performance at New Houston and the North Pole. At VL-1 production was lowest with an average of 3.8 kg/sol produced with the highest specific energy of 55 kW-hr/kg. Viking Lander 2 had similar results with an average of 4.3 kg/sol produced and a specific energy of 53 kW-hr/kg. New Houston, with its increased concentration, had more favorable production with an average of 14 kg/sol and a reduced specific energy of 17 kW-hr/kg. Finally, at the North Pole, with a concentration during summer of nearly 10 times that of Viking Lander 1, average daily production was 25.8 kg/sol at the best overall specific energy of 10.5 kW-hr/kg.

CONCLUSION

Results from this study demonstrate that the WAVAR concept is a feasible method for replacing regenerative water losses from the life support system with indigenous water in NASA's Mars Reference Mission. The WAVAR design presented integrates into the Reference Mission in a configuration mounted on the top of the existing laboratory habitat module and has a total dry mass of 885 kg. Simulations show that the design has varying capability under different conditions on Mars. The four conditions simulated in the study are water vapor content and temperature fluctuations at the Viking Lander 1 site, the Viking Lander 2 site, the North Pole based upon Viking Orbiter data, and a hypothetical site called "New Houston" which is a site with daily fluctuation trends based upon those seen at the Viking Lander 1 site but with a average yearly water vapor concentration equivalent to the global average. The results of the simulations show that the WAVAR satisfies the requirements under VL-1, New Houston, and North Pole conditions. The low average power required for each of the successful cases suggests that an increase in fan power may be used to increase flow velocity and the rate of water adsorption without exceeding the power constraint of 5% of Reference Mission power. This would be especially beneficial at the North Pole, where an increase in average power to 15 kW increases the daily net gain of water to 42 kg/sol. Low vapor concentrations at the VL-2 site preclude efficient use of the WAVAR. The most efficient and productive site for WAVAR use is the northern polar region, where the vapor concentration is almost an order of magnitude higher than that measured at the VL-1 site during the summer. Under the New Houston and North Pole conditions, WAVAR performance is good enough to consider the complete replacement of seed hydrogen by the WAVAR in the NASA Reference Mission.

In addition to the simulation results, use of ambient winds to augment water production was examined. The results of analysis indicate the addition to the flow through the WAVAR unit from ambient wind power is negligible when compared the flow provided by fan actuation.

The results presented here were obtained using data of varying degrees of reliability. In order to perform a more rigorous simulation, the properties of zeolite 3A under the low pressures and temperatures on the Martian surface need to be determined experimentally. Tests are also needed to determine to what extent, if any, CO_2 blocks the adsorption of water in zeolite under runtime conditions. In addition, direct measurements of diurnal changes in the atmospheric water content at various sites on Mars are needed in order to determine the extent by which fog increases the water available for adsorption during periods of low ambient temperature.

ACKNOWLEDGEMENTS

The authors would like to extend their thanks to Seung Chung, Ben Diedrich, John Liptac, Doug MacSparran, Arti Nadkarni, Ryan Schwab, Andrew Shell, and Susana Quintana, who participated in early stages of the project. The authors are also indebted to Dr. Eckart W. Schmidt of Hazmat, Inc. for his encouragement and many helpful suggestions.

REFERENCES

1. Hoffmann, S. J, *et al.*, *"Mars Reference Mission,"* NASA Johnson Space Center, Summer 1997.
2. Baker, D. A and Zubrin, R. M., "Mars Direct: Combining Near-Term Technologies to Achieve a Two-Launch Manned Mars Mission," *Journal of British Interplanetary Society*, 1990. Also published as AAS 97-366, in *From Imagination to Reality: Mars Exploration Studies of the Journal of the British Interplanetary Society* (Part I: Precursors and Early Piloted Exploration Missions), ed. R. M. Zubrin, Vol. 91, AAS *Science and Technology Series*, 1997, pp. 225-239.
3. Ash, R. L., Dowler, W. L. and Varsi, G., "Feasibility of Rocket Propellant Production on Mars," *Acta Astronomica*, Vol. 5, 1978, pp. 705-724.
4. Connolly, J. C., Personal communication, Johnson Space Center, Jan 22, 1998.
5. Williams, J. D., Coons, S. C. and Bruckner, A. P., "Design of a Water Vapor Adsorption Reactor for Martian In Situ Resource Utilization," *Journal of British Interplanetary Society*, Vol. 48, pp. 347-354, 1995. Also published as AAS 97-355, in *From Imagination to Reality: Mars Exploration Studies of the Journal of the British Interplanetary Society* (Part I: Precursors and Early Piloted Exploration Missions), ed. R. M. Zubrin, Vol. 91, AAS *Science and Technology Series*, 1997, pp. 59-73.
6. Carr, M. H., *Water on Mars*, Oxford University Press, Oxford, UK, 1996, pp. 3-46.
7. Jakosky, B. M., and Haberle, R. M., "The Seasonal Behavior of Water on Mars," in *Mars*, Kieffer, H. H., *et al.*, eds., The University of Arizona Press, Tucson, 1992, pp. 969-1016.
8. Jakosky, B. M, and Farmer, C. B., "The Seasonal and global Behavior of Water Vapor in the Mars Atmosphere: Complete Global Results of the Viking Atmospheric Water Detector Experiment," *J. Geophys. Res.*, Vol. 87, 1982, pp. 2999-3019.
9. Ryan, J. A., Sharman, R. D., and Lucich, R. D., "Mars Water Vapor, Near Surface," *J. Geophys. Res.*, Vol. 87, 1982, pp. 7279-7284.
10. Smith, P. H., *et al.*, "Results from the Mars Pathfinder Camera," *Science*, Vol. 278, 1997, pp.1758-1765.
11. Davies, D. W., "The Vertical Distribution of Mars Water Vapor," *J. Geophys. Res.*, Vol. 84, 1979, pp. 2875-2879.

12. McKay, C. P., "Living and Working on Mars," *The NASA Mars Conference*, Reiber, D .B., ed., AAS Publications, *Science and Technology Series*, Vol. 71, San Diego, CA, 1988, pp. 516.
13. Ruthven, D. M., Shamshuzzaman, F., and Knaebel, K. S., *Pressure Swing Adsorption*, VCM Publishers, Inc., New York, NY, 1994, pp. 1-65.
14. Roussy, G., Zoulalian, A., Charreyre, M., and Thiebaut, J. M., "How Microwaves Dehydrate Zeolites," *J. Phys. Chem.*, Vol. 88, 1984, pp. 5702-5708.
15. Whittington, B. I., and Milestone, N. B., "The Microwave Heating of Zeolites," *Zeolites*, Vol. 12, 1992, pp. 815-818.
16. Coons, S. C., Williams, J. D., and Bruckner, A. P., "In Situ Propellant Production Strategies and Applications for a Low-Cost Mars Sample Return Mission," AIAA 95-2796, 31st AIAA/ASME/SAE/ASEE *Joint Propulsion Conference and Exhibit*, San Diego, CA July 10-12, 1995.
17. Coons, S. C., Williams, J. D., and Bruckner, A. P., "Feasibility Study of Water Vapor Adsorption on Mars for In Situ Resource Utilization," Paper AIAA 97-2765, 33rd AIAA/ASME/SAE/ASEE Joint Propulsion Conference, Seattle, WA, July 6-9, 1997.
18. Adan-Plaza, S., *et al.*, "Extraction of Atmospheric Water on Mars for the Mars Reference Mission," *Proceedings of Mars Exploration Forum*, Houston, TX, May 5-7, 1998, in press.
19. Grover, M. R., and Bruckner, A. P., "Water Vapor Extraction from the Martian Atmosphere by Adsorption in Molecular Sieves," Paper AIAA 98-3301, 34th AIAA/ASME/SAE/ASEE Joint Propulsion Conference, Cleveland, OH, July 13-15, 1998.
20. Grover, M. R., Odell, E. H., Smith-Brito, S. L., Warwick, R. W., and Bruckner, A. P., "Ares Explore: A Study of Human Mars Exploration Alternatives Using *In Situ* Propellant Production and Current Technology," AAS 96-332 in *The Case for Mars VI*, McMillen, K. R., ed., AAS Publications, *Science and Technology Series*, Vol. 98, San Diego, CA, 1999, pp. 309-339.
21. Ferall, J. F., *et al.*, "Life Support Systems Analysis and technical Trades for a Lunar Outpost", NASA-TM-109927, NASA Jet Propulsion Laboratory, Pasadena, CA, 1994.
22. Breck, D. W., *Zeolite Molecular Sieves*, Wiley-Interscience, New York, 1974.
23. Dyer, A., *An Introduction to Zeolite Molecular Sieves*, John Wiley & Sons Ltd., Chistester, Great Britain, 1988, pp. 1-3.
24. "GRACE Davison Molecular Sieves Brochure for Zeolite 3A," W.R. Grace & Co., Baltimore, MD, (no date).
25. Roque-Malherbe, R., Personal communication, Instituto de Tecnologia Quimica UPV-CSIC, Universidad Politecnica de Valencia, Valencia, Spain, May 8, 1998.
26. UOP Technical Brochure UOP Molecular Sieves, UOP Molecular Sieve Adsobents, Garden Grove, CA, p. 10.
27. Marsh, K. N., "Recommended Reference Materials for the Realization of Physicochemical Properties," Blackwell Scientific Publications, Oxford, 1987, pg 256, Table 9.2.14.2.
28. Ishimaru, *Electromagnetic Wave Propagation, Radiation and Scattering*, John Wiley & Sons, New York, 1991, pp. 90-110.
29. Niu, M.C-Y., *Composite Airframe Structures*, Conmilit Press Ltd, Hong Kong, 1992, pg. 131.
30. Purchas, D., *Handbook of Filter Media*, Elsevier Advanced Technology, Oxford, UK, 1996, pp. 214-216.
31. NASA Commercial Technology Network, http://technology.ksc.nasa.gov/WWWaccess/Opport/aerogel.html.
32. Laitone, E. V., "Aerodynamic Lift at Reynolds Numbers Below 7×10^{4}," *AIAA Journal*, Vol. 34, No. 9, Sept. 1996, pp. 1941-1942.
33. Ruthven, D. M., *Principles of Adsorption and Adsorption Processes*, John Wiley & Sons, New York, 1984, pp. 206-207.
34. Hirschfelder, J. O., Curtiss, C. F., and Bird, R. B., *Molecular Theory of Gases and Liquids*, John Wiley & Sons, New York, 1954, pp. 1122, 1126.
35. Svehla, R. A., "Estimated Viscosities and Thermal Conductivities of Gases at High Temperatures," Technical Report R-132, Washington National Aeronautics and Space Administration, 1963, pp. 21, 68, B2.
36. Zurek, R. W., *et al.*, "Dynamics of the Atmosphere of Mars", in *Mars*, Kieffer, H.H., Jakosky, B.M., Snyder, C.W. and Matthews, M.S. eds., The University of Arizona Press, Tucson, AZ, 1992, p. 852.

37. Ryan, J. A., and Sharman, R. D., "H$_2$O Frost Point Detection on Mars?," *J. Geophys. Res.*, Vol. 86, 1981, pp. 503-511.
38. Moelwyn-Hughes, E. A., *Physical Chemistry*, 2nd ed., Pergamon Press, Oxford, UK, 1961, pp. 275-279.
39. Kieffer, H. H., Jakosky, B. M., Snyder, C. W. and Matthews, M. S. editors, *Mars*, University of Arizona Press, Tucson, AZ, 1992, pg. 842.

A COMPARISON OF IN SITU RESOURCE UTILIZATION OPTIONS FOR THE FIRST HUMAN MARS MISSIONS

Kristian Pauly*

The current plans of NASA for the first missions to Mars involve in situ resource utilization (ISRU). Goal of the six month study carried out at the Johnson Space Center (Ref. 1) was to obtain an independent assessment about whether ISRU can provide the advantages in mass and cost that are claimed by the approvers. Furthermore it was tried to obtain a non-biased comparison of the different ISRU options proposed for a human Mars mission in order to find the optimum option for such a mission. The study is based as far as possible on actually built production units and less on previous theoretical studies.

It is shown that estimations of the approvers of ISRU are very often too optimistic and that many options, which look good at first sight, have to be ruled out due to practical reasons during detailed review. Nevertheless, some options remain very promising and actually have the potential of decreasing mass and cost of such a mission.

BACKGROUND

The Design Reference Mission (DRM) (Ref. 2) of the NASA Mars Exploration Study Team is the current "best idea" of NASA to fly humans to Mars. The DRM is not a final result - it is a documentation of the present research status and changes permanently as the development continues. If new ideas substantiate that they can improve the mission performance, the DRM is altered accordingly (Ref. 3).

The current design involves a split mission approach with predeployment of the hardware that is needed for the surface stay (about 550 days) and the return to Earth in the launch window prior to the launch of the first piloted flight. By means of the split mission approach and new technologies, such as aerocapturing, the size of the ships that have to be launched to Mars decrease significantly. Thus, excessive assembly operations in low Earth orbit can be avoided, which reduces risk and cost. Each flight can be lifted to orbit with not more than two launches of a Shuttle derived launcher with a payload of about 85 metric tons.

Another technology that is used to reduce risk and cost is the use of indigenous resources on Mars. This technology was already suggested in the sixties and in the seventies (by Ash & al) (Ref. 4), but was implemented in the planning as a mission critical element not until Zubrin's Mars Direct plans (Ref. 5).

GOAL

Goal of the in situ propellant production (ISPP) on Mars is to fuel up an ascent vehicle that is able to deliver a payload of 6359 kg into a 1 sol Martian orbit (see Figure 1). This num-

---

* Exploration Office, NASA Johnson Space Center. Contact Address: Dipl.-Ing. Kristian Pauly, Division of Astronautics, Technische Universität München, D-85747 Garching, Germany.
E-mail: pauly@warr.lrt.mw.tu-muenchen.de.

ber is taken out of the current Design Reference Mission (Ref. 3). For an ascent vehicle that uses a liquid methane / liquid oxygen propellant combination, this results in a propellant mass in an order of magnitude of 50 tons. For different propellants the engines, the specific impulse, the tank sizes, and basically the whole Mars Ascent Vehicle vary. All these parameters have to be taken into account to obtain a fair comparison between the options.

Figure 1   Mars Ascent Vehicle with ISRU Plant (Ref. 2).

Apart from the in situ production of propellants, the DRM suggests a life support system cache. The idea is that in the case of the breakdown of primary and secondary life support system the survival of the crew can be assured with the help of an open loop system. The consumables in this case would be provided by the ISRU plant. This option has also be taken into account.

Furthermore, it was tried to estimate the effect of an in situ rover fuel production. During their stay on the Moon, the Apollo crews had only a very limited sortie radius and always stayed in walking distance from the ascent vehicle. For a human Mars mission with surface stays of over 500 days the mobility and the crew's radius of action has to be increased significantly. Battery powered rovers cannot provide this range, and the use of plutonium powered rovers is dwarfed by political issues. It is tried in this study to find out whether ISRU could be a way out of the dilemma.

CANDIDATE ISRU PROCESSES

A plethora of different processes has been suggested for ISRU. In this study options have been considered, which involve the utilization of the Martian atmosphere. The acquisition of atmosphere is relatively simple compared to drilling or collecting regolith. This was a handicap given by the Mars Exploration Study Team, because the feasibility of other options can hardly be estimated. But it must be kept in mind that the usage of indigenous resources is not limited to the atmosphere; ground ice and other resources could once be very useful, too.

The processes that were considered in this study are:

- Sabatier: $4H_2 + CO_2 \rightarrow CH_4 + 2H_2O$
- Water Electrolysis: $2H_2O \rightarrow 2H_2 + O_2$
- Carbon Dioxide Electrolysis: $2CO_2 \rightarrow O_2 + 2CO$

- Reversed Water Gas Shift: $H_2 + CO_2 \rightarrow H_2O + CO$
- Methane Pyrolysis: $CH_4 \rightarrow 2H_2 + C$
- Fischer / Tropsch: $CO + \left(1 + \frac{m}{2n}\right)H_2 \rightarrow \frac{1}{n}C_nH_m + H_2O$
- Methanol Synthesis: $2H_2 + CO \rightarrow CH_3OH$
- Ethylene Synthesis: $4H_2 + 2CO \rightarrow C_2H_4 + 2H_2O$
- Bosch: $CO_2 + 2H_2 \rightarrow 2H_2O + C$

The list is not complete; other promising ideas are currently under investigation (e.g. synthesis of higher hydrocarbons). But only little information about these new ideas is available up to now, so it was decided to limit the computer simulation to these processes.

With these processes, the following fuels can be produced on Mars:
- Methane (CH_4)
- Methanol (CH_3OH)
- Carbon Monoxide (CO)
- Amorphous carbon (C)
- Ethylene (C_2H_4)
- Hydrogen (H_2)

The oxidizer in all cases is liquid oxygen (O_2).

ACQUISITION OF DATA

To estimate the mass, power and volume properties of the mission elements, it was tried to base the estimation formulae as far as possible on actually built machinery and less on previous studies. Ref. 6 to Ref. 10 give only a very limited overview of the references used.

To obtain a fair comparison, the theoretical specific impulses of the different propellants were calculated with computer codes and then multiplied with the same engine efficiency for all options.

COMPUTER MODEL

The model consists out of the following modules, as shown in Figure 2:
- Mars Ascent Vehicle
- Atmosphere Acquisition
- Power Supply
- Processing Data
- Combination of Processes
- Propellant Data
- Liquefaction
- Comparison

The elements of the computer model are described in the following paragraphs.

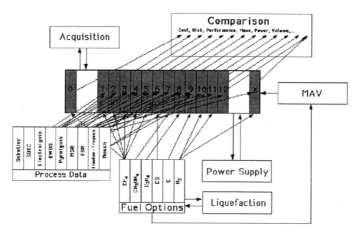

Figure 2 Structure of Computer Model.

MARS ASCENT VEHICLE

As mentioned before, the modeling of the Mars ascent vehicle is of crucial importance for the overall simulation. Different fuels lead not only to different propellant masses, but also to different tank masses, cooling requirements, etc. Thus, an accurate vehicle modeling that takes into account all these aspects is essential. The numbers of other ascent vehicles (like the Apollo Lunar ascent stage) are not automatically transferable to a Mars ascent stage.

The ascent vehicle was modeled by means of assumptions, taking in account the results of sizers used by Johnson Space Center and the Marshall Spaceflight Center. The results of the modeling of ascent vehicles that use different propellant combinations is summarized in Figure 3. The calculations consider a 20% dry mass margin and residual propellants at burnout.

ATMOSPHERE ACQUISITION

For small missions, a new atmosphere acquisition technology proved its usefulness: the sorption pump. Instead of compressing the Martian atmosphere by means of mechanical compressors this technique uses molecular sieves to adsorb the carbon dioxide. Unlike a mechanical compressor, a sorption pump does not contain moving parts, but achieves the densification of carbon dioxide by alternately cooling and heating a sorbent bed material. Because of the lack of moving parts, it has a significant potential for long lifetime, reliability and robustness. This material (for example zeolite) adsorbs low pressure gas at low temperatures and releases high pressure gas at higher temperatures. Accordingly, the Mars ISPP Precursor (MIP) experiment onboard the 2001 lander uses this technology to acquire the Martian atmosphere. Nevertheless, the simulation shows that the system mass of a mechanical compressor increases less with the mass flow than the system mass of a sorption pump. At a production rate of about 50 kg per day the masses are about equal. For human missions with almost one order of magnitude higher mass flows, the sorption pumps are heavier than mechanical compressors. As promising as the sorption pumps may appear for small missions, they are probably the wrong way for human missions.

Other technologies, like for example batch freezers that try to acquire the carbon dioxide by cooling it down below its freezing point, have still to prove their viability.

Ascent Payload				
w/o Margin	5299	kg		
w Margin	6359	kg	with	20% Margin

Achieved deltaV

deltaV	5625	m/s
(losses and control included)		

Isp and O/F ratios of different fuels with LOX:
Theoretical values based on pc=1200psia / exp. ratio 150:1

efficiency of engine	0,9487					
	CH4	CH3OH	C2H4	CO	C	H2
Isp[s] theoretical value	390	362	388	300	335	475
Isp[s]	370	343	368	285	318	451
O/F[-]	3,6	1,6	2,8	0,6	2,1	6,0
Dry Mass Margin	20,00%	20,00%	20,00%	20,00%	20,00%	20,00%
Str. Factor St. 1 w Marg	9,88%	10,80%	9,88%	11,61%	11,61%	13,35%
Str. Factor St. 2 w Marg	14,34%	14,99%	14,34%	17,47%	17,47%	20,60%

Def.: The structure factor is the dry mass of a stage without the mass of its payload (its upper stages) divided by the ignitio
The unused fuel is included in the structure, mass not in the propellant mass

Percentage of propellant that remains in tanks:	2,0%

Mass Breakdown for different Fuels

1 Stage

	CH4	CH3OH	C2H4	CO	C	H2
Payload w Margin	6359	6359	6359	6359	6359	6359
Str. Mass w Margin	6445	10772	6561	-1617098	23500	6433
Dry Mass w Margin	12804	17131	12920	-1610739	29859	12792
Propellant total	52446	82610	53505	-12313559	172492	35396
Propellant used	51397	82610	53505	-12313559	172492	35396
MAV Total Launch Ma	65249	99741	66424	-13924298	202351	48188
Fuel	11401	31773	14080	-7944231	55643	5057
O2	41044	50837	39424	-4369327	116849	30340

2 Stages assuming optimum staging

	CH4	CH3OH	C2H4	CO	C	H2
Payload w Margin	6359	6359	6359	6359	6359	6359
1st Stage deltaV1	1919	1920	1921	1922	1923	2280
Str. Mass w Margin	6393	9033	6474	25925	15358	7387
Dry Mass w Margin	37608	46571	37956	109900	70138	32577
Propellant	27114	37069	27587	113329	62106	22753
Propellant used	26572	36328	27035	111062	60864	22298
2nd Stage deltaV2	3707	3706	3705	3704	3703	3345
Str. Mass w Margin	4476	5627	4515	14671	9570	5189
Dry Mass w Margin	10835	11986	10874	21030	15929	11548
Propellant	20380	25552	20609	62946	38851	13642
Propellant used	19973	25041	20197	61687	38074	13369
Ignition Mass	31216	37538	31482	83976	54780	25190
Total						
Str. Mass w Margin	10869	14660	10988	40595	24928	12576
Struc. Factor w Margin	16,8%	17,5%	16,8%	18,2%	18,9%	22,7%
Propellant	47494	62621	48196	176275	100957	36395
Propellant used	46544	61369	47232	172749	98938	35667
MAV Total Launch Ma	64722	83640	65543	223229	132244	55330
Fuel	10325	24085	12683	113726	32567	5199
O2	37169	38536	35513	62549	68390	31196

Comparison

							DRM 3.0
Str. Mass w Margin							
1 Stage	6445	10772	6561	-1617098	23500	6433	4069
2 Stages	10869	14660	10988	40595	24928	12576	
Propellant Mass							
1 Stage	52446	82610	53505	-12313559	172492	35396	38446
2 Stages	47494	62621	48196	176275	100957	36395	

Figure 3 Mars Ascent Vehicle Model.

The baseline of the modeling of the pumps was hardware built for the International Space Station respectively for the 2001 Mars Lander.

POWER SUPPLY

The Design Reference Mission assumes a nuclear power plant on the surface of Mars. The reactor design is based on the work done in the fifties, the sixties and the seventies during the NERVA project (Nuclear Engine for Rocket Vehicle Application). A reactor of the required size of some 100 kW_e was designed by the NASA Lewis Research Center. The results of this research work is the baseline for the modeling of the nuclear power source. The graphs shown in Figure 4 are an example of how the masses of the power source were estimated. Similar relations are derived for all subsystems.

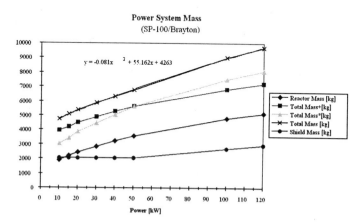

Figure 4   Mass of Power Supply.

Figure 4 shows, that the mass of the nuclear reactor changes only little with the increasing power demand. Thus, saving of energy is not as crucial as it is with a solar power source. Political issues (like the question of ground testing of these nuclear devices on Earth) and the issue of site contamination on Mars have still to be answered.

Preliminary results of an ongoing study of the Exploration Office show, that only for keeping the crew alive, some 3000 m^2 of solar array area are needed. It is questionable whether the power need of the outpost could be satisfied with solar energy. That is especially a problem for the first cargo mission, where the solar arrays would have to be deployed and operated automatically. For later missions, humans could assist the deployment and cleaning of the solar arrays, but the needed areas are still huge.

Other resources like wind energy or geothermal energy were suggested by James & al (Ref. 11) and Zubrin & al (Ref. 5), but the feasibility of these options has still to be proven. Thus, the computer model only involved the simulation of nuclear and solar energy.

PROCESS DATA / COMBINATIONS OF PROCESSES

With the nine processes listed above, a plethora of process combinations can be obtained. Not every combination makes sense; the combination must be able to produce fuel, oxygen and water. The selection of the process combinations was based on the following two requirements:

- The combination must be able to produce fuel, oxygen and water (for life support system cache). This is necessary only for the first mission - for the follow-on missions cache production is not required.
- The number of processes must not exceed three, to limit the complexity of the ISRU-plant.

The combinations considered in this study are:
0. No ISRU (as reference)
1. Sabatier & Solid Oxide Electrolysis Cell
2. Sabatier & Water Electrolysis
3. Sabatier & Electrolysis & Pyrolysis
4. Sabatier & Electrolysis & Reversed Water Gas Shift
5. SOEC & Fischer-Tropsch & Pyrolysis
6. SOEC & Reversed Water Gas Shift (CO)
7. RWGS & Methanol Synthesis Reactor & Electrolysis
8. RWGS & Ethylene Synthesis Reactor & Electrolysis
9. Sabatier & Electrolysis & Pyrolysis (Hybrid C)
10. Solid Oxide Electrolysis Cell & Bosch (No ISRU Fuel)
11. Water Acquisition & Electrolysis (WAVAR)
12. Water Acquisition & Electrolysis & Sabatier
13. Water Acquisition & Electrolysis on Phobos

Most of the processes do not produce their products in the ratio that is needed for later applications (e.g. combustion). For the first mission, three substances are needed: the fuel (e.g. methane), water for the LSS and oxygen as oxidizer for the engines and for the LSS-cache. In this case, three different amounts must be produced. Therefore, three different variables - three "regulators" are needed. The first variable is the reactant input mass. The other variables have to be provided by the processes.

For the follow-on missions (the second and the third mission to Mars) things get easier, because no water has to be produced and therefore the number of variables is reduced from three to two.

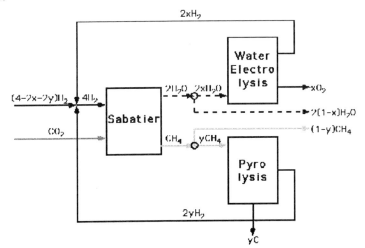

Figure 5   Sabatier / Electrolysis / Pyrolysis Flowchart.

For a first estimation of mass, power and volume properties of the plant, the mass flows through the different subsystems of the plant must be calculated. This is done as shown in Figure 5 at the example of the option 9 ("Sabatier & Pyrolysis & Water Electrolysis"). For a detailed calculation of the overall mass, power and volume properties, a more detailed design (see Figure 6) of the ISRU plant is necessary.

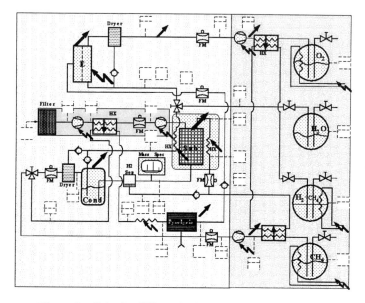

Figure 6   Sabatier / Electrolysis / Pyrolysis Plant Design.

LIQUEFACTION AND STORAGE

Apart from the amorphous carbon all propellants must be liquefied. In the case of methanol, this is no big technical problem. Other fuels like hydrogen for example require more finesse, not only in terms of liquefaction but also in terms of storage (active cooling). Reliquefaction of hydrogen is difficult, currently it is preferred to reduce boiloff as far as possible instead of trying to reliquefy the hydrogen. New technologies like gelling hydrogen or nanotubes could have the potential of reducing the overall mass.

EVALUATION

For the downselection of options the following metrics come into question:
- Cost
- Mass
- Power
- Volume
- Mission Performance
- Human Health and Safety
- Risk
- Level of Technology Readiness
- Complexity
- Political / Public Appearance

As shown previously, the power requirement of the ISRU plant plays no role if the power source is a nuclear one. For a solar power source the required power levels are probably too high for all options. Thus, power seems not to be a metric suited for the downselection process.

The volumes of the ISRU plant and the feedstock are only of importance regarding the accommodation into the cargo lander shroud.

The mission performance is the same for all options. The human health is not effected by any method.

To estimate the risk level respectively the level of safety is difficult, since the amount of long-term experience is very diverse for the different options. A good benchmark in this situation is a comparison of the level of technology readiness (LTR) and the complexity of the different options (see Figure 12). The political / public appearance of the different option does not vary very much the options.

At the beginning of the thesis it was planned to downselect the best mission(s) by means of the metric cost. Unfortunately, the NASA cost estimation database was not available until the end of the six months study. Since mass is the most important input for a cost estimation program, it seems to be appropriate to use this variable instead for downselecting. However, it must be said that mass does not automatically equal cost.

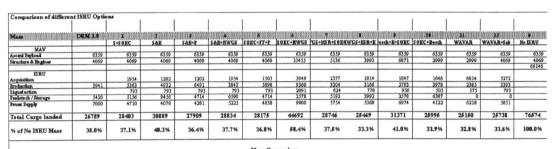

Mass	DRM 3.0	1 S+SOEC	2 S&E	3 S&E+P	4 S&E+RWGS	5 SOEC+IT+P	6 SOEC+RWGS	7 GS+MSR+SOEWGS+ESR+E	8 orch+E+SOEC	9 SOEC+Bosch	10	11 WAVAR	12 WAVAR+Sab	No ISRU
MAV														
Ascent Payload	6359	6359	6359	6359	6359	6359	6359	6359	6359	6359	6359	6359	6359	6359
Structure & Engines	4069	4069	4069	4069	4069	4069	10455	5136	3993	6871	2999	2999	4069	4069
														66146
ISRU														
Acquisition		1954	1202	1202	1954	1503	3049	2377	1814	1847	1646	6614	5272	
Production	3941	5363	4932	6491	3843	5898	9360	3304	3166	5793	3978	2385	3393	
Liquefaction		793	793	793	793	793	2091	624	778	950	505	575	793	
Feedstock / Storage	5420	5156	9456	4714	6590	4714	2578	5192	3992	2578	6387	0	0	
Power Supply	7000	4710	4078	4281	5225	4838	9900	5754	5368	6974	4122	6218	5851	
Total Cargo landed	26789	28403	30889	27909	28834	28175	44692	28746	25469	31371	25995	25150	25738	76574
% of No ISRU Mass	35.0%	37.1%	40.3%	36.4%	37.7%	36.8%	58.4%	37.5%	33.3%	41.0%	33.9%	32.8%	33.6%	100.0%

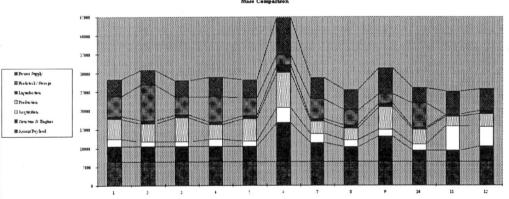

Figure 7   Preliminary Comparison of ISRU Options.

After two months of investigation, the number of options was reduced to six by means of the preliminary results shown in Figure 7. This figure shows the „ticket home mass" for the first mission. This is the sum of the masses of everything that is needed to return crew and payload to the Earth Return Vehicle in Mars orbit. This includes not only the ISRU plant, but also the Mars ascent vehicle, the power supply and the mass of the imported hydrogen. A compari-

son of only the ISRU plant mass alone definitely would not be fair since for example a lightweight plant that produces propellants with a bad I_{sp} can still debase the mission mass budget. The remaining options after the downselection were:
- Sabatier & Solid Oxide Electrolyte Cell
- Sabatier & Water Electrolysis
- Sabatier & Water Electrolysis & Pyrolysis
- Reversed Water Gas Shift & Methanol Synthesis & Water Electrolysis
- Reversed Water Gas Shift & Ethylene Synthesis & Water Electrolysis
- Solid Oxide Electrolyte Cell & Bosch

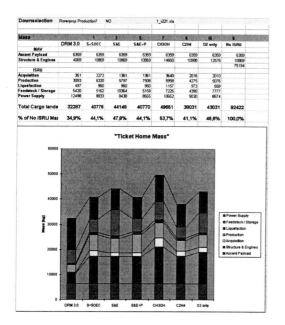

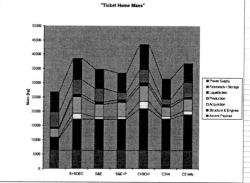

Figure 8   Mass with Cache / without Roverfuel.
Figure 9   Mass without Cache / without Roverfuel.

RESULTS

In the further progress of the study, the models of the different subsystems became more and more detailed. The final results of the mass estimates after the downselection are shown in Figure 8 to Figure 11. For a human Mars mission, carbon monoxide and amorphous carbon are not feasible options, since the specific impulse is too low and would result in a huge mars ascent vehicle.

Methanol was seen as a promising fuel in many sample return studies. Its advantages are that it is not cryogenic and can be stored easily. The disadvantage though is a comparatively low I_{sp}. For small missions (such as sample return) with solar power production, the reduction of power requirements is of vital importance. For missions with a nuclear reactor as power source, the mass can only be reduced very slightly by reducing the power requirement. The computer simulation has shown that the power consumption has a very small effect whereas the specific impulse is a very important player.

Methane seems to be a very good choice. Methane engines are a relatively simple compared to for example hydrogen engines. Russians are currently testing their RD-169 methane engine which will by the engine for the new Riksha-1 launcher that will have its maiden flight within the next two years (Ref. 12).

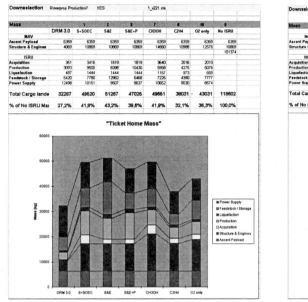
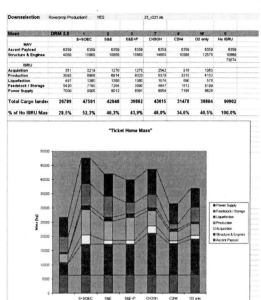

Figure 10   Mass with Cache / with Roverfuel.
Figure 11   Mass without Cache / with Roverfuel.

An ethylene option was also considered in this study. It seems that the advantages of methane over methanol are outmatched by this fuel; the amount of hydrogen that has to be imported is considerably decreased, the specific impulse is about the same as for methane. The problem is, however, that this is a comparatively young idea; the application of ethylene fuel in rocket engines has still to be proven.

	Weight	Battlestar	Sab+SOEC	Sab+E	Sab+E+P	Sab+E+RWGS	SOEC+F/T+P	SOEC+RWGS
LTR	1	Launcher 6.5 Aerobrake 7 big Lander 7 **7**	Sab 7 SOEC 6 **6**	Sab 7 E 8 Zubrin **7.5**	Sab 7 E 8 P 6 **6**	Sab 7 E 8 RWGS 5 **5**	SOEC 6 F/T 5 P 6 **5**	big MAV SOEC 6 RWGS 5 **5**
Complexity	1	3 **5**	2 **6**	3 **6**	2 **6**	3 **5**	3 Filter **4.5**	3 Mix **4**	3 Filter **4.5**
Others	?	LSS-Cache? Launcher??	Nuclear	Nuclear	Nuclear	Nuclear	Nuclear	Nuclear

	Weight	RWGS+MSR+E	RWGS+ESR+E	Bosch+E+SOEC	SOEC+Bosch	WAVAR+E	Phobos E
LTR	1	RWGS 5 MSR 6 E 8 Zubrin **7**	RWGS 5 ESR 5 E 8 Meyer **5.5**	Bosch 5 E 8 SOEC 6 Amorphous C 2 **3**	SOEC 6 BOSCH 5 LH$_2$ 6 **6**	WAVAR 4 E 8 LH$_2$ 6 **4.5**	big Lander 7 E 8 Phobos 3 LH$_2$ 5 **4**
Complexity	1	3 **5**	3 Filter **4.5**	4 **4**	3 **6**	3 **5**	4 **4.5**
Others	?	Nuclear	Nuclear	Nuclear	Nuclear	Nuclear	LSS-Cache on Phobos Non-Nuclear

Figure 12   Level of Technology Readiness / Complexity.

CONCLUSION

One conclusion of the study is that the ratio of plant mass to produced propellant mass decreases with increasing mass flows. Thus, for a human mission it makes more sense to invest in a heavier plant to produce more efficient propellants - even if they are cryogenic. The importance of the specific impulse increases compared to small missions. The optimum choice of ISRU options for a small mission (like Mars sample return missions) is very likely different from the optimum options for large missions (human Mars mission). This rises the question whether it makes sense to see sample return missions isolated or whether it could be more reasonable to see them as dress rehearsals for human missions and to choose the processes that are optimal for the big missions.

Redundancy and margins are very likely to be underestimated. Three redundant units in every subsystem with each providing 50% of the total required mass flow with an opportunity of cross-linking seems to be a reasonable approach.

It is questionable whether the idea of a life support system cache out of water and oxygen on the Martian surface instead of a second backup system as proposed in the Design Reference Mission makes sense.

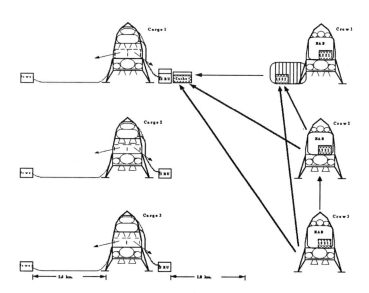

Figure 13 Backup Strategy of DRM.

Three redundant systems are not only needed on the Martian surface, but also on the way towards Mars. The production of rover fuel could avoid the need of plutonium powered rovers as well as the need of huge amounts of rover fuel. In case of a major malfunction the rover fuel depot could provide the crew with oxygen, water, heat, and electrical current.

Another open issue is the storage of hydrogen over extended periods of time and on the surface of Mars. Storing cryogenic fuels will never be easy, but progress has been made during recent years and new technologies, e.g. gelled hydrogen or nanotubes, seem to be promising. If it turns out that hydrogen storage is to difficult to realize, then the propellant production will probably be limited to oxygen.

OUTLOOK

All the testing that has been done up to now must not give the impression that this is already sufficient. No testing on Earth, no matter how sophisticated, can ultimately assure that the systems will work in the Martian environment. Thus, before human missions can be launched, several robotic precursors will be sent towards the Red Planet, to prove the reliability of new technologies such as ISRU. NASA is planning to send a number of different probes with ISRU-experiments to Mars, starting with the MIP experiment onboard the 2001 Lander. In this experiment the feasibility of oxygen production out of the Martian atmosphere will be demonstrated. In the next launch window (2003), the first end-to-end in situ propellant production unit will be established where fuel is not only produced, put also stored and used. The 2003 Lander will also be the first lander of a new generation of landers which are larger than the landers of previous years. This type of lander will also be the basis for the first 2005 Mars Sample Return (MSR) Mission. It is very likely that this mission (and a similar mission in 2007) will be based on the ISRU concept. If the new technologies that are involved by the design reference mission prove their reliability in these missions, a first human Mars mission could take place in 2014.

One reason to go to Mars is the fact that with results obtained with the help of comparative planetology can help us to solve our problems here on Earth. The greenhouse effect was discovered on Venus by NASA scientists - the Martian atmosphere consists almost purely of a greenhouse gas (CO_2) and Mars has a "global ozone hole". Mars was once a planet with less harsh conditions, and it could be of crucial importance to understand why that changed.

The Design Reference Mission also comprises the research in the use of solar energy and the storage of hydrogen over long periods of time. Most processes that were under investigation in this work convert a greenhouse gas into fuel and oxygen. All these techniques could be very well used here on Earth, too.

It seems likely that the human exploration of deep space is not feasible without in situ resource utilization. This is especially true for permanent outposts. If humans want to explore deep space, they have to learn to use indigenous resources. Mars seems to be the break-even point, where the ISRU approach becomes more advantageous than the classical approach.

The question whether ISRU will be used for the first human Mars mission has still to be answered. But there is no question that humans will explore the Red Planet in the next century. Humans have never stopped to push back the boundaries. Mars is the next step.

Figure 14   Inevitable Descent (Pat Rawlings).

REFERENCES

1. Pauly, K. 1998, Mars Mission Scenarios Involving In Situ Resource Utilization, Diplomarbeit at the Division of Astronautics, Technische Universität München.
2. Hoffmann S. & Kaplan, D., 1997, Human Exploration of Mars: The Reference Mission of the NASA Mars Exploration Study Team, NASA Special Publications 6107.
3. Drake, B. G., June 1998, Reference Mission Version 3.0 (Addendum), NASA Johnson Space Center Publications EX13-98-036.
4. Ash, R. L., Dowler, W. L., Vars, G., 1978, "Feasibility of Rocket Propellant Production on Mars," *Acta Astronautica*, Vol. 5, pp.705-724.
5. Zubrin, R. & Wagner, R., 1996, *The Case for Mars*.
6. Zubrin, R., Frankie, B., Kito, T., 1997, "Mars Methanol In Situ Propellant Production SBIR I Study," NAS 9-97082, Pioneer Astronautics, Lakewood.
7. Crow, S. C., 1997, "The MOXCE project: New Cells for Producing Oxygen on Mars," AIAA 97-2766, AIAAA/SAE/ASME/ASEE 33rd Joint Propulsion Conference.
8. Zubrin, R., Price, S., Mason, L., Clark, L., Phase III Final Report: An End-to-End Demonstration of a full scale Mars In Situ Propellant Production Unit, Contract NAS 9-19145, Martin Marietta Astronautics, Denver.
9. Zubrin, R., Burnside-Clapp, M., Meyer, T., 1997, "New Approaches for Mars In Situ Resource Utilization based on the Reversed Water Gas Shift," AIAA-97-0895, 35th Aerospace Sciences Meeting & Exhibit.
10. Deffenbaugh, D. M., Green, S. T., Miller, M. A., Treuhaft, M. B., 1998, *Mars In Situ Propellant Production*, Southwest Research Institute, San Antonio.
11. James, G., Chamitoff, G., Barker, D., 1998, Resource Utilization and Site Selection for a self-sufficient Martian Outpost, NASA/TP-98-206538.
12. Karnozov, V., 1998, "Russian Cryogenic Projects," Article in *Launchspace*, Vol. 3 (June / July), 44-46.

PRODUCING A BRICK FROM A MARTIAN SOIL SIMULATE

David Seymour

INTRODUCTION

Producing bricks from the Martian soil has been a topic floating around for a while now. The idea is to use the local Martian resources to form underground vaulted chambers that could help store an atmosphere and protect humans from cosmic rays and solar radiation.[1] Through some researching, I have found some scientific research already performed on producing objects from Martian soil. The idea is becoming popular, I believe, due to the fact that it is much cheaper to produce habitats on Mars using Martian resources than to produce the habitats on Earth and send them to Mars.[2]

The purpose of this paper is to see if it would be possible to use some of the fine regolith that covers Mars, mix it with the correct amount of water, and form the mixture into usable bricks for construction. There is a severe limitation to be worked when dealing with this subject though. There is no Martian soil on Earth since no missions to Mars have returned soil samples. Another limitation is that it is still unknown if clays are present on Mars. The major purpose for clays in brick manufacturing is the green strength they provide for the bricks while they are being produced.

It was once believed that clays were abundant on Mars but now it is believed that only a few clays that contain iron might be found on Mars3 (p. 581). Spectroscopic observations of the surface of Mars show that many forms of clays are not observed. Nontronite (an iron rich clay) could be a minor constituent of the soil but the authors also add that spectroscopic evidence for clay minerals remains inconclusive

Salts are also thought to be found on Mars (p. 622).[3] The highly oxidizing atmosphere of Mars is thought to contribute to salt formations. Salts thought to dominate on the surface are $MgSO_4$, Na_2SO_4, and $NaCl$.

Boyd, Thompson and Clark have done some research on producing flat circles out of Martian material (p. 542).[4] They produced a material they called "Duricrete". They used a mixture of 12% $MgSO_4$, 1% $NaCl$, 2% Fe_2O_3, and 85% clay to simulate a Martian soil solids content. The two different types of clay used were, bentonite and Pennsylvania nontronite.

Boyd, Thompson, and Clark experienced cracking and warping from their samples (p. 542).[4] To correct these problems, they tried a number of solutions. Slowing down the evaporation rate helped lessen the cracking problem. Another was to add a matrix material such as nylon mesh, rayon cloth, kevlar fiber, or glass wool.

Replacing the water with pure sulfur was also effective. With this method they heated the solids to 150°C where the sulfur polymerized. Upon cooling, they also tried compressing the polymerized material which produced the strongest of the sulfur cemented duricretes.

Boyd, Thompson, and Clark said that the most effective method of producing duricrete was to use compaction to squeeze out the water used to make the mud. Their strength results show that compressed samples with a matrix material were indeed higher then the rest of the results.

DISCUSSION ON MAKING THE BRICKS

The Materials for the Bricks

I plan on using the same mixture as Boyd, Thompson, and Clark to begin with. It seems to represent the Martian soil as best as we know it presently to be. I will be doing more research into whether salts (such as $MgSO_4$ and $NaCl$) can actually be found on the surfaces of Mars. I will also be doing more research into how salts affect brickmaking. I may adjust the mixture if the results of this research disagree with the mixture used by Boyd, Thompson, and Clark.

I would like to run some characterization tests on the clay I use for a more detailed look at what I am using. Such a test might be atomic characterization.

I will try adding matrix materials such as those Boyd, Thompson, and Clark used. I am also interested in how they used sulfur instead of water to hold the bricks together. I would like to try a few test bricks using just sulfur instead of water. These sulfur bricks will be hand molded (more information on hand molding below).

Forming the Bricks

I would like to try extruding the bricks to begin with. This is how most bricks are produce on Earth and would seem to be the easiest way to producing them on Mars5.(p. 12) Unfortunately, I might not have the equipment needed to produce them in this fashion. My next option would be to try producing them using a hand mold. Hand molding would consist of pieces of wood held together by clamps to form a mold into which the stiff mud will be placed. Then the wood can be removed from the sides of the brick. In either case, I plan on making the bricks about 8x4x2 inches.

The next step would be to compress the bricks. I am unsure how this will be done at this time and will be consulting with some professors about the set-up. I would like to see the differences between a compressed brick and an uncompressed brick so some bricks will by-pass this step.

Drying the Bricks

I plan on having to dry the bricks rather slowly but yet controlled because cracks will form otherwise. I will try using an oven first with no air movement. If this is too much, my next plan is to dry the bricks out in the open with a fan blowing gently on them.

If I produce the sulfur bricks I will not have to worry about the drying step. I will heat them to a temperature above 119°C so that they can polymerize. If I have a system for compressing the bricks I will compress a few polymerized sulfur bricks.

Firing the Bricks

I plan on firing the bricks to a temperature where they are well burnt. Swain states that well burned bricks for construction have similar characteristics such as uniform color, free from cracks, and ring when struck with a hammer (p. 143).[6] I believe that these characteristics would also apply to the bricks I hope to make.

Testing

I plan on performing compression tests on the bricks. Swain describes some standard methods for compression tests on bricks (p. 144)[6] Different strengths are measured by how the brick is tested. If the brick is laying down on its large face, then the strength values are larger than when it is standing up on its smallest face. I will test bricks both ways but believe that the strength values for the bricks laying down are more useful because that will be the position that they are used in when constructing a habitat.

I would also like to perform an efflorescence test on the bricks. Efflorescence is the process of soluble salts in the bricks leaching out onto the surface of the brick because of water migrating to the surface to evaporate (p. 268).[7] The crystallization of the salts deposited on the surface of the brick cause deterioration. I expect that the $MgSO_4$ used in creating the brick will show up during this test.

CONCLUSIONS

The Martian soil might be formed into bricks to use in constructing habitats on Mars. It is possible that clays and salts can be found on Mars to use in brick manufacturing. The research done by Boyd, Thompson and Clark show that they were able to produce shapes from a simulate to the Martian soil with good results using compaction and matrix materials.

Further research and testing on producing a brick from a simulate Martian material will be preformed. Some possible tests that will be performed on the bricks are compression strengths and efflorescence.

This experiments will be different from ones that use actual Martian soil. Currently, it is not know if clays and salts are present on Mars or what form they are in if they are present. Further exploration of Mars will be able to provide more accurate information for this topic.

REFERENCES

1. B. Mackenzie, "Building Mars Habitats Using Local Materials," AAS 87-216, in C Stoker, ed., *The Case for Mars III*, Volume 74, *Science and Technology Series* of the American Astronautical Society, Univelt, San Diego, CA, 1989.
2. Robert Zubrin and Richard Wagner. *The Case for Mars*. New York City: Touchstone, 1997.
3. A. Banin and B. Clark. "Chapter 18-Surface Chemistry and Mineralogy." *Mars*. ISBN# 0-8165-1257-4. Library of Congress call letters: QB641.M35 1992. University of Arizona, 1992.
4. Robert C. Boyd, Patrick S. Thompson, and Benton C. Clark, "Duricrete and Composites Construction on Mars," AAS 87-213, in C. Stoker, ed., *The Case for Mars III*, Volume 74, *Science and Technology Series* of the American Astronautical Society, Univelt, San Diego, CA, 1989.
5. W. B. McKay. *Brickwork*. London: Longman Group Ltd., 1974.
6. George Fillmore Swain. *Structural Engineering: Fundamental Properties of Materials*. New York, McGraw-Hill Book Company, Inc., 1924.
7. F. H. Clews. *Heavy Clay Technology*. England, William Clowes and Sons, Limited, 1955.

Suggested Reading

John S. Lewis. *Mining the Sky*. Addison Wesley, 1996.

THE CASE FOR A MARS BASE ISRU REFINERY

Kelly R. McMillen[*] and Thomas R. Meyer[*]

During the early phases of human Mars exploration, in-situ resource utilization (ISRU) will lower costs, expand capabilities, and serve as an enabling technology for establishing permanent colonies. Martian atmospheric resources can be used to provide consumables such as fuel, oxidant, breathable air, and water that are critical for early human missions. Martian atmospheric carbon dioxide and imported hydrogen can be used, for example, as feedstock for the catalytic production of oxygen, methane, methanol, water and other propellants (Zubrin, 1991, 1996, 1997, 1998, Zubrin, Meyer, McMillen 1998, Meyer 1989, 1981). These processes utilize catalytic reactors containing small amounts of iron, nickel and other suitable catalysts, plus gas selective membranes, electrolysis, and other easily implemented gas separation techniques. Waste carbon monoxide from carbon dioxide reduction processes together with hydrogen can be combined to produce other liquid and gaseous fuels and chemical compounds. Excess heat from an exothermic Sabatier reaction can be diverted to minimize heat requirements in endothermic processes such as the reverse water-gas shift reaction. Valuable synergies can be realized by integrating various processes. Oxygen and fuel production processes can be combined so the thermal and material wastes of one process can be utilized by the other thus forming a unique Martian "chemical refinery" that features internal hydrogen recycling and production of a purified carbon monoxide intermediate by-product. Turbines can also be used to recover mechanical energy from high-pressure waste gas and systems can share common hardware and feedstock systems. Chemical feedstock, power, heat and mechanical energy are utilized efficiently and conserved in the design of these robust Martian atmospheric refineries whose technologies may also find applications in industrial waste utilization technology on Earth.

INTRODUCTION

There are several factors which when taken together make a compelling case for establishing a Mars base refinery. These include first and foremost, the ongoing need for large quantities of consumables to support human exploration, the need to minimize the high cost of transporting bulk consumables from Earth, safety considerations that require the maintenance of adequate reservoirs of vital life support compounds, the convenience of valuable readily available feedstocks obtainable from the Martian atmosphere, and options to use innovative materials processing technologies that can be integrated to optimize system performance. In addition, refineries designed to primarily utilize atmospheric gases as feedstock will have a lower environmental impact than surface mining. Finally, this technology is the basis for establishing self-sufficiency on Mars that will make eventual permanent human habitation feasible. Let us elaborate on these points.

MARS BASE ISRU REFINERY:
AN ENABLING TECHNOLOGY FOR HUMAN EXPLORATION

The Mars Base In-Situ Refinery builds on NASA's technological concepts for habitat physico-chemical control features (MSFC 1997) and the original Mars Gas Extractor (Meyer,

---

[*] University of Colorado/Boulder Center for Science and Policy, P.O. Box 4883, Boulder, Colorado 80306.

1981- Figures 2, 3). It provides the basis for an integrated systems approach to surface exploration by combining consumables production with habitat design. A Mars base can conceivably be constructed piece-by-piece as energy production, reliability and power issues are tested and resolved. Would you rather drive across the Australian Outback with a gas can or build reliability into your chances for survival by constructing a gas station? When asked when he believed we would see the first relay station on Mars, Sir Arthur C. Clarke, replied, "it's a good idea and necessary for extended human missions, I would guess around the year 2030!" This hypothetical schedule requires a long-term plan, implementing a strong synergy between associated processes (oxygen, water, hydrogen and propellant production), high conversion ratios, conservation of mechanical and heat energy and high H_2 leveraging.

CONVENIENCE - ATMOSPHERIC RESOURCES ARE BOTH USEFUL AND UBIQUITOUS

As a major component in the Martian atmosphere (95.3%), carbon dioxide is a candidate feedstock for the manufacture of oxygen, propellants, and a whole range of Fischer-Tropsch products (Frankie, this volume). Compounds manufactured from CO_2 and hydrogen can be stored easily in ambient Mars conditions (+ 15°C to - 100°C).

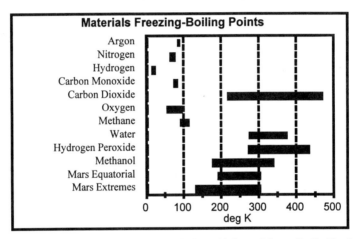

Figure 1   Materials Freezing and Boiling Points. (Adapted from B. C. Clark, 1991).

A Pulse Tube Refrigeration system (PTR) that contains no movable parts and in which heat transfer between the gas and the tube wall follows linearized conservation equations, could suffice for liquid storage of hydrogen, methane, and oxygen on the surface of Mars. Oxygen can be stored in either a gaseous (GOX) or liquid (LOX) form under these conditions. Compression of carbon dioxide for use as a cryogen may also be feasible on Mars' surface.

SAFETY - A MARS BASE REFINERY INCREASES SAFETY AND ROBUSTNESS

A key function of a Mars Base Refinery will be to accumulate and maintain reservoirs of consumables. The amount of each compound held in reserve will depend on safety and expected system-loading requirements. These represent an important safety reserve that can be used during refinery maintenance downtime and in the event of emergencies.

NEED: SCIENTIFIC EXPLORATION REQUIRES CONSUMABLES

The primary purpose for developing a Mars Base refinery is to supply consumables such as air, water, and fuel for scientific exploration. Mars holds a vast trove of scientific knowledge that will help us better understand and protect our own planet, understand the origin and evolution of our solar system, and gain important insights into origin of life. Mars is also a laboratory for habitability research where we can appraise the feasibility of modifying the Martian climate and adapting terrestrial species that might facilitate eventual human habitation of Mars. Such studies will merit the establishment of manned bases and laboratories on Mars that are comparable to those in Antarctica and whose research operations could easily span many decades. NASA's life support systems will assist in attaining basic self-sufficiency on Mars.

Table 1
ESTIMATE OF CONSUMABLES USE BY MARS BASE ELEMENTS

Mars Base Elements*	Life Support			Facility / Vehicle		Water Recycling
	O_2	N_2	H_2O	Fuel	Oxidant	Potential
Biosphere / Habitat	M	M	L	M	M	Yes
Ascent Vehicle	S	S	S	L	L	No
Rover / Excursion Vehicle	S	S	S	M	M	Yes
EVA / Field Activities	S	S	S	-	-	No
Fuel Cell / Power System	-	-	-	L	L	Yes
ISRU Facility	S	S	S	M	M	Yes

Table indicates relative usage estimates: S,M,L = Small, moderate or large volume use.

ISRU AS AN ENABLING TECHNOLOGY FOR EXPLORATION: LOWERS COSTS & MASS

ISRU is an enabling technology for exploration because it drastically lowers the transportation costs of importing consumables, particularly rocket propellant needed for ascent vehicles for return to Earth. Since it takes on the order of 100 tons of liquid oxygen and propellant to return a crew of five persons from Mars surface back to Earth, and since the transportation cost to deliver material to the surface of Mars is in the range of $20-$50,000 per kilogram, if this fuel can be produced from local resources on Mars, then a an enormous savings in transportation costs can be achieved. In fact, this savings could be the critical factor that determines whether a human exploration program is feasible at all.

In the early stages of exploration, the logistical difficulty of obtaining water on Mars suggests a strategy where the hydrogen component of the propellant would be imported from Earth while the remainder of the constituents are obtained from Mars resources as proposed in the Mars Direct scenario (Zubrin and Baker 1984, 1991).

As an illustration consider a rocket that uses oxygen and methanol propellant. From the combustion equation we see that

$$CH_3OH + 1.5\ O_2 \rightarrow CO_2 + 2\ H_2O \quad \Delta H = -675.72 \text{ kJ/mol of } CH_3OH$$

If we add the mass in atomic weight units of each element, we have 12 units of carbon, 40 units of oxygen, and 4 units of hydrogen for a total of 56 units for each molecule of methanol burned. Of this 52 units represent material that could be obtained by processing Martian atmospheric carbon dioxide while the remaining 4 units consist of hydrogen that would initially be imported from Earth. Since the hydrogen represents only about 7% of the entire mass of the fuel, this translates into very significant savings over shipping the entire mass of fuel at a cost of several billion dollars for each crew of 5 persons to be returned home from Mars. As another example, if ethylene, a more powerful propellant could be synthesized on Mars, the amount of hydrogen required drops to only 3.2%.

$$C_2H_4 + 3 O_2 \rightarrow 2 CO_2 + 2 H_2O \quad \Delta H = -1322.49 \text{ kJ/mol } C_2H_4$$

Thus a Mars Base Refinery that is able to produce water, oxygen and ethylene would be a tremendous asset for human exploration. Table 2 summarizes the mass leverage of in-situ propellant for Earth-derived hydrogen.

Table 2
IN-SITU PROPELLANT MASS LEVERAGE[a]
FOR EARTH-DERIVED HYDROGEN RELATIVE TO TERRESTRIAL H_2/O_2 FUEL

Fuel	Reaction Stoichiometry	I_{sp}[b]	Leverage[c]	Effective I_{sp}[d]
Hydrogen (baseline)	$H_2 + 1/2\ O_2$ (Earth)	460	1	1
Hydrogen	$H_2 + 1/2\ O_2$ (Mars)	460	9	4,140
Methane	$CH_4 + 2\ O_2$	380	20	7,600
Ethane	$C_2H_6 + 7/2\ O_2$	365	24	8,760
Ethylene	$C_2H_4 + 3\ O_2$	373	31	11,563
Acetylene	$C_2H_2 + 5/2\ O_2$	410	53	21,730
Methanol	$CH_3OH + 3/2\ O_2$	315	20	6,300
Ethanol	$C_2H_5OH + 3\ O_2$	330	24	7,920
Carbon Monoxide	$CO + 1/2\ O_2$	259	--	-------

[a] After Hepp et al., 1991.
[b] Isp values Zubrin
[c] Physical Mass Leverage = Mass Fuel/Mass Hydrogen
[d] Effective Isp of the Earth-derived H2 when leveraged by the Mars-derived components (Leverage x Isp).

REVERSE WATER-GAS SHIFT (RWGS) AS AN EXAMPLE OF AN INTEGRATED SYSTEM

The reverse water-gas shift (RWGS) reaction coupled with water electrolysis and ethylene production can be configured to produce oxygen from Mars atmospheric carbon dioxide using a continuous flow process that operates at moderate temperatures and pressures and is easily scalable (Zubrin, Meyer, McMillen 1997) (figure 4). An adsorption-based separator is recommended for separating out the component gases and filtering the aerosols from the Mars atmosphere. The reason for this is that at the low temperatures of the Martian night, CO2, loading can approach theoretical limits for certain adsorbents. Operating on the change in temperature from day to night, this diurnal cycle provides an effective optimization for use of adsorbent mass (Flynn, McKay, Sridhar, 1996). Additionally, under the low-pressure conditions of Mars, adsorption pumps are often able to produce pressure ratios much larger than single-stage mechanical compressors. They use fewer moving parts and are more easily scaled to small dimensions than are mechanical methods.

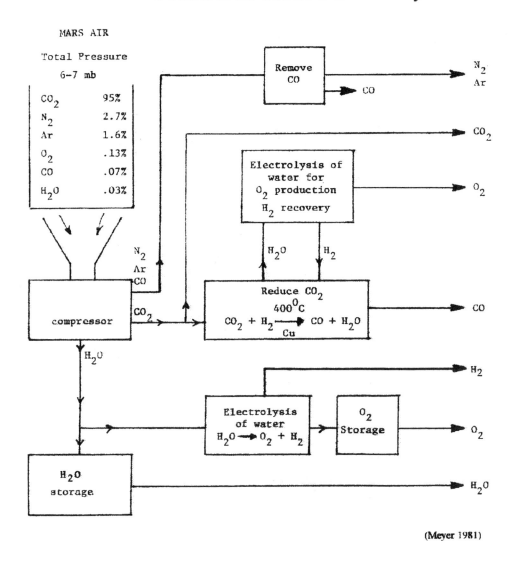

Figure 2   The Gas Extractor (Meyer, 1981).

The feedstock for this process is ubiquitous carbon dioxide and an initial charge of imported hydrogen. Since hydrogen is not consumed and can be recycled, this makes possible the design of a continuous flow oxygen production system where the only other feedstock requirement is for additional hydrogen (or water) to compensate for system losses.

Figure 3   Carter Emmart's Concept of the Gas Extractor (Meyer, 1981).

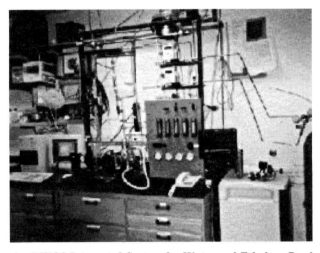

Figure 4   RWGS Integrated System for Water and Ethylene Production.

Table 3
MATERIALS THAT CAN BE MADE USING MARS AIR

Material	Representative Process
CO	Direct or liquefaction
N	Liquefaction, fractional distillation
Ar	Liquefaction, fractional distillation
H_2O	Dehumidification of Mars air
O_2	Reduction of CO_2, Sabatier
CO	Reduction of CO_2, Sabatier
C	Reduction of CO_2, Bosch
H_2	Electrolysis of H_2O
H_2O_2	Auto-oxidation, electrolysis
NH_3	Electrosynthesis
N_2H_4	Raschig
HNO_3	Oswald
NO	Reduction of HNO_3, other
NO	Oxidation of NO
N_2O_2	Polymerization of NO_2
HCOOH	Electrochemical reduction of CO_2
CH_4	Catalytic hydrogenation of CO
CH_3OH	Catalytic hydrogenation of CO

* for a summary of higher hydrocarbons manufacturable on Mars, see Brian Frankie, this volume.

CONCLUSION

On Mars your footprints are forever. Atmospheric refineries will have less impact on the Martian environment than surface or subsurface mining. Optimal use of Martian natural resources is assumed during the atmospheric refining process in which carbon dioxide is filtered from the air. All carbon that is taken out of the atmosphere is returned upon combustion or upon reoxidation of the carbon monoxide. The CO is recycled inside the catalytic reactor to produce fuel or propellant. So our net usage is zero. Nothing is removed from the environment.

An integrated ISRU Refinery offers redundancy, energy savings, conservation of feedstock, and recycling of waste products: heat, materials, and mechanical energy. Examples: energy savings - RWGS can use waste heat from SE, conservation of feedstock - waste CO from RWGS is feedstock for In-Situ Propellant Production (ISPP). Ways to improve the efficiency of various processes by integrating them into a Mars Base Refinery include using waste heat from one process to supply another efficiency is improved by using the waste product form one process as feedstock in another (figure 5). Likewise, recovery of mechanical energy from the high-pressure exhaust of one process enables optimization of mechanical efficiency that improves the efficiency of the entire refinery system.

Future - ISRU is the basis for self-sufficiency, and the foundation for future expansion and eventual colonies. On Earth, the atmospheric refinery is an experimental testbed for design of energy efficient integrated systems with industrial applications and in extreme environmental outposts such as the Antarctic, Arctic or desert communities.

REFERENCES

1. Ash, R. L., Dowler, W. L., Varsi, G., 1978. "Feasibility of Rocket Propellant Production on Mars." *Acta Astronautica*. 5: 705-724.
2. Biological Contamination of Mars: Issues and Recommendations. 1992. Space Studies Board, National Research Council. National Academy Press, Washington, D.C.
3. Clark, B. C., 1979. "The Viking Results - The Case for Man on Mars." AAS 78-156, in *The Future United States Space Program*, R. S. Johnson et al., eds., AAS *Adv. Astronaut. Sci.*, 38-I: 263-278.
4. Clark, B. C., J. L. Kalkstein, S. Meyer., 1991. "The Case for the Methanol Powered Planetary Rover." 42nd Congress of the Intl. Astronautical Federation, Oct. 5-11, 1991, Montreal Canada.
5. Finn, J. E., Sridhar, K. R., McKay, C. P., 1996. "Utilization of Martian Atmosphere Constituents by Temperature-Swing Adsorption." J.B.I.S. 49:423-430. Also published as AAS 97-360, in *From Imagination to Reality: Mars Exploration Studies of the Journal of the British Interplanetary Society* (Part I: Precursors and Early Piloted Exploration Missions), ed. R. M. Zubrin, Vol. 91, AAS *Science and Technology Series*, 1997, pp. 127-142.
6. Finn, J. E., McKay, C. P., Sridhar, K. R., 1996. "Martian Atmospheric Utilization by Temperature-Swing Adsorption." 26th Intl. Conf. on Environmental Systems, Monterey, CA July 8-11, 1996. SAE #961597.
7. Frankie, Brian. "Production of Higher Hydrocarbons on Mars." This Volume.
8. Haberle, Robert. 1986. "The Climate of Mars." *Scientific American*, 254(5): 54-63.
9. Halmann, M. M., 1993. *Chemical Fixation of Carbon Dioxide: Methods for Recycling CO_2 into Useful Products*. CRC Press, Boca Raton, FL.
10. Hepp, A. F., G. Landis and D. Linne. 1991. Material processing with hydrogen and carbon monoxide on Mars. NASA Tech Memorandum 104405.
11. Lee, J. M., Kittel, P., Timmerhaus, K. D., Radebaugh, R., 1995. "Steady Secondary Momentum and Enthalpy Streaming in the Pulse Tube Refrigerator." In *Cryocoolers 8*. R. G. Ross, Jr., Ed., Plenum Press, NY. 359-369.
12. Mars Environmental Document, NASA document, JSC. 1988.
13. McMillen, K. R., Meyer, T. R., Clark, B. C., 1999. "Methanol as a Dual-use Fuel for Earth and Mars." In, *The Case for Mars V*, ed. P. J. Boston. Vol. 97, AAS *Science and Technology Series*, Univelt, Inc., San Diego, CA.
14. Meyer, T. R., 1981. "Extraction of Martian Resources for a Manned Research Station." *J. British Interplanetary Soc.* 34:285-288.
15. Meyer, T. R., McKay, C. P., 1981. "The Atmosphere of Mars: Resources for the Exploration and Settlement of Mars." AAS 81-244, in *The Case for Mars I*, ed. P. J. Boston. Vol. 57, AAS *Science and Technology Series*, Univelt, Inc., San Diego, CA. 209-232.
16. Meyer, T. R., McKay, C. P., 1989. "The Resources of Mars for Human Settlement." *J. British Interplanetary Soc.* 42:147-160. Also published as AAS 97-373, in *From Imagination to Reality: Mars Exploration Studies of the Journal of the British Interplanetary Society* (Part II: Base Building, Colonization and Terraformation), ed. R. M. Zubrin, Vol. 92, AAS *Science and Technology Series*, 1997, pp. 3-29.
17. Meyer, T. R., McKay, C. P., 1995. "Using the Resources of Mars for Human Settlement." AAS 89-489, in *Strategies for the Exploration of Mars: A Guide for Human Exploration*, eds. C. R. Stoker, C. Emmart. Vol. 86, AAS *Science and Technology Series*, Univelt, Inc., San Diego, CA. 393-442.
18. Zubrin, R., T. R. Meyer, K. R. McMillen. 1998. Mars In-Situ Resource Utilization Research: MIRUR. General Analysis of the Reverse Water-Gas Shift Reaction and Scaling Relationships for Mars ISSP Systems. Report to NASA JSC, Jan. 1998. ACRP NCC9-45.
19. Zubrin, R., T. R. Meyer, K. R. McMillen. 1998. "Mars In-Situ Resource Utilization: Reverse Water-Gas Shift Reaction." Space 98 Conference, Albuquerque, NM April 30, 1998.
20. Zubrin, R., Baker, D., Gwynne, O. 1991. "Mars Direct: A Simple, Robust and Cost-Effective Architecture for the Space Exploration Initiative." 29th Aerospace Sciences Meeting, Reno, NV Jan. 1991. Also published as AAS 90-168, in *The Case for Mars IV: The International Exploration of Mars* (Mission Strategy and Architectures), ed. T. R. Meyer, Vol. 89, AAS *Science and Technology Series*, 1997, pp. 275-314. Presented in summary in R. Zubrin, D. Baker, "Humans to Mars in 1999," Aerospace America, Aug. 1990. AIAA-91-0326.

21. Zubrin, R. Price, S., 1995. Mars Sample Return Utilizing In-Situ Propellant Production. Lockheed Martin Marietta Final Report on NASA JSC Contract NAS-9-19359, March 1995.
22. Zubrin, R. Price, S., Mason, L., 1994. "Report on the Construction and Operation of a Mars In-Situ Propellant Plant." 30th AIAA/ASME Joint Propulsion Conference, Indianapolis, IN, June 1994. AIAA-94-2844.
23. Zubrin, R. Price, S., Mason, L., 1995. Mars Sample Return With In-Situ Resource Utilization: An End-to-End Demonstration of a Full Scale Mars In-Situ Propellant Production Unit. Lockheed Martin Marietta Final Report to NASA JSC Jan. 1995. NAS-9-19145.

Chapter 14
HUMAN FACTORS

Shuttle astronaut Scott Horowitz gives the Pilot's view of human Mars exploration.

Johnson Space Center's John Connolly presents NASA's concepts for human Mars missions.

COPING WITH EFFECTS OF ENFORCED INTIMACY ON LONG DURATION SPACE FLIGHT[*]

Lara Battles[†]

Mars awaits us. By public or private means, humankind is going to Mars. The technology necessary to get there is not the problem; it exists today. There is considerable discussion as to which technologies will work best and provide for the safety of the crew when balanced against the cost of the mission.

No matter how we approach the challenge of "getting there," doing so will take a long time. To justify expense, research time on Mars must be allowed, and a return trip will take into consideration optimal timing for the launch. All this amounts to an extended period during which the first Mars crews will come to know each other, undoubtedly, better than they might wish. Dealing with intense relationships in close quarters for extended periods will require much forebearance of these spacefarers, and the success of these missions, the quality of their research, even the development and survival of neophyte Martian culture will be permanently influenced by the personality, gender, and psychological well-being of their crew members.

Historically, Earth-based ventures of discovery of heroic proportions involved short-term efforts with built-in redundancy (the Columbus model) in which an intensive effort would be made to accomplish a particular goal with one grand effort. When it was successful, a relatively short burst of stressful, goal-directed activity would be rewarded with the accomplishment of a specific goal and promises of riches and status upon return home. The longer voyages and ventures which succeeded allowed for short bursts of intensive effort followed by periods of tranquility, refreshment, and socializing, as in Cook's voyages of the Pacific. Island visits broke the monotony and fearfulness of the long voyages of discovery.

The opening of the West in our own country was largely reliant on the military model, which to a fair degree describes our present space program. Traditional Army discipline shaped itself around the exigencies of the wilderness landscape, and successful units valued skills and knowledge of the land and its inhabitants more than pure chain of command. While shipboard life in a hostile environment requires a clear chain of command, well-drilled survival routines, and reliable discipline, long-term, high-level maintenance of that discipline is prone to the minor mutinies of fatigue, disinterest, and boredom. As in the Old West, enthusiasm for spit and polish appears to vary with the inverse of the distance from the mission's control. The acuity of discipline is renewed following downtime, but with the route to Mars painfully short of tropical islands or friendly natives; the stressbusters of the Martian crew must be brought along or incorporated into the design of the voyage.

---

[*] Copyright, Lara Battles, 1998. All Rights Reserved.
[†] Lara Battles, M.A., M.F.T., 405 East Branch Street, Suite H, Arroyo Grande, California 93420.

With projected mission lengths running from eighteen months to three years, traditional crews of single, "Top-Gun," male or male-acting astronauts may not be the best choice. Certainly all those heading for Mars will have optimal and redundant training; the best the mission's financiers can obtain. The nature of long term space flight itself dictates what will make for crew compatibility for the long haul. The physical, emotional, and social side effects of space flight demand special concessions to the ship's environment in order to provide adequately for mission personnel's comfort and composure. The expenses of providing for genuine crew member comfort ought not be the target for economizing. This extraordinary first group of extraterrestrial human settlers deserves the best resources. While much of what makes for group comfort will have to do with shipboard amenities, some of it will necessarily depend on the makeup of the crew, itself.

The impacts of long duration space flight are addressed in depth in other resources (Connors, Harrison & Akins, 1985; Harrison, Clearwater & McKay, 1990), but a consensus exists that psychological stressors are continuous and significant. The primate's reaction to the unknown, fear, is a constant companion of the crew, and while it may be moderated by good personal intrapsychic hygiene, by astronautic competency, by distractions, and even denial, still the continual presence of the high stakes for any error tends to wear on personnel. Historically, NASA has kept its crews busy far longer than an eight hour workday, but this strategy, if employed on a Mars mission, might create work stresses and resentment due to fatigue. Indeed, an "Us vs. Them" mentality has been found to develop during longer missions, with Mission Control being perceived as unsympathetic or even out of touch with the real issues of shipboard life.

Crowded living space is always an issue because of liftoff cost considerations. Designers of ships, shuttles, and space stations aim for maximal crew comfort in a minimum of space, not always to great effect. This issue will likely be exacerbated on a Mars mission because of its duration. If "Cabin Fever" may be experienced by persons shut in only by the snowy winters here on Earth; how much more might it affect a tightly packed crew whose only relief from the crowded quarters is a spacewalk? From this perspective, one can empathize with those who have succumbed to the "Call of Space," and attempted to become "one with the universe" by going out there unprotected.

The machinery that makes space livable for humans by recycling air and wastes does not do so silently. Noise and noisome aromas must be tolerated where there are no alternatives. These contribute to a generally stressed state common to spacefarers. Further, once on course, the long voyage may take its toll in boredom, loneliness, and isolation. And if the voyage goes well, it *will* be boring.

Enforced comradery and a militaristic chain of command tend to promote social withdrawal. While withdrawing permits the individual to refresh his sense of self and personal boundaries, loneliness may be the outcome. Separation from loved ones is likely to be felt more acutely. As shipmates grow more aware of their disengagement from home life and the social roles that maintained their sense of personal identify and worth, self esteem tends to fall. This separation from family and friends can exacerbate to a generalized discomfort and estrangement...the edge of an anxiety disorder or the anomie of the refugee.

Traditional solutions for these psychological and social stressors of space flight have involved keeping the astronauts extremely busy (perhaps to prevent having time for reflective thought) and exercising. The exercising is certainly a necessary part of any mission that involves weightlessness, and it helps with stress reduction. However, for the very long duration flight, it may be wise to pack a healthy supply of antidepressant medications...checked out preflight for tolerance...to minimize the unavoidable effects of reduced mobility and crowding over the long run. The suggestion that full spectrum lighting be used to make up for loss of exposure to sunlight (and thus prevent incidental development of seasonal affective disorder depression caused by lack of exposure to sunlight has been addressed by Harrison (Harrison, 1990).

Privacy, then, must be seen as more important in space than on Earth because it must substitute for getting off ship. Ship structure has historically limited or prohibited private socializing, while shipmates complain of an overwhelming feeling of not being able to get away from crewmates or off the ship. Where it is not possible to get away in reality, one might effectively do so virtually. It would appear an economy to provide crew with the best personal computers can offer in leisure-supporting software. Ideally, such programs would encourage in-depth use of the creative mind (to replicate the state of "deep play" found in children fully engaged in imaginary play). This kind of recreation would permit shipmates to refresh themselves and get away from it all within their own quarters. The use of large, wall-mounted screens and virtual reality may further enhance the sense of "getting away from it all." New software might even be streamed in should the need arise. Virtual reality programs ought, also, to become a standard to help with the boring prospect of three hours of daily exercising. Ordinary avocations such as journalling, letter writing, playing musical instruments, pursuing space-sparing hobbies, and using a large CD library are to be encouraged. "Outdoor sports" of stargazing and spacewalking may help when the sense of overwhelming togetherness is extreme, but must be limited due to radiation effects.

Choosing the right crew for such a mission is critical. "Right Stuff" competencies are essential, but social skills, sensitivity, creativity, and cooperation—regardless of rank or chain of command—ultimately will set the culture for shipboard life. Ideal contenders would have social characteristics of high sociability and liking for people, good manners, playfulness, honesty, tolerance, a strong work ethic, and the appropriate use of power. Intrapsychic qualities of resilience, creativity, empathy, cheerfulness, high "emotional intelligence," independent thinking, and the ability to endure extended inconvenience will all be desirable. Communications skills which are engaged persistently and patiently to solve the toughest problems are vital, because in so small a space, rancor cannot be permitted to take root. Strong candidates would also be capable of daydreaming and passing time without supportive stimulus. Crews report that sharing empathetic humor makes a great difference in the quality of the flight, as does the absence of seriously annoying habits, and the ability to sleep without snoring loudly. Crew cohesiveness is enhanced by shared interests and passions. Pre-flight shakedowns or extended practice solving problems together help identify and resolve potential interpersonal trouble zones. The matching of crew members' personalities on a long flight is of much greater importance than it was on the Apollo missions. Last minute substitutions of personnel would be ill-advised.

Space manners in general would tend to the conservative: one must live with the consequences of everything one does, as well as what one's shipmates do. The basics of clear communication as well as a consensually shared set of rules or manners will be necessary, and spelling it out may be critical if crews do not come from the same cultures on Earth. Issues that will need regulating through social manners and safety regulations will include those affecting noise, smell, use of air, water, and interior space of the ship. Social issues are apt to arise if failings occur with regard to the management of tempers, or even the "emotional bolts" of the unspoken feelings. A profound respect for privacy is the cornerstone of shipboard harmony. Irritating habits might be of no concern on Earth would potentially need special handling and intervention to prevent ongoing animosities. Always a question will be the use of alcohol or other intoxicants. It would seem churlish to deny the occasional use of mild intoxicants, while it would be folly to permit their abuse.

Many of the negative social effects of long duration space flight appear to be those that affect any group living in crowded conditions and isolation. It is interesting to note that Shannon Lucid's stay on the Mir station was characterized as quite harmonic, and she, herself speaks of the importance of compatible shipmates (Lucid, 1998). One might view her gender as a strong contributing factor to the social compatibility of the flight. Male-only groups tend to be more competitive and physical while female-only groups tend to be more social and nurturing, but also occasionally ribald. Mixed gender groups tend to bring out the most polite behavior in both males and females, particularly where pair-bonding is clear and firm. While there were no pairings on Ms. Lucid's flight, the presence of a mature female may well have been the stabilizing effect the made for so socially agreeable a mission. On an earlier mission of exploration, Sacajawea's presence smoothed the way for Lewis and Clark, and created a different kind of comradery among the explorers. It was, and is, the woman's touch.

What place, then, does sex have in long duration space flight? One might as well ask what place it has on Earth. While sex, viewed in a narrow sense, is the vehicle for reproduction, it is not in this light that it is considered here; with space radiation and microgravity effects on the developing fetus unknown, it would be recklessly unethical to conceive a child in spaceflight at this point in our scientific development.

Sex, however, has many other beneficial effects. For compatible, bonded couples, it's a normal activity which is generally considered to be enjoyable. Sex is a basic human need, and one that has been heretofore politely ignored in the realm of space flight. Skin hunger and sexual deprivation are among the primary complaints of pair bonded astronauts who fly without their wives, and this very lack is one of the primary causes of loneliness and depression. Emotional intimacy is enhanced through sexual contact. Affectional needs are met, and the biochemical side effect of the orgasmic experience is a diminishment of depression and of obsessing, and a general release of "feel good" and psychological bonding chemicals, thus enhancing couple and crew cohesion.. It reduces stress and tension. It provides a basis for a good deal of humor. Sex by oneself is simply not the same, nor nearly as effective in combating the negative psychological side effects of space flight, as it is with one's partner.

Including women in the crew, or, more aptly, pair-bonded couples (and, in the USA, married ones) just makes sense. An all-male or all-female crew is not representative of how humans prefer to live in the long term. If Mars settlement is to be normalized, it will need to include women as well as men, and the culture established there will effectively be set by the culture imported on the initial settlement and exploration ships. To have a balanced culture, one must provide that balance.

Clearly, issues of jealousy and loyalty come with sex. For this reason, it is suggested that older couples initially form the crews, optimally with the females past the age of menses...both because of the elimination of the issue of pregnancy and because older couples tend to have worked out their concerns of commitment. Couples tend to provide strong emotional resources for each other, being able to confide in each other and foil one another's moods. At worst, they can fight with each other instead of the rest of the crew, trusting the relationship to endure when a less tested crew-based relationship might not. Besides, they can make up.

Microgravity effects may, ultimately, dictate many of the answers to the question of sex in space. There is evidence (Grindeland, 1998) that in microgravity the leuteinizing hormone production is reduced, affecting the production of testosterone in males and reducing it. If a similar reduction is seen in females, even long term pair-bonded couples may have less interest in sex, per se. There is also a reduction of the immune system, and in females, and this may permit ordinary bacteria inhabiting the female's sexual tract to develop a nasty attitude. This would certainly diminish one's interest in contact, but the affectional aspects of having one's partner along...and the desire to protect and help that partner, may yet have a strong salient effect on crew morale and mission success.

Of a less captivating but more primitive nature is the possibility of using scent as a stimulus for mood management and anomie reduction. Air on space stations is known to be stale, verging on rank. While the human nose is not our best sense organ, and, fortunately, it will usually give up sending us the same signal over the long haul, the major complaint from spacefarers is not unpleasant aroma, but the lack of novel aroma. In particular, the loss of the "smell of green" has troubled Russians on the Mir. Scent can evoke powerful memories as well as cue the mind to alertness, cheerfulness or even irritability. In the author's personal experience, the scent of mildew is extremely irritating, but it is the remnants of this scent that is so hard to clear from a closed air system. While it would be inappropriate to release "better" scents into the ship's air supply in a wholesale fashion, judicious use of fast-decaying scent packets may help regularize the environment by reducing monotony and creating a welcoming, supportive, and pleasant setting. Perhaps the low level irritability that comes of breathing the same old air could be relieved, temporarily, by the personal use of scent inhalers that really are the breath of springtime.

Adding women to the crew and improving the ship's aroma are small enhancements when one considers the overall trip to Mars. One hundred years from now, when commercial spaceliners will be transmitting families and businessmen and women between Earth and Mars, staff positions of ship's recreation leader, counselor, and possibly even ship's fool will make the long transit more endurable, enjoyable, and healthy. For our intrepid Mars crew, including in the crew a member trained to intervene when crew members clash without resolution may be a critical factor in preventing a mission failure. The ship's counselor could watch for isolating behaviors, signs of depression and intervene with counsel or psychotherapy, and recommend psychotropic medications when appropriate. In a secondary role as ship's recreation leader, this mission specialist would initiate social cohesiveness activities for recreation (song, drama, dance, acrobatics, general amusement) and for problem solving (crew meetings, group problem solving). And lastly, as ship's fool, s/he could encourage and cajole crew members to perform the mandatory exercises upon which their physical and mental health depends.

Space may be the final frontier, but humans have just started addressing how to approach it. Normalizing our efforts is critical if the cultures we export from Earth are to have the elements necessary for balance and renewal. Including women is a vital step in that direction, and acknowledging our nature as sexual beings is another. Moving away from scientific and military asceticism to a respect for our human need for a culture which includes history, art, individual pursuits and novelty will go a long way towards creating the spacefaring culture that will be able to travel not only to the reaches of our solar system, but beyond.

REFERENCES

Connors, M. M., Harrison, A. A., and Akins, F. R.: *Living Aloft, Human Requirements for Extended Spaceflight.* NASA SP-483. Washington, D.C.: National Aeronautics and Space Administration, 1985.

Grindeland, Richard E., "Microgravity Effects," speech given at Space Down to Earth conference, NASA-Ames Research Center, Moffett Field, CA, August 5, 1998.

Harrison, A. A., Marsflight Human Factors: The Martin Marietta Manned Mars Mission Study, unpublished report, 1990.

Harrison, A. A., Clearwater, Y. A., McKay, C. P., Eds.: *From Antarctica to Outer Space: Life in Isolation and Confinement.* NY: Springer-Verlag, 1991.

Lucid, Shannon W.; "Six Months on Mir", *Scientific American*, vol. 278, no. 5, 1998, pp. 46-55.

ON OUR BEST BEHAVIOR: OPTIMIZATION OF GROUP FUNCTIONING ON THE EARLY MARS MISSIONS

Vadim I. Gushin[*] and Marilyn Dudley-Rowley[†]

It is something of an irony that a planet named for the God of War challenges mortal humans to psychosocial behaviors which may far surpass what is seen normally. Mars cannot be explored without a human presence on that planet. There is no robotic substitute for processing complicated information, making hard decisions, and solving non-standard problems. Many phenomena will be unknown and will need the intuition and the multiple perspectives of human cognition to discern them.

With a manned mission, come all the foibles of what it means to be human. Unfortunately, dysfunctional acts and events could jeopardize the autonomous, long-duration missions as Mars exploration requires. Space and other extraterrestrial environments require of people behaviors closer to zero-tolerance for deviance. This unavoidable fact calls for optimization of group functioning.

Understanding optimization calls for comprehending the occurrences and frequencies of deviance in extreme environments among teams living and working there; for baselines of optimal standards of performance still yet to be drawn; and for the diagnosis, prevention, and correction of less-than-optimal behaviors, not only on an individual basis, but especially on a group basis. Several perspectives and methods are discussed. Finally, it is contended, optimizing group functioning for the "first Martians" could lead to a higher order of behaviors conducive to international cooperation on Earth. In the conquest of Mars, we conquer ourselves.

Much about long-duration spaceflight is unknown, and the journey to Mars is *the* mission which is on the minds of all who concern themselves with flights of long-duration.

In spite of the American space program's success to date with unmanned probes, Mars cannot be properly explored without a human presence on that planet. There is no robotic substitute for processing complicated information, making hard decisions, and solving non-standard problems. A Mars flight would involve complicated situations with many variables; decisions which need to be made quickly; the operation of a number of complicated systems; and the need to maintain the reliability of those systems. Moreover, many phenomena encountered during the mission will be unknown and will need the intuition and the multiple perspectives of human cognition to discern them.

With a manned mission to Mars, come all the foibles of what it means to be human. Added to this, functions will be performed under increased information stress and responsibility due to a high degree of autonomy of flight. Unfortunately, deviant or dysfunctional acts and events could jeopardize the autonomous, long-duration missions as Mars exploration requires. They will exacerbate the detrimental effects of spaceflight on humans and could lead to disas-

---

[*] Institute for Biomedical Problems, Moscow, Russia.
[†] OPS-Alaska, 2664 Montana Road, Fairbanks, Alaska 99709.

ter. Even a decision which is somewhat less than optimal could result in a failure to fulfill the mission program.

Comprehending the occurrences and frequencies of these acts is essential to optimizing crew training and safety; making inferences about long-term individual and group responses to extreme environments; identifying critical factors and underlying mechanisms affecting those responses; assessing the psychosocial contributions to the optimal human habitability of space; designing equipment and systems; and providing insight into conflict between human capabilities and system engineering methodologies which can inform spacecraft design, mission planning, and related group operations, and lead to new processes and procedures.

Two traditions have emerged corresponding to the American and Russian space programs, respectively. The American approach essentially has been to ignore psychosocial concerns, and indeed most of the spectrum of human factors issues. This arose from the American focus on short-term flight. The Russian approach was to study psychosocial concerns substantially under the rubric of biomedical support to the Russian space program. This was in response to the Russian focus on long-term flight. Social and behavioral scientists have not taken the American tradition lying down, and have been crying "folly" for years on this issue which is even apparent to engineers. "Mad" Don Arabian, Director, Mission Evaluation Room, Johnson Space Center, defined every component of a spacecraft in terms of systems, even the astronauts (Lovell and Kluger 1994, pp. 287-288). To borrow the analogy, it can be said that NASA has neglected the systems which are human. Three and a half decades after the American entrée into space, this is still true, although a turnabout is in the making (Dudley-Rowley and Patrick Nolan 1997, pp. C3-C4).

One clear example are the various life support test projects and facilities operating under the Advanced Life Support program at Johnson Space Center (Maass 1998). In the meantime, a Russian team at the Institute for Biomedical Problems in Moscow are pressing ahead with a high-fidelity International Space Station (ISS) Simulation. Unlike the American experiments which focus more on biological recycling in a contained environment, the Russian study examines all biomedical concerns including how Simulation team-mates who comprise multinational crews interact as they go about working and living in containment. The ISS program is an important rehearsal for the Mars venture. International Space Station Simulation project scientists have considered the requirements of a Mars mission. The most important factor of biomedical support on such a mission is ensuring "reliability in performance of each cosmonaut and of the whole crew at all mission stages", with timely and accurate fulfillment of all components of the flight program (Ilyin, Kholin, Gushin, and Ivanovsky 1992).

The Russian work aside, much still needs to be done. While people are certainly capable of adapting to a variety of extreme, isolated environments for long periods of time, physical, psychological, social, and cultural research still needs to be conducted to further facilitate human adaptation to hostile, confined, and isolated conditions, because how people adjust and adapt to these conditions can affect not only their mental health and social cohesion but also their performance of assigned duties (Levesque 1991). For, after all, "duties" on both a space station and on a Mars mission will be much more complex than on short duration flights where many functions were or could be handled by ground controllers. The physical infrastructures of these long-duration ventures require built-in, on-board diagnostic capabilities wired into the spacecraft and equipment and not to the ground. This will require of the crew a level of maintenance, component replacement, redundancies, and techniques that have not been invented yet (McCurdy 1993, pp. 153-154).

The main concern in any manned space mission is condition of the craft and life support. Unlike some other extreme environmental settings here on Earth, where a certain amount of time may be available to remedy a situation, damage to life support systems in space is an immediate concern, i.e., breachment of environmental containment. Some degree of human dysfunction is tolerated and even expected in terrestrial analogs, but space and other extraterrestrial environments require of us behaviors closer to zero-tolerance for deviance (Dudley-Rowley and Nolan 1997, p. C4). Similarly, Richard Jennings, former Chief of Flight Medicine at Johnson Space Center, has called for "fastidiousness in high-risk environments (Jennings 1997)."

One way to optimize group functioning of Mars crews is to promote active use of self-control and self-regulation methods among individual members (Ilyin et al. 1992, p. 271). These methods alone will not suffice, however. Diagnosis, prevention, and correction must not only be performed on an individual basis, but on a group basis as well. "[The] success in the mission will depend not only on the professional skill and motivation of each crewmember but to the same (and probably, greater) extent on the quality of their joint activity (p. 279)." Former NASA psychiatrist, Patricia Santy, has said that crewmembers must be trained to comprehend the role of social interaction. "Crew psychological training for the Mars venture must allow the crew an understanding of group dynamics; they must learn how to respond to stress and prevent its impact on performance; they must know the impact of leadership style; and to know how to cope appropriately with interpersonal tensions (Santy 1997)." Russian researchers add that training measures need also include practicing group interaction in extraordinary situations (Ilyin et al., p. 278). This even may not be enough to achieve a high degree of group functioning. From his experience with the Biosphere II project, Roy Walford remarked that the breakdown in group cohesion which occurred in containment came as a surprise because of a great deal of prior teamwork focusing on extreme circumstances (Walford 1997). Other Biosphere II crew have made similar observations, how group cohesion disintegrated in containment after having literally held the lives of crewmates in one's hand in training experiences preceding enclosure (Sheddan 1995). A deep water caving expedition is another case of interest to space researchers where serious interpersonal divisions occurred. The expedition's team had trained together for four years without any group dysfunction. In fact, the team leader was an astronaut candidate program finalist who had not been selected upon further consideration (Bishop 1997).

It is becoming clear that along with prior training of any kind, Mars crews must have a system of monitoring and diagnosis on board to determine psychophysiological status, physical, professional, and psychological performance; and to be able to correct any undesirable changes (Ilyin et al. 1992, p. 277). An on-board multi-functional Complex, doubling as a medical and physical training center, could provide this system. Instrumentation could generate tests to determine psychological, physiological, and professional skills fitness of operators (p. 278). Results could be compared against a computer-generated baseline.

A problem presents itself at the outset and Patricia Santy has phrased the question succinctly: "What performance standards will be allowed before intervention is made (Santy 1997)?" This is part of the dilemma in the study of psychosocial aspects of long-duration spaceflight. We barely have begun to move beyond the results of a number of containment and prolonged bed-rest experiments (Gushin and Efimov). Results of these studies seem to maintain that interventions can be made based on a dynamic diagnosis of the functional state, current results, and crewmembers' styles (Ilyin et al. 1992, p. 272). Results are also beginning to lend themselves to quantification and point to a second generation of rigorous studies. This can help us in the realm of diagnosis, where a psychosocial index or indices would be desirable.

The authors, along with our experimental psychologist colleague, Tom Gorry, are seeking such an index. We propose to call this instrument, the Altman Scale, after Irwin Altman, the social psychologist who first conceptualized such an instrument. We will videotape the interactions of the multinational crews in the International Space Station Simulation in Moscow next year. Our overarching twin ideas are to define a baseline about group functioning in space and space-like environments in order to define social states; and to develop the instrument to predict experiences groups have in those environments, for application as a measure of team social states, so that undesirable social states may be avoided and optimal ones maintained. Special foci are: 1) how the physical environment is actively used to cope with interpersonal compatibility and incompatibility and 2) precursor events of off-nominal, sociological deviant functioning. Acts and events which indicate off-nominal, sociological deviant functioning of crew have already been determined from expeditionary records (Dudley-Rowley and Nolan 1997, p. C6; Dudley-Rowley 1997). These are: 1) Unusual, bizarre, or puzzling behaviors (such as withdrawal and life-threatening behaviors); II) Acts of aggression (verbal and physical); and III)Acts of deliberation (such as resources theft, hoarding, or hogging resources, not doing one's work, and violating safety rules).

We will observe chronological trajectories and distributions of group functioning along six dimensions which may be couched as testable hypotheses. These are:

1. As mission duration increases, team members use more communicative modes corresponding to their developing more ways to convey information about how they feel to one another as well as cognitive information; and their increasing self-disclosure and addressing personal topical areas without the need to probe or with any hesitancy.

2. As mission duration increases, the more unique or idiosyncratic to the team their communications become.

3. As mission duration increases, the more efficient the team's communications become in terms of speed and back-and-forth understanding.

4. As mission duration increases, fewer communicative cues need to be exchanged to synchronize activities.

5. As mission duration increases, team members increase their movement in and out of discussion areas in facile ways, their exhibitions of intimate, touching gestures and body positions, their exchange patterns, and their movement into others' personal space without discomfort and much preamble.

6. As mission duration increases, team members increase their expressed judgements and their expression of feelings about one another.

We will videograph from the beginning of the ISS Simulation to the end of it. From this record, we will draw quantifiable data from random and systematic samples. Investigators' expectations are that communicative modes, unique communication, informalities, and evaluations will increase over time; and that queries for elaborative explanations and verbiage relating to synchronous activities will decrease over time. We will chart curves for each dimension over time. If the expected increases and reductions in the dimensions are not happening as expected as mission duration increases, it may be correlated with psychosocially off-nominal behaviors, and this is a clue of how off-nominal the social state is at any given time along the chronological trajectory. As dimensional increases and reductions run counter to expectations, there is probably going to be a correlation with increase in volume of unusual behaviors, aggressive actions, and deliberate negative acts. Each of the Altman dimensions will be graphed on profile using stanines. The acceptable degree of variation for each dimension can only be assessed after the Simulation is run and gauged for further robustness by looking at records from prior simulations or through use of the Altman Scale in analogs and actual flights. We expect our study over the course of the ISS Simulation to provide a level of quantitative rigor not yet available in the area of psychosocial aspects of long-duration missions.

In conclusion, we would like to point out that long-duration flights, like Mars missions, are more than just groups of people flying around in space. They are emergent societies. Roy Walford has remarked that isolated groups resemble primitive societies, not quasi-military bases (Walford 1997). This is consistent with Dudley-Rowley's observation that contained environments have the dynamics of hunting-gathering societies (Dudley-Rowley 1995). Patricia Santy has said that the Mars crews will develop a separate culture fairly rapidly (Santy 1997). What a culture it will be, too! The missions to Mars undoubtedly will be multicultural. National space programs will have to work together much like the Russian and American crews in Arthur C. Clarke's 2010 to save themselves on the dawn of a new relationship with the Cosmos. The peopling of Mars will rapidly draw humanity together in ways that diplomatic alliances have not. Optimizing group functioning for the "first Martians" could lead to a higher order of behaviors conducive to international cooperation on Earth. In the conquest of Mars, we conquer ourselves.

REFERENCES

Bishop, Sheryl. Sep 1997. Presentation at the Medicine on Mars Conference. Center for Advanced Space Studies, Houston, Texas.

Dudley-Rowley, Marilyn. 1995. "Modeling Social Interaction for the Nauvik Project, A Closed Ecological Life Support System," Fairbanks, Alaska. American Institute of Aeronautics and Astronautics Publication.

Dudley-Rowley, Marilyn. 1997. "Deviance Among Expeditioners: Defining the Off-Nominal Act Through Space and Polar Field Analogs," *Human Performance in Extreme Environments*, Vol. 2, No. 1: pp. 119-127.

Dudley-Rowley, Marilyn and Patrick Nolan. 1997. "Deviance Among Team Personnel in Space and Analog Polar Field Environments: the Effects of Size and Heterogeneity of Crew and Mission Duration on Behavior and Performance," a proposal to the Sociology Program, National Science Foundation.

Gushin, Vadim I. and Vladimir A. Efimov. Http://www.Geocities.com/CapeCanaveral/Launchpad/1033.

Ilyin, Eugene A., Sergei F. Kholin, Vadim I. Gushin, and Yuri R. Ivanovsky. 1992. "The Human Factor in a Manned Mars Mission," *Advanced Space Research*, Vol. 12: pp. 271-279.

Jennings, Richard. Sept. 1997. Presentation at the Medicine on Mars Conference. Center for Advanced Space Studies, Houston, Texas.

Levesque, Marc. 1991. "An Experiential Perspective on Conducting Social and Behavioral Research at Antarctic Research Stations." pp. 16-19 in *From Antarctica to Outer Space: Life in Isolation and Confinement*, edited by A. A. Harrison, Y. A. Clearwater, and C. P. McKay. New York: Springer-Verlag.

Lovell, Jim and Jeffrey Kluger. 1994. *Apollo 13 [Lost Moon: The Perilous Voyage of Apollo 13]*. New York: Pocket Books.

Maass, Peter. June 1998. "Mission to Mars, 2008?" *Wired*.

McCurdy, Howard. 1993. *Inside NASA: High Technology and Organizational Change in the U.S. Space Program*. Baltimore: Johns Hopkins University Press.

Santy, Patricia. 1997. Presentation at the Medicine on Mars Conference. Center for Advanced Space Studies, Houston, Texas.

Sheddan, Marylin K. 1995. "Role Changes During Long-Term Missions: An Anecdotal Assessment." American Institute of Aeronautics and Astronautics Publication.

Walford, Roy. Sep 1997. Keynote Presentation at the Medicine on Mars Conference. Center for Advanced Space Studies, Houston, Texas.

MARS MISSION OPERATIONS

Kenneth E. Peek

Considerable thought has been applied to the idea of going to Mars with regard to the hardware and the Astronauts/Cosmonauts who would make the journeys. However, not as much attention has been directed to how such voyages would be supported from the ground on the blue, not the red planet. Earthbound mission operations are the subject of this paper. It will present issues relating to ground support for manned Mars expeditions and address some of the concerns that will have to be overcome.

If NASA support is obtained, existing infrastructure can be utilized to varying degrees. However, conflicts with other programs, namely the International Space Station, Space Shuttle, and various JPL interplanetary missions should be expected. If a private venture is planned, this mission support apparatus must be created or leased. Using the NASA paragon, the list would include at a minimum:

- A Mission Control Center (MCC) including all supporting material and staffing
- The Deep Space Network (DSN) and/or Shuttle Tracking and Data Relay Satellite System (TDRSS) equivalent for voice/telemetry routing from the spacecraft and Martian ground sites to the Earth.
- Launch services and ground support equipment/personnel for the launch vehicles.

Each of the above items would bring a unique set of challenges to the mission. Some of these would include:

MCC

---

* The views expressed are those of the author and not necessarily those of United Space Alliance or NASA. The author is a Shuttle/ISS flight controller at the Johnson Space Center in Houston.

The MCC would be the nerve center of the flight on Earth. All information to and from the vehicle would nominally go through the control center for safety and efficiency reasons. This would necessitate redundancy in the control center for both personnel and equipment.

Specific flight operational parameters would include establishing a set chain of command in the flight control room, especially if the missions are international. Timely sharing of information between controllers and centers would be another issue that needs to be addressed pre-flight. The Shuttle/*Mir* program has had relatively good success with this, but there have been times during critical situations when information has not been disseminated in a timely manner. This situation must be avoided as much as possible. The International Space Station will be a good proving ground for additional ground control/commanding across several different control centers. If the flight were mostly an American mission, Shuttle protocol (which has evolved from programs such as *Skylab*, Apollo, etc.) would be an appropriate model.

Console support for the missions would parallel present day Mir or planned ISS experience. It may not be necessary to have the entire team in the control room for the duration of the flight. Indeed, complete staffing would not be desirable for personnel and cost considerations, especially during non-critical flight phases. Minimal staffing for telemetry and other critical positions would be required, as would an acting flight director. The rest of the team would be on standby and readily accessible if need arose. For major events such as launch, checkout, etc. the entire flight control team would be on-console. As the mission progressed, and the communications "lag" became more pronounced, the spacecraft crew would have to be more autonomous for real-time anomalies. The MCC teams would shift into a support role providing technical expertise for scientific data collected, chronic issues related to vehicle/crew performance, and other non-immediate concerns. For the return leg of the voyage, the ground support process would likely be reversed. It should be expected that spacecraft and crew performance would almost inevitably degrade the longer the mission went on, and that would necessitate increased support from the ground for problems the crew could not resolve.

Depending on the Mission goals and the Mars vehicle design, the number of positions in the control room would vary. Using the Shuttle program as a model, these would include a flight director, telemetry/communications, navigation, propulsion, environmental control, ground control (for MCC links and hardware) maintenance and others. Other positions such as a payload coordinator, robotics experts, etc. would be needed only if the mission is structured in such a way as to require those disciplines. There could also be new positions created or older ones from the Apollo era brought back for the entire mission or specific flight phases. A field geologist backroom would almost certainly be one of these "reincarnated" spots, whereas a "telepresence" operator (analogous to the one staffed for the *Mars Pathfinder* mission) may be needed for manned missions.

Although the Martian day is similar to Earth's (Mars' day is approximately 37 minutes longer), over an extended period of time this difference can be pronounced. For a crew in transit between Earth and Mars, a gradual adjustment can be phased in. For the ground crew, shifts and sleep cycles will vary to support the mission. This is currently done with many manned space flight missions to coincide with launch windows, orbital rendezvous and other mission dependent events.

Attrition of the control team during the flights will have to be addressed. For a long mission, it is almost inevitable that ground controllers will leave the mission for various reasons. That expertise will have to be replaced. Lessons learned from JPL interplanetary missions as well as *Skylab*, *Mir* and ISS should be expanded upon. This can be lessened somewhat with attention to morale. An espirit de corps should be reinforced. It is doubtful that there will be a shortage of qualified personnel willing to contribute their skills to this historical endeavor.

For MCC operations, issues pertaining to voice loops/protocol, a common language, measurement systems, etc. would need to be addressed pre-flight. Again, ISS experience should be used as a model here. English would be a good choice for a common language, due to the large number of applications in which it is already used (i.e. commercial aviation). If the spacecraft is largely US built, the English system of units should be used (due to industry in America not adopting the SI system as fast as the rest of the world) for issues relating to hardware. Other measurements for flight parameters such as distance traveled, velocity and others could be SI, depending on flight planning preferences. Regardless, conversion programs should be readily available. International participation would modify these standards to meet the needs of the participants.

Ground support hardware (workstations, display screens, servers, etc.) will need to be chosen. Workstations with readily expandable software and hardware that can be easily upgraded should be selected. Support personnel for such equipment will need to be readily accessible at all hours.

Due to the weight and volume that paper documentation would entail, the majority of procedures, reference material, etc. for the crew to use on the flight must be electronic. However, certain critical checklists should have a paper backup (as is currently used with the Space Shuttle). This would ensure that checklists would always be accessible to the crew and any "bitflips" or computer resets that will inevitably occur (due to radiation events, software/hardware problems, etc.) would not adversely affect the crew in time critical situations. ISS experience will prove valuable here, as this is what is currently planned.

Training details for ground controllers and crew would have to be worked out. It has been suggested that the research stations on the Antarctic continent would be a good analog for Mars crew training (remoteness, climate, etc.). Another possibility is the Devon Island site in the Arctic currently being developed by the Mars Society. These sites as well as other worldwide facilities need to be linked to Mission Control. Communications, transportation, resupply and other logistics issues will need to be ironed out. Training time would also have to be negotiated if using NASA facilities. Heavy training loads from the Shuttle and ISS programs should be expected in addition to actual flight operations.

Integrated simulations including corresponding communication time "lags" will be needed to realistically portray the flights. Malfunction scenarios need to be planned and worked out "real-time". Shift scheduling, days or week on console at a time, cross training, on-the-job-training and other items are all issues that need to be addressed. The ground training and mission design processes should begin as early as feasible to work these issues.

Other training issues need to be evaluated. To simulate Martian gravity, KC-135 parabolic flights or neutrally buoyant Helium balloons which are used to manipulate Shuttle Payloads for robotic arm (RMS) training should be expanded on (both of which are currently being done at JSC). A similar balloon technique was also used in filming the HBO television series *From the Earth to the Moon* to portray lunar gravity for the actor/astronauts with realistic results.

Public awareness of the mission from the MCC Public Affairs Officer (PAO) will need to be planned. Should the air to ground links be open to the public (NASA custom) or more reserved (Russian)? These activities should not be overlooked, as this will be the missions' window to the outside world. Events to celebrate milestones should be planned for both crew and ground controller morale as well as public awareness. Hi-profile celebrities at the Capcom/PAO console position, with reasonable supervision, should be encouraged. Viewing rooms should be accessible for the general public and private guests. Websites patterned after the ones created for *Mars Pathfinder* and *Mars Global Surveyor* should be established, including the practice of making data almost instantaneously available to the public.

For many, if not all of the above items, Apollo experience will prove invaluable. It would not be unrealistic to expect many veterans of that program to readily render their knowledge to a Mars mission, including those in retirement. This should be utilized to the fullest extent possible.

COMMAND AND TELEMETRY

Commands and telemetry would need to be received, formatted and directed between any Mars vehicle and the MCC. Presumably, this would be performed by a combination of the DSN, TDRSS, and domestic satellite infrastructure. A system of network time sharing between the different users would need to be negotiated. Additional expansion of the Deep Space Network, depending on the need for continuous communication, may be required. This would tie into the Mars vehicle design. The more autonomous the spacecraft, the less oversight the mission would need from the ground. However, experience has shown the value of an "extra set of eyes" at the MCC to monitor the large number of parameters all spacecraft contain.

Going outside of the DSN to include other available systems should be explored. This has been done for the *Galileo* project with the *Parkes* radio telescope in Australia. It is not part of the Deep Space Network, but *Galileo* has been granted station coverage in the past to help boost performance.

The TDRSS system would be applicable for a Mars spacecraft in low earth orbit at the beginning and ending of each mission. The capability for DSN and TDRSS to work together through the same vehicle has been proved through Shuttle experience. A similar telemetry scheme should be employed for the Mars vehicle. The potential conflict with Shuttle operations could be avoided by negotiating the (assumed) small number of launches per year of the Mars vehicle to not overlap the Shuttle fleet. ISS operations using TDRSS are planned continuously however, and this will necessitate time-sharing.

Continuous ground coverage from the surface of Mars would only be available with orbiting satellites, and even then there will be times that the Mars crew will be out of touch with Earth due to solar conjunction. Whether or not constant communications from the Martian surface is even necessary will have to be worked out in mission design. However, for emergency reasons the design should afford maximum opportunity for said telemetry (this would be analogous to the Space Shuttle having priority on the TDRSS network for an on-orbit emergency)? The planned use the *Mars Global Surveyor*/Mars Relay system as a communications platform for the future Martian surface vehicles such as *Mars Polar Lander* should serve as an appropriate test bed.

LAUNCH SERVICES

With the ever-increasing number of launches from the world's spaceports, conflicts are bound to occur. This could be due to pad availability, range conflicts, etc. VAB (Vertical Assembly Processing) accessibility would also factor into this for a launch vehicle based on current Space Shuttle technology and launched from Florida. Competition for resources at the

Kennedy Space Center (or any other potential launch site) will have to be managed. This would include both material and manpower.

Other options for launch sites should be explored. The potential for using the completed (but never used) Shuttle launch site at Vandenberg Air Force Base in California should be evaluated. Orbital dynamics, refurbishment and other issues may prove to be too difficult to overcome for that site however. The Russian launch sites at Biakonur with their *Energia* heavy-lift infrastructure should also be investigated (as of this writing, there are reportedly one assembled *Energia* booster in stowage with parts for 2 more). With the current state of the Russian economy, and the difficulties in procuring Russian equipment for ISS, the latter possibility must be carefully thought out however.

NASA currently operates manned launch vehicles in a "handoff" mode. When the Shuttle launches, the Firing room at KSC is in charge until the Solid Rocket Boosters ignite. At that moment, MCC in Houston takes over for the duration of the mission until Orbiter wheelstop. Consideration should be given to integrating these two operations if a private venture is planned (for potential cost reduction). If leasing/timesharing of NASA facilities were pursued, the current system would likely be the appropriate choice due to its proven functionality.

SUMMARY

Regardless of whether or not NASA facilities (or ESA, RSA, NASDA, etc.) are used, many of the issues addressed above will arise. Some of these problems have been addressed in other programs, most noticeably Apollo, *Skylab*, Shuttle/*Mir* and the ISS. Variations must be expected however, and solutions will need to be re-tailored for the upcoming Mars missions.

ACKNOWLEDGMENTS

Special thanks to Ronald Sostaric for technical input and presentation of the paper to The Mars Society.

REFERENCE SOURCES

VAFB for Shuttle launch site: http://www.vafb.af.mil/30swmain.html (History at a Glance - Chronology).

Deep Space Network (DSN): http://deepspace.jpl.nasa.gov/dsn/.

Jet Propulsion Laboratory: http://www.jpl.nasa.gov/ (Mars Pathfinder, Global Surveyor, Polar Lander, project Galileo information and DSN).

Energia Russian/Ukrainian boosters: http://solar.rtd.utk.edu/~mwade/Ivfam/energia.hm.

Mission Control at JSC and Shuttle/Mir/ISS protocol and TDRSS: http://www.jsc.nasa.gov/ (Additional information provided by various flight controllers at the MCC).

KSC launch info: http://www.ksc.nasa.gov
United Space Alliance
NASA
Personal observations

MAR 98-068

AT WHAT RISK IS IT ACCEPTABLE TO COMMIT TO A MANNED MARS MISSION?

Dennis G. Pelaccio* and Joseph R. Fragola†

This paper identifies key factors and issues one must consider to commit to a manned Mars mission. Discussion emphasizes the critical issue of technical risk to commit and perform such a mission successfully. An assessment is provided which examines the current status of these key factors and considerations to support a commitment to undertake a manned Mars mission. The concept of qualitative risk management is introduced, and its potential to manage risk associated with a future manned Mars mission is considered. As part of the paper's discussion, the technical risk issue associated with man's last space exploration venture, the landings on the moon, is examined. The lessons learned from the manned Moon exploration program are then considered in the context of providing guidance in the decision to undertake a manned mission to Mars.

I. INTRODUCTION

A number of critical sociological, economic, and technical factors must come together for a serious commitment to be undertaken for a manned Mars mission by a nation, or by the world of nations. The three required factors to commit to such a mission are: 1.) strong political leadership; 2.) the availability of financial resources; and, lastly, 3.) the technical capability to pursue this goal. Other major factors that are not necessarily required for a commitment to a manned Mars mission, but could provide additional rational for such a mission, are: to investigate and exploit potential commercial and/or scientific payoff(s) by colonizing Mars; or the need to colonize Mars for the survival of civilization if there is a catastrophic event that takes place on Earth, such as a collision from a large asteroid.

If one examines past great exploration initiatives, such as Columbus discovering America or the manned moon mission landings, the three major factors to support these historic initiatives did come together successfully. In the case of the Columbus exploration initiative, Queen Isabelle of Spain provided the strong leadership and financial resources of her country to take on this high-risk effort. For her foresight, her country reaped many rewards, including access to riches of the then considered "Far East", claims on a new land, and enhanced predominance in the "World-Order of Nations" at that time. On the other hand, the United States (US) provided the strong political direction and financial support required to ensure that the initial phase of the manned Moon program was successful. For this large focused investment of political will and national resources, the US reaped major benefits, both in terms of establishing, without a doubt, its technical and government superiority in the bi-polar "Cold War World-Order," and in developing numerous technical spin-offs, which are still commonly prevalent in society today. The "true, measurable," planetary scientific payoff from the manned Moon program, if one considers its impact on society today, can still not be justified. This does not mean that the

---

* Chief Engineer, Pioneer Astronautics, 445 Union Boulevard, Suite 125, Lakewood, Colorado 80228. E-mail: strcspace@aol.com.
† Vice President, Advanced Technology Division, Science Applications International Corporation, 7th West, 36st Street, New York, New York 10018. E-mail: joseph.r.fragola@cpmx.saic.com.

planetary science findings of the manned Moon program will not increase in value to society, as time goes on. The manned Moon program also is a good example of highlighting the importance of the three required factors to perform such an undertaken. When interest in the manned Moon program began to subside in the late 1960's and early 1970's, due to other society interest (the Vietnam War, civil rights, and civil unrest in nations cities) and the competition of resources brought on by these issues, the program was substantially reduced in scope and cancelled abruptly. The last Moon landing took place just three years after the first, in 1972.

By examining these and other similar exploration initiatives, extensive insight can be gained in understanding the levels of technical risk that were considered acceptable to commit to such historic endeavors. Of notable interest, is that when these initiatives were undertaken, their technical risk levels were well above those typically considered acceptable for current state-of-the-art aerospace systems. Hence, many of the technologies and systems required to perform such initiatives were not available to engineers when they were approved to go forward. This fact raises many technical risk questions that must be addressed before commitment of a manned Mars mission should be undertaken. One key question that must be addressed is: What is an acceptable technical risk level to commit to a manned Mars mission? This can translate into another related critical question such as: What and how much technology is required to meet the technical risk level to successfully undertake a manned Mars mission? Because of the technical complexity and large investment required to perform a successful manned Mars mission, it is likely that levels of technical risk comparable to past historic exploration endeavors will have to be accepted, if such a mission is to be undertaken in the next 10 to 30 years. Additionally, other critical questions that also need to be answered are: How does one identify and track the technical maturity progress of technologies required for a manned Mars mission? and, Is there an optimum (minimum cost and/or time) investment strategy that can be identified to best use the limited resources available, to develop the necessary technologies?

This paper is an initial attempt to address the critical questions, just mentioned. To address these questions, as well as to better understand the current environment from which to consider a manned Mars mission at this time, critical sociological and technical factors are first examined. The concept of quantitative risk assessment is then introduced to address the critical technology maturation issues that must be quantitatively considered, to commit to a successful manned Mars mission. The paper then examines the manned moon exploration experience. This program, by its nature, provides useful insight in risk management of such a high-risk program. Finally, recommendations are provided that will ensure that the infrastructure is in place to make the "best-use" of limited technology development resources required to support such an endeavor, as well as provide the necessary insight in what the acceptable risk is to commit to a manned Mars mission.

Again, it should be noted, that if a manned Mars mission is to be undertaken in the next 10 to 30 years, there more than likely will be a relatively high level of risk associated with it. The key question remains: What is the acceptable level risk to pursue (commit) to such a worthy endeavor for man-kind? Even after a serious commitment is made to undertake a manned Mars mission, in all likelihood it will take an additionally 5 to 15 years to perform such an endeavor.

II. CURRENT ENVIRONMENT TO CONSIDER COMMITTING TO A MANNED MARS MISSION

The three required factors to commit to such a mission, as previously mentioned, are: 1.) strong political leadership; 2.) the availability of the proper level of financial resources; and, lastly, 3.) the technical capability to pursue this goal. Other major factors that are not necessarily required for a commitment to a manned Mars mission, but could provide additional rational for such a mission, are: to investigate and exploit potential commercial and/or scientific payoff(s) by colonizing Mars; or the need to colonize Mars for the survival of civilization if there is a catastrophic event that takes place on Earth, such as a collision from a large asteroid. Figure 1 depicts the relationship of these critical sociological, economic, and technical factors on the commitment to undertake a manned Mars mission. By examining these decision influencing factors, the current overall environment can be defined, at least qualitatively, to establish a starting point, from which a decision to commit to a future manned Mars mission will be made.

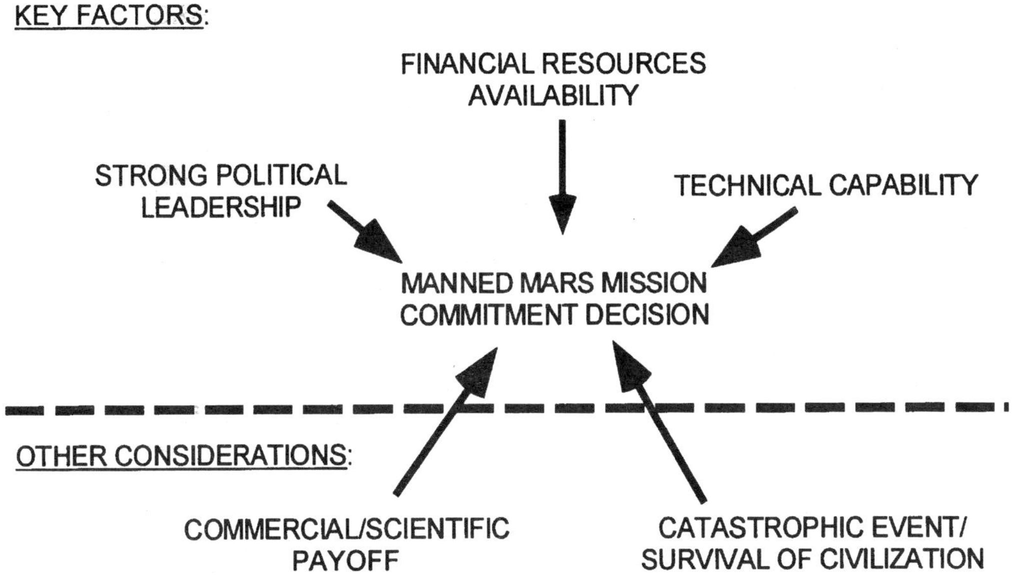

KEY FACTORS:

FINANCIAL RESOURCES AVAILABILITY

STRONG POLITICAL LEADERSHIP

TECHNICAL CAPABILITY

MANNED MARS MISSION COMMITMENT DECISION

OTHER CONSIDERATIONS:

COMMERCIAL/SCIENTIFIC PAYOFF

CATASTROPHIC EVENT/ SURVIVAL OF CIVILIZATION

Figure 1 The Relationship of the Various Decision Influencing Factors and Considerations to Support a Commitment Decision of a Manned Mars Mission.

The remainder of this section, assesses the current status of each decision influencing factor/consideration to support the commitment of a manned Mars mission today (August 1998). A simple color code guide is used to support the assessment. Red signifies that this influencing factor can not support commitment to a manned Mars mission at this time. There are significant deficiencies to support such a mission. Much work needs to be accomplished in this area. A yellow color rating indicates that there is "some reasonable support and/or maturity" to commitment of a manned Mars mission. More investment and/or effort to generate more political will needs to be expended in this area. The last assessment color code indicator, green, indicates that factors associated with this each decision influencing factor/consideration area is currently in proper order (degree of necessary maturity) to support a commitment to undertake a manned Mars mission. Before a serious commitment can take place, to ensure success, all the key decision influencing factors should be rated as green.

An assessment of key manned Mars mission commitment decision influencing factors is summarized in Table 1. This assessment is based on producing a serious commitment decision in the next five years. As one can see by reviewing Table 1, at present (August 1998), the environment to support a commitment decision to undertake a manned Mars mission, is not very good. None of the influencing decision factors are rated green. Even with rather negative assessment, not all is lost, because there is some reason for hope. This hope is due to the ground swell of grass roots for such an effort, and other factors such as robust world economy, and the technology base from which to draw upon for such an endeavor. The following discussion summarizes the various decision factors, independently, as well as other major decision influence assessment considerations.

Table 1
DECISION INFLUENCING FACTOR/CONSIDERATION ASSESSMENT SUMMARY
- August 1998 -

Decision Influencing Factor	Red	Yellow	Green
Strong Political Leadership	√√*	√	
Financial Resources Availability	√√	√	
Technical Capability	√	√√	

* a "√√" notation mark indicates a predominate assessment position.

Strong Political Leadership: This influencing factor is currently assessed as red, but there is optimism that with a little organization and work by the public, a yellow rating is possible soon. The current red rating is primarily based on the fact that no focused direction and support has been provided by the political leadership of any country in the world, to undertake a manned Mars mission. There has been a lot of discussion on this topic in recent times, but only very limited resources have been made available by the various spacefaring nations of the world. Examples of such recent efforts include President Bush's Space Exploration Initiative (SEI) program, and the recent unsuccessful France/Russian Mars balloon surface exploration effort. Even the current low-cost, US Mars robotic exploration program, which has been a success and has demonstrated its value, has only marginal resources allocated to it. The positive element associated with decision factor is that there is intense grass-roots public interest in a manned Mars mission. Public opinion polls have consistently shown over the years, that there is extensive support by the public for space exploration, and in particular for a manned mission to Mars. The recent interest in the US Mars robotic Pathfinder mission, when it landed on July 4, 1997, attests to this fact. The entertainment industry (movie and television media) has invested large amounts of funding and technical resources to provide an extensive amount of high-quality programming in this area. They truly understand the intense public interest in this area.

Financial Resources Availability: Like the previous decision factor discussion, this influencing factor is currently assessed as red, but there is some hope for optimism that this rating could improve. One major negative aspect associated with the current assessment of this factor, is that only little or marginal financial support is being provided by the world community, as previously mentioned. Little support is being given by the Europeans, Russians and Japanese to the Mars exploration area at this time, while the financial support associated with the current US Mars robotic exploration program is marginal at best. Additionally, no focused investment in manned Mars exploration is currently being provided by any the world's spacefaring countries at this time.

This is in contrast to the current overall world economic situation, which is in relatively good shape (most recent events in Russia and Asia not withstanding). Most industrial country economies are in the beat shape in years. Presently, the two largest world economies, the US and European economies are performing in a growth mode, while the other two major spacefaring countries, Russia and Japan, are having economic difficulties. Even with the current economic difficulties in Russia and Asia, the financially stronger nations, such as the US and the European Community, could use the space system development and launch assets in these nations to maximize the geographical participation and world-wide investment in a manned Mars program. With no major world conflicts (wars) or Cold War stalemates currently being contested, one would think that adequate financial resources could be acquired at this time to undertake man's next grand exploration challenge, a manned mission to Mars.

Technical Capability: Presently, there are many mission architecture ideas and technologies emerging that can support manned Mars mission, if a reasonable acceptable level of technical risk is assumed.[1-6] Hence, the yellow classification rating is selected for this particular decision influencing factor. Mission architectures range from the grandiose SEI architecture[1,2] to the much more simpler, realistic, near-term Mars Direct mission architecture proposed by Zubrin and Baker.[3,4] There is absolutely no shortage of ideas on how to perform a manned mission to Mars.[1-6] Key technologies such as those associated with nuclear propulsion and power systems, which have successfully been demonstrated in the past, provide a critical technology data bases from which to draw upon.[7,8] Even the current International Space Station (ISS) program can address some of the critical health and operational issues associated with sending a manned mission to Mars. As a minimum, though with much pain, the ISS program is establishing the necessary organization and working relationship framework required to undertake a large international supported future manned Mars mission. The great technical and intellectual resources that were dedicated to Cold War, over the past 50 years, may now be available as input to support a manned Mars mission.

Even with the reasonably well established technology base from which to draw, the necessary infrastructure is still lacking in many key required technology areas. Some examples of these areas include low-cost earth-to-orbit transportation systems, and nuclear propulsion and power systems. In some ways, recent US government policy may have set back development of these key technical areas. To perform a manned Mars mission, especially on a sustained basis, a low-cost, large payload capability, earth-to-orbit transportation system is essential. To date the US government commitment to field such a system is marginal, at best. Even the current reusable launch vehicle program is not funded at the high level required on a consistent yearly basis. Currently, the US government also does not provide the necessary incentives, such as in the form of tax breaks, initial government payload launch commitments, and government provided liability insurance, to attract the proper level of commercial investment required to develop such systems. Additionally, the recent US government policy has let the technical expertise base erode in high-leverage technology areas for a manned Mars mission. An example of this is in the nuclear propulsion area, which would have a positive impact on any future manned Mars mission. In the early 1990's the US government made a decision to essentially stop all research and development in this critical Mars exploration technology area. This decision has produced an extensive void, in terms of manufacturing and test infrastructure, and technical expertise, which is required to develop this technology. The irreplaceable, hands-on expertise gained during the successful nuclear propulsion programs in the 1960's and early 1970's has been lost for good. Developing this expertise now is possible, but would require a substantial undertaking.

Other Decision Influence Considerations: As of today, no major quantitative (dollars and cents) commercial and/or scientific payoff benefit to colonize Mars has been identified. Hence, it is likely that the true commercial and/or scientific payoff benefits of Mars will not be established until more exploration (robotic and/or manned) of Mars is performed. The recent Hale-Bopp Jupiter/asteroid impact event has raised some interest within the world community about using Mars as a possible Earth civilization safe haven. Just recently, the National Aeronautics and Space Administration (NASA) has initiated an effort to seriously assess the potential of asteroid impact on Earth. Lastly, the issue associated with the lack of major competitive pressures between the countries of world needs to be considered. This is a unique time in the history of the world, and performing a grand exploration undertaking in such an environment, will be breaking new ground. Thus, a key question will have to be addressed: Can high-risk exploration prosper in such an environment? Addressing this decision influencing consideration may be the critical factor that dictates if a commitment to a manned Mars mission in the 10 to 30 years, is undertaken. Discussion now focuses on quantitative risk assessment methods, which have the potential to address many of the outstanding manned Mars mission risk questions that have previously been presented.

III. QUANTITATIVE RISK ASSESSMENT

The Quantitative Risk Management (QRM) process provides a comprehensive structured assessment methodology to access the risk of emerging systems. In the simplest of explanations, this process can identify and quantify risk of key system parameters by mathematically representing them with appropriate weight or probabilities, and system element interrelationships. Many examples of applications of this assessment methodology are reported in the literature.[9-14] It is a proven risk assessment process which is well established (considered as everyday engineering practice) in the nuclear power industry, but is only used on an "as needed basis" in the aerospace industry. NASA only modestly started to implement the use of QRM after the Shuttle Challenger incident. Various elements of QRM, which are listed in Table 2, can be used throughout the system development process. That is it can be used from the initial concept assessment definition/comparison stage, to the final design and system operation phase. This assessment process can address and quantify key system issues and risk, such as that associated with mission success, design robustness, cost, schedule, logistics and maintenance requirements and operational support.

Table 2

VARIOUS ELEMENTS OF THE QRM PROCESS THAT CAN BE USED THROUGHOUT THE SYSTEM DEVELOPMENT PROCESS

- RISK APPORTIONMENT MODELING
- PROBABILISTIC DECISION ANALYSIS
- UNCERTAINTY ANALYSIS
- PROBABILISTIC DESIGN ANALYSIS
- FAILURE MODE AND EFFECTS ANALYSIS
- HAZARDS ANALYSIS
- COMMON CAUSE FAILURE IDENTIFICATION
- STATIC VERSUS DYNAMIC MODELING
- BAYESIAN UPDATING
- DATA BASE MATURITY/DEVELOPMENT

The QRM process elements listed in Table 2, can be used in identifying and developing a promising manned Mars mission system. Risk apportionment, probabilistic decision and uncertainty analysis can provide critical guidance in the initial concept assessment definition/comparison development stage of a manned Mars mission. Numerous mission/system design options can be examined at this stage of development. System and support technology requirements can be defined (including supporting technology development maturation plans), and their impact in terms of cost and schedule, can be assessed. From these study results, optimum mission/system design option(s) can be identified for further examination. As the promising system design option(s) mature in terms of its development, failure mode and effects, and hazards analysis, as well as common cause and dynamic modeling analysis provide insight into the reliability, maturity and robustness of the system. From this analysis a critical items list is identified, which will be composed of potential single point failures. This list is continuously tracked throughout the development of the system. Dynamic modeling of the system can quantify the influence of system state knowledge and response (the ability of the system to reconfigure itself and/or perform in a degraded state) to determine its robustness to function. During the development process, the maturity state of the system design is continuous compared to the technical risk, cost and schedule goals initially defined at the beginning of its development. Probabilistic uncertainty and design analysis is applied as needed in the design process to address the weak points in the system, as it is being developed. Additionally, estimates of system maturity throughout its development process, are characterized by applying Bayesian updating techniques. That is the knowledge/experience gained through system development is applied to the QRM analysis of the system to provide a better (less uncertain) estimate of system maturity. To ensure accuracy in applying the QRM process to manned Mars mission, the supporting data base must be representative of the system being accessed. Hence, development of such a data base to support the assess of manned Mars mission/system design option is critical. Typically, when any new advanced system option is assessed using the QRM process, such as is the case for the manned Mars mission, "surrogate" data is first used. That is the current supporting data available to the analyst is adjusted accordingly, to account for unique operating characteristics and environments associated with the new system being assessed. As the system matures, supporting data becomes available that is more representative of the system. This data then replaces the surrogate data, in the analysis, where possible.

In addition, QRM, by its nature, provides a structured framework from which to communicate system issues and risk to all levels management and support personnel. The process provides an orderly framework from which to communicate the mission design approach and selection, as well as the associate assumed risk and project progress to the highest level Government policy decision maker, to program management, technical support personnel, and even the public.

Typically, in the past, implementation of this assessment methodology to complex, realistic, system has been difficult because of the computational resources required. This is not the case today because of the recent advancements in computational technology. Hence, the idea of applying QRM process to a complex real life system, such as that necessary to perform a manned Mars mission, is now a feasible option to consider.

Due to the complexity, scope, and likely international aspects of a future manned Mars mission, and the inherent capabilities associated with the QRM process to properly access, track and communicate technical risk of such a mission, is it is essential that it be used as soon as possible. In this respect, review of many of the past, as well as the current NASA manned mission concept assessment studies fail to "quantify" technical risk, and its impact mission architecture and supporting technology selections/ratings. Hence, one needs to question the validity of past and present manned Mars planning study results. The QRM process, by its nature, also provides policy and technical decision makers necessary insight required to address another critical question if we are to forward on a manned Mars mission: Even if we have established what the acceptable level of technical risk is, how do we know when we have achieved it? Being able just to accurately address such a question is critical in managing the related technology development and financial resource allocation issues associated with such a mission. As an historical note, the next section which follows examines how the issue of technical risk was managed in the US manned lunar landing exploration program, also known as the Apollo program.

IV. THE MANNED MOON EXPLORATION PROGRAM EXPERIENCE

The theoretical basis for quantitative risk assessment was firmly established in time for the birth of NASA and the onset of the Apollo program. During the Apollo program, NASA shied away from using quantitative risk approaches despite evidence of interest early on. In fact, in the months following Kennedy's announcement of the lunar program, the founders of NASA decided that they had to have quantitative numerical risk goals for the Apollo mission. These goals, after much discussion, were decided to be 1 out of 100 for mission completion (0.990), and that 1 out of 1000 for returning the crew safely (0.999). They also understood that setting a risk of failure goal was not enough, but rather that "identification of potential failures and their risks is essential to a successful design and thus to a successful mission."[15] Further, NASA managers knew for this program that: "Risk is the basic common denominator for decision measurement."[15] This early reasoning led to the development of quantitative risk models which were initiated for all the Apollo program elements. The development proceeded along with the program so that by the mid-1960's models or modeling approaches existed at least for the Apollo Command and Service Module (CSM), the Lunar Module (LM), and the Saturn V launch vehicle.[16-18]

Despite the availability of these tools and the recognized need to deal with risk in a quantitative fashion, NASA soured to quantitative approaches as the program progressed and fell back on a decision making approach based upon five qualitative factors: 1.) review of all significant equipment modifications incorporated since the last design review and all anticipated modifications not as yet approved; 2.) the identification of and the determination of the qualification status of any system component whose failure, by itself, could cause loss of life, stage, or space vehicle (single failure point(s) (SFP)); 3.) a review of all vehicle and special system test results; 4.) a review of all significant failures and subsequent corrective actions; and, 5.) a review of unsolved problems, plans for corrective action, and estimated completion dates.

The identification of single failure points was predominantly accomplished by the performance of Failure Mode and Effects Analyses (FMEA). During the course of such analyses, each constituent part of a system was reviewed to determine its potential modes of failure and the subsequent effects which that mode of failure would have upon the component itself, as well as the assembly of which it was a part, (its impact on the subsystem, system, vehicle, mission, and crew). This bottom-up analysis was thereby intended to identify individual components whose failure might put the mission at risk. The analysis also indicated potential approaches that could be implemented within the existing design or, alternatively, possible design changes which might be made either to eliminate the failure mode, reduce its frequency to a acceptably low level, or mitigate its consequences. In this way, an FMEA exhibited some of the features expected of a risk analysis. Single failures which could not be eliminated or mitigated were collected across the design along with the rationale as to why they were retained and a list of all those identified SFPs were included in a Critical Items List (CIL). This list allowed the items to receive special attention in development, manufacture, installation, and test. Since the FMEAs and their associated CILs were critical determinants in the five factor decision process previously mentioned, the process as a whole is often referred to as the "FMEA/CIL" process. The FMEA/CIL process therefore was a static qualitative, bottom-up approach oriented toward assessing and reducing the risk of single independent component failures causing the loss of crew, vehicle, or mission.

While the FMEA/CIL approach certainly proved to be successful in producing reliable spacecraft and launch vehicles (based upon the success of the Apollo program), each of its characteristic features carried with them some drawback. An extended discussion of these drawbacks has been provided elsewhere,[19] however, the following list provides a summary of the problems: 1.) no natural probability cutoff; 2.) no risk focus; 3.) directed at single independent failures ignoring correlated failure or common cause failure impact; 4.) difficulties in incorporating human and software errors; 5.) difficulties in dealing with dynamic situations; 6.) no systematic approach for identifying and dealing with uncertainties; and 7.) significant financial cost in terms of test and personnel resources.

Given the drawbacks of the FMEA/CIL process and the original intent of NASA to obtain quantitative assessments of risk, it is logical to ask why NASA would turn its back on quantitative assessment and so tightly embrace a qualitative approach with so many known problems. The answer to such a question is of course, at least to some degree, somewhat speculative, but the second author's (J. Fragola) experience and the available historical evidence provide support to one possible answer. The evidence is as follows: 1.) many of the defects of the FMEA/CIL process were not that serious given the environment extant during the Apollo program era; 2.) those that were serious were not adequately addressed by the quantitative approaches available at the time anyway; 3.) the predictions available from existing quantitative models were completely unacceptable and inaccurate as forecasts of the risk "to be incurred" during an actual mission; and 4.) abundant personnel and test facility resources were provided to establish the needed confidence that each critical item had been properly addressed. Additionally, while rather primitive human error quantification approaches did exist at the time[20,21] they were developed for process oriented and not control oriented tasks and would not be ready to address control oriented tasks for at least another decade and a half. Further, if human error quantification techniques were lacking, software techniques were essentially nonexistent. The first symposium on the subject was not even held until 1973![22] Several years after the first manned lunar landing. As for the problems in dealing with common cause failures, dynamic situations and uncertainties, the available quantitative techniques of the day (i.e., fault trees, and reliability block diagrams (RBDs)) included no common cause failure approach, were also basically static when applied, and rarely, if ever, addressed uncertainties.

As previously noted, the quantitative risk assessment approaches available in the 1960s appeared not to offer very much above the qualitative to recommend them. That alone might have been enough to doom quantitative analysis. However, in the second author's opinion, what assured the demise of quantitative analysis was that its predictions were bad. Bad in the sense that the predicted failure probabilities were so high that they appeared to be obviously inconsistent with the test and early unmanned flight experience. The exact predicted values for the entire mission were not able to be resurrected for this paper (they were actually one of the few things classified in the entire lunar program at the time). However, various sources place the Saturn V launch vehicle mission success point estimate at about 0.88,[15] and the second author's recollection was that the CSM and LM estimates were 0.90 and 0.95, respectively. These estimates would indicate an overall mission failure rate or risk of one in four missions (or a mission success probability of 0.75). Additionally, from the second author's recollection, these final point values resulted after review and update of initial estimates which were significantly worse! Why did these estimates appear so wrong to the project team (and hindsight appears to have borne out their viewpoint since there was only one mission failure, Apollo 13, in the entire program and the mean mission risk demonstrated was about 4 times lower)? More importantly, why were the predictions developed by the reliability analysts such poor forecasts? Again, although any answer has to be considered speculative, the following rationale seems supported by the evidence.

The decision makers at NASA (and by the way not the traditional NACA types and the Huntsville "Germans"; at least not at first according to one source23), seemed to fully recognize that there was no possibility of flying enough test missions to be confident in the system because of the number of times it worked. They seemed to know instinctively that the only way they could build up the "infrastructure of confidence"24 required to give the go-ahead for a manned lunar flight was by structuring it from the engineering insights gained in development and the growth observed in the testing conducted. Thus they knew, because of the significant learning process expected and planned for, that significant growth in reliability would occur in the system throughout development. They reasoned that the history of past programs and early failures in Apollo only indicated what the reliability of the system had been in the past, not what its performance was forecasted to be over the spectrum of actual manned missions. In this way, they heuristically structured their design and testing processes not so much to be investigative, but rather to be confirmatory. They would expect the design to perform at flight levels so they would test at levels well above those expected during flight to provide them confidence that performance at flight levels would be assured. Thus, designs were robust, failures at flight levels were few, and the root cause analysis and corrective action programs ensured that those that did occur would tend to occur early enough in the test program so that their root cause could be eliminated. From this perspective, the decision makers had great confidence that even though estimates made from the scant early program history available and the history of past programs and equipment might indicate a mission risk of one in four launches, the actual risk was much lower.

On the other hand, the reliability analysts of the day were just beginning to address the issues of reliability growth and other alternatives for forecasting the risk in developing systems. While the theoretical basis was well established, practical applications for approaches such as the Bayesian combination of history, test and flight data were lacking. Since these approaches were considered experimental at the time, it is likely that the analysts based their predictions on a classical estimate derived directly from the test data and the history that was available, without growth considerations. For the reasons mentioned above, the test data available were extremely pessimistic. Further, the past flight experience was not at all good. A recently declassified report indicates just how bad it was.25 The report issued just one month prior to the Kennedy moon program announcement indicated that US ballistic missiles had only a 70% success rate and that only 50% of US spacecraft had reached successful orbits. Given these considerations, it should not be at all surprising that quantitative estimates, even when properly performed, would significantly underestimate the actual in-service reliability. From such a pessimistic viewpoint, the forecasts produced were likely to be unrealistically unflattering to the program. Additionally, they were available so late in the program as to be of little use in design improvements, even in the relative sense, where they might have had at least some value. Therefore, the entire exercise might well have been viewed as counter-productive.

Whatever the actual reason might have been for quantitative assessment's fall from grace, it was certainly the case that these estimates were not widely circulated even within NASA and were sparingly, if ever, released to the public at large. With the quantitative exercise too late to be of aid as a design tool and counter-productive when eventually available, it is not surprising that the qualitative FMEA/CIL process (which had been seen as widely useful and, though somewhat ex post facto, much more timely) was therefore seen as the far superior approach. While the above scenario might not be entirely accurate it does provide one plausible explanation as to why the FMEA/CIL process received full endorsement in the subsequent Shuttle era, and why quantitative approaches, at least comprehensive system level ones, were not employed again by NASA and its contractors, until after the 51L mission (the ill-fated Challenger mission of January 28, 1986).

Early attempts at reducing probabilistic models to allow for quantification had been tried at Grumman Aerospace Corporation[17] and North American Aviation[18] during the Apollo program, as was mentioned. These attempts had represented the Boolean equations derived from RBDs in terms of event sets, which approximated the probability of success rather well in the case where each event had a reasonably high probability. Soon thereafter, techniques surfaced in the nuclear industry which took advantage of the network dualism of RBDs and fault trees to represent the failure to perform the top event of a fault tree in terms of its minimal cutsets (i.e., cutsets with no repeated events). This examination on how the issue of technical risk was managed in the US manned Moon landing exploration program, provides a good real life experience data point from which policy decision makers, technologist and designers can draw upon in defining a plan for committing to, and developing a future manned Mars mission.

V. CONCLUDING REMARKS AND RECOMMENDATIONS

A brief review of some past grand exploration initiatives shows that many sociological, economic, and technical factors must come together for a commitment to undertake a manned Mars mission. This initial study effort indicates that much can be learned by examining such past exploration initiative. Examining such initiative, in detail, would provide extremely useful insight in the understanding and setting of risk goals for a manned Mars mission. It is recommended that a world renown committee of experts be assembled, immediately, to study in detail, the sociological, economic, technical, and risk taking aspects associated with many of history's grand exploration initiatives, that would likely have relevance to future manned Mars mission. As a minimum, this committee should be composed of experts, which have appropriate expertise in the areas of history, sociology, engineering, and risk analysis. Results from such a project could provide much needed insight and information to current manned Mars mission planning study efforts.

Based on the brief review of past exploration initiatives considered in this study, it is likely that a high level risk will have to be accepted if a commitment to a manned Mars mission is to be forthcoming in the next 10 to 30 years. The past manned Moon landing program experience shows that a mission probability of success risk of less than 0.75 was acceptable to commit to such a program. This risk level is much less than that considered acceptable for manned space flight today (a mission probability of success risk greater than 0.90). In the past manned Moon landing program, the use of QRM methods were applied in a somewhat limited fashion, but the adoption of engineering design and system development approaches such as those associated with reliability growth (learning through the development program), and confirmatory development testing at the component and systems level, were essential in making the program a success. This experience indicates that much positive can be said about undertaking a high risk project, such the Moon landing program, even without having all the design solutions and technologies required to perform it, at the time of commitment.

Due to the complexity, scope, and likely international aspects of a future manned mission to Mars, adopting QRM to properly access, track and communicate technical risk of such a mission is essential. QRM provides a useful organized way to address the technical risk issue associated with a complex manned Mars mission. This risk management technology is well proven, and with advent of emerging high-speed computational technologies, it is a viable option to consider. Implementation of QRM methods can address issues such as what, when, and how we do meet the technical risk goal(s) of such a mission.

It is strongly recommended that the QRM process be adopted immediately by the international community, for all planning and development efforts. Adoption of this assessment approach can be applied to all phases of development, and will provide a provide a rational basis for the architecture /design of the mission, establishing the technology requirements to support such a mission, and the sequent commitment to proceed. Additionally, data from ongoing and future programs related to a manned Mars mission should be gathered and organized accordingly, as a minimum, to support QRM assessment efforts. QRM should also be adopted immediately, as the "international engineering language" for this mission. Adoption of international QRM process standards, assessment tool and data base development and distribution, and training of all level of appropriate personnel should be undertaken by all spacefaring nations that will be involved with the manned Mars mission. A future manned Mars mission will be neither cost-free nor risk-free, but will be well within the gasp of groups of men and women, and nation-states who are bold enough to venture forth from the world and to continue mankind's destiny to explore.[26] Onward to Mars and beyond!

ACKNOWLEDGEMENTS

The authors would like to acknowledge the support of their respective companies for support in preparation of this paper.

REFERENCES

1. "Report of the 90-Day Study on Human Exploration of the Moon and Mars," NASA Headquarters, Washington, DC, November 1989.
2. "America at the Threshold: Report of the Synthesis Group on America's Space Exploration Initiative," U.S. Government Printing Office, Washington, DC, May 1991.
3. Zubrin, R., and R. Wagner, *The Case for Mars: The Plan to Settle the Red Planet and Why We Must*, The Free Press, New York, NY, 1996.
4. Zubrin, R., D. Baker, and O. Gwynne, "Mars Direct: A Simple, Robust, and Cost Effective Architecture for the Space Exploration Initiative," 29th Aerospace Sciences Meeting, AIAA 91-0326, Reno, NV, January 7-10, 1991. Also published as AAS 90-168, in *The Case for Mars IV: The International Exploration of Mars* (Mission Strategy and Architectures), ed. T. R. Meyer, Vol. 89, AAS *Science and Technology Series*, 1997, pp. 275-314.
5. Hoffman, S. J., and D. I. Kaplan (eds.), "Human Exploration of Mars: The Reference Mission of the NASA Mars Exploration Team," NASA Special Publication 6107, July 1997.
6. Drake, B. G (ed.), "Reference Mission Version 3.0 Addendum to the Human Exploration to Mars: The Reference Mission of the NASA Mars Exploration Team," NASA Johnson Space Flight Center Publication No. EX13-98-036, June 1998.
7. Koenig, D. R., "Experience Gained form the Space Nuclear Rocket Program (Rover)," LA-10062-H, Los Alamos National Laboratory, Los Alamos, NM, 1986.
8. Robbins, W. H., and H. B. Finger, "An Historical Perspective of the NERVA Nuclear Rocket Engine Technology Program," AIAA 91-3451, AIAA/NASA/OAI Conference on Advanced SEI Technologies, Cleveland, OH, September 4-6, 1991.

9. Apostolakis, G. (ed.), *Probabilistic Safety Assessment and Management*, Volumes I and II, Elservier, New York, NY, 1991.
10. Haugen, E. B., *Probabilistic Mechanical Design*, John Wiley & Sons, New York, NY, 1980.
11. Henley, E. J., and H. Kumanoto, *Probabilistic Risk Assessment, Reliability Engineering, Design, and Analysis*, IEEE Press, Piscataway, NJ, 1992.
12. Ryan, R. S., and J. S. Townsend, "Application of Probabilistic Analysis and Design Methods in Space Programs," *Journal of Spacecraft and Rockets*, Vol. 31, No. 6, November-December 1994, pp. 1038-1043.
13. Saaty, T. L., *The Analytic Hierarchy Process*, McGraw-Hill Book Company, Inc., New York, NY, 1980.
14. Siddall, J. N., *Probabilistic Engineering Design - Principles and Applications*, Mercel Deckker, Inc., New York, NY, 1983.
15. Cato, R. E. Jr. and Wheadon, W. C., "The Impact of Failure Data on Management of a Launch Operations Reliability Program," *Annals of Assurance Sciences*, 8th Reliability and Maintainability Conference Proceedings, 7-9 July 1969, Gordon & Breach, New York, 1969. LCN64-22868.
16. McKnight, C. W. *et al.*, "Automatic Reliability Mathematical Model," North American Aviation, Inc., Downey, CA, NA66-838, 1966.
17. Weisburg, S. A. and J. H. Schmidt, "Computer Technique for Estimating System Reliability," *Proceedings 1966 Annual Symposium on Reliability*, pp. 87-97.
18. "Saturn V Reliability Analysis Model Summary," SA-502, MSFC Drawing No. 10M30570, August 1967, NASA/MSFC, Huntsville, AL.
19. Fragola, J. R., "Space Shuttle Program Risk Management," Reliability Availability Maintainability Symposium (RAMS) 96, Las Vegas, NV, January 1996.
20. Swain, A. D., J. W. Altman, and L. W Rook, *Human Error Quantification: A Symposium*, SCR-610, Sandia National Laboratories, Albuquerque, NM, April 1963.
21. Meister, D., "Methods of Predicting Human Reliability in Man-Machine Systems," *Human Factors*, 1964, pp. 621-646.
22. *Proceedings of the 1973 IEEE Symposium on Computer Software Reliability*, Cat No. 73CH0741-9CSR, IEEE, New York, 30 April - 2 May, 1993.
23. Murray, C. and C. B. Cox, *Apollo: The Race to the Moon*, Simon and Schuster, NY, 1989.
24. Fragola, J. R., "Reliability and Risk Analysis Data Base Development, An Historical Perspective," *Reliability Engineering and System Safety*, Special Issue on Reliability Data Bases, Elsevier - North Holland, Amsterdam, The Netherlands.
25. Moody, J. W., "Reliability of Ballistic Missiles and Space Vehicles," Working Paper, Reliability Office, George C. Marshall Space Center, Huntsville, AL, April 15, 1961.
26. Fragola, J. R., "Engineering the Risk in Space Systems," Christian Huygens Lecture, TU Delft, Delft, The Netherlands, February 1997.

THE CASE FOR NURSES
AS KEY CONTRIBUTORS TO MARS EXPLORATION TEAMS

Mary Ellen Symanski*

A group of people who may be overlooked as valuable contributors to Mars exploration mission are nurses. It is my contention that they would be valuable members of Mars exploration teams not only to care for injured team members, but to offer assistance to others in coping with physical and mental challenges in the environment. Zubrin contends that the engineer is the most valuable member of the Mars exploration team due to the his or her ability to manage and repair equipment. Nurses can "fix" many "people problems" in non-pretentious and practical ways similar to the manner in which engineers work out problems with equipment. Nurses are broadly educated yet practically trained. They are natural multi-taskers accustomed to working long hours in stressful environments and thinking on their feet. Characteristics of nurses that would be valuable on a Mars expedition include flexibility, ability to handle many medical and psychological predicaments, good "people skills," and the capacity to improvise supplies in whatever settings they find themselves. Nurse-practitioners, who have additional education, are trained to handle basic health problems, and are used in health practices as "physician extenders." It is my contention that the human cargo going up to space is very valuable and as complex as the sophisticated equipment. Ingrained in the traditions of the nursing profession are the values of helping others and promoting well being and health. Nurses are "doers" as well as thinkers and problem solvers, and would readily pitch in with whatever other tasks need to be done. I believe that the contributions of nurses on Mars expeditions would make these people worth their weight many times over.

Nurses represent a group of people who could potentially be very useful and make low-profile yet extremely valuable contributions to Mars exploration teams. One of the revered leaders of modern nursing, the late Dr. Martha Rogers, predicted that at the turn of the century people would be living in "moon villages and space towns," and that the discipline of nursing should focus attention on how people will live in the future.[1] Unfortunately, our presence in space is not as great as Dr. Rogers envisioned it would be, and her suggestion that nurses should consider a role in space health care has not yet struck much of a cord within the mainstream of the nursing profession. While the average nurse may be too down-to-earth to think much about space, the characteristics of being sensible and realistic will help make nurses valuable additions to long term space exploration missions.

BACKGROUND ON THE NURSING PROFESSION

For those who have only passing acquaintance with nurses, I will give a little background information about the profession. Nurses are not simply a lesser educated subset of the medical profession nor are they merely health care technicians. The nursing profession has its own unique history, traditions, and educational system. While there is some overlap with medicine in terms of the goals of health and healing and treatment approaches, the nursing role is very

---

* Mary Ellen Symanski, Ph.D., R.N., Associate Professor of Nursing, University of Maine, Orono, Maine 04469. E-mail: mesphd@aol.com.

different from that of a physician. Some aspects of this nursing role enable very valuable contributions to space expedition teams. While the expert skills and knowledge of physicians will be important for health care in space, nurses could make great contributions in their own right that will be explained in this paper.

Nursing has a rich and complex history dating back to the earliest records of human life. Nursing has roots that are intertwined with the development of medical knowledge but the discipline has separate shoots and branches, for example the work of the Hospitaliers, a religious order of men who provided excellent nursing care during the Crusades.[2] In terms of modern nursing, the Crimean War is often cited as a starting point, with a focus on the work or Florence Nightingale. Miss Nightingale was an educated woman from a wealthy family, but followed a calling to serve the sick, and ultimately cared for diseased and wounded soldiers during the Crimean war.[3] Nurses have contribute countless hours of service behind the scenes during all of the modern wars. They served as symbols of light and hope in miserable circumstances, while courageously and competently going about their work.[2]

Nursing traditions provide a foundation for those entering the profession. These traditions are passed on to new nurses by experienced nurses in educational programs and in the workplace. Part of the indoctrination takes place in the form of rites of passage such as "pinning" ceremonies, where nursing students hear speeches about the values of nursing and receive a pin as a symbol of their membership in the group. Nurses and nursing students who are of high caliber in terms of scholarship and service may choose to be inducted into Sigma Theta Tau, an international honor society for nurses.

The core values of honesty, integrity and service to others are parts of the nursing tradition that will serve Mars expeditions well. The American Nurses' Association Code of Ethics[4] lists eleven responsibilities of nurses, including the duty to respect the human dignity of people and the obligation to take individual responsibility for actions. A nurse is called upon to live the value of honesty each time he or she makes a mistake and needs to call the physician and nursing supervisor to report the error. Nurses live the value of honesty when they administer medications and treatments only within ordered protocols and candidly tell patients when they do not know something.

Unethical behavior can be observed in the fictional character "Dr. Smith" on the 1960's television series "Lost in Space." This stowaway who claimed to be a professor, illustrated the damage that can be done to a mission when a member operates with selfish motives and does not consistently tell the truth. I am not proposing that nurses are the only people who are honest, nor can I claim that all nurses are impeccably honest. As a group however, nurses tend to have a trust-based relationship with the public.[5]

Nurses could contribute positively to a tone of good character and honest interactions on a Mars expedition. Nurses generally enter the profession to help people and are not usually motivated by fortune or fame. Issues of honesty versus fraud, and greed for profit versus the advancement of science and the human condition, will be themes that take on more and more importance as space travel becomes common place.

NURSING EDUCATION AND EXPERIENCE: SUITABLE FOR MARS SERVICE

The traditional education and experience of nurses combined with newer theoretical approaches introduced in the later part of this century, contribute to the suitability of nurses to space expeditions. I will elaborate upon these qualifications starting with the more recent developments in the science of nursing.

A cohort of notable nursing theorists was educated at the doctoral level in the 1960's with support of the Federal Nurse Scientist program of 1962. This program was intended to stimulate and develop research in nursing by providing financial resources to institutions and scholarships to nursing graduate students.[2] The theorists approached the science of nursing from different angles. Their frameworks for nursing practice are well suited to health care in space, perhaps even better than health care on earth, currently characterized by downsizing, downstaffing, and bureaucratic demands. The original group of nursing theorists, including Orem, King, Leininger, Rogers, Watson and Roy had a considerable influence on nursing education in the United States and Canada, and to some extent, abroad. The ideas and world views of these major nursing theorists have permeated nursing education from the 1970's to the present day.[6] The theories and conceptual frameworks emphasize building positive, healthy communities within a diverse and multi-cultural world.

Martha Rogers, DSc introduced a theory of "unitary human beings" in the 1960's at New York University, contending that humans are energy fields in constant interaction with the environment.[7] The late Dr. Rogers, a well read woman who studied physics and philosophy as well as nursing, encouraged nurses to stop viewing health and illness in the usual mechanistic and compartmentalized way, and to instead look at human physiological patterns such as blood pressure and activity from a broader environmental perspective. Rogers was one of the early theorists who encouraged nurses to think about their work far into the future, in an era when life in space would be a reality.[8]

Madeleine Leininger is a nurse with a doctorate in anthropology who served as Dean and Professor of Nursing at the Universities of Washington and Utah. She spearheaded a movement to make nurses more cognizant of the cultural issues that need to be considered in health care. She also conducted research on the nature of human caring across the lifespan among people of different cultures.[9] Nurses in space will be well served by sensitivity to cultural issues, as it is likely that teams will include people from diverse backgrounds. An appreciation of the many nuances of how people care for each other will be worthwhile too, in a remote environment with people living in close proximity.

Another nurse viewed as a leader in the area of human care is Dr. Jean Watson. Watson is a nurse with a background in mental health nursing and a doctorate in educational psychology. She established the Center for Human Caring at the University of Colorado.[10] Her belief is that care is a major component of the nursing role, and that the human touch of caring takes on additional importance in a highly technological society.

Dr. Dorothea Orem, formerly of Catholic University, Washington, D.C., emphasized the role of nurses in helping people care for themselves.[11] Self-care will be imperative in the space environment, where there will not be a plethora of health professionals around, nor will there be facilities such as hospitals and doctors' offices. On NASA missions the self-care philosophy has been successfully employed, with astronauts taught first aid and other skills associated with health and hygiene in space. Orem's background in the discipline of education contributed to her focus on the teaching role of nurses, and most nursing programs include a great deal of content about instructing people in health practices. The skill of teaching people to care for themselves in various ways will be important in space, for example people will need to be taught radiation avoidance techniques, how to use maintain good hygiene practices, and how to care for injuries in field conditions.

Sister Callista Roy, with background that includes a doctorate in sociology, took principles from Harry Helson's general adaptation theory and other sources to develop a theoretical framework for nursing that emphasizes people's physical and psychological adaptive capabilities.[12] Adaptation to the new space milieu will be a key theme for health in space, and nurses educated from this perspective will have a positive rather than pathologic framework from which to view the health needs of people in the new environment.

While the nursing theorists have contributed more recent additions to nursing curricula, basic education for nurses continues to include a heavy emphasis on biomedical sciences. Many of these topics have clear implications for health care in space. For example nurses are educated in care of burns and traumatic injuries, maintaining fluid and electrolyte balance, infectious disease prevention and hygiene, dealing with aches and pains, skin and foot care, bowel and bladder issues, and the care of new mothers and infants. Nurse-practitioners have additional education and training in primary health care, including the ability to diagnose and treat many health conditions.

ADDING SPECIALIZED SKILLS FOR SPACE CONDITIONS TO A SOLID BASE

The knowledge and skills of nurses form a sound basis on which to add specific training on health needs in microgravity situations and other space health challenges. Members of the Space Nursing Society discuss applications of earthbound nursing to space settings, for example the similarity of microgravity conditions to bedrest.[13] The Space Nursing Society (SNS) is an international organization established in 1991 to promote the interest and participation of nurses in space endeavors. At a recent conference, Linda Plush, current executive director and past president of the SNS, enumerated potential areas where nurses could contribute in space settings. Examples of such areas include motion sickness and balance problems, delayed wound healing, and the treatment or prevention of exposure to toxic chemicals or radiation.[13] Plush asserts that nurses have often been in the forefront providing health care in extreme and hostile settings, and it is only logical that they expand their practice to the space environment (personal communication, June, 1998).

Speakers at SNS conferences have put forth a variety of ideas about nursing roles in space. John C. Proctor a British nurse who works in the offshore North Sea oil platforms suggests that there are commonalties to providing health care in remote and hazardous conditions, and reminds the public that health care takes place wherever there are human beings working and living.[14] Dr. Barbara Czerwinski sees potential for nursing research in the area of facilitating hygiene practices for women in space.[15]

Nurses also receive a great deal of education and experience that contributes to positive "people skills." Nurses have background in human development and sociology. Examples of the types of situations they handle in a day's work include helping patients to set goals in rehabilitation, teaching people how to manage various health conditions, reassuring confused older adults, comforting dying patients, and handling complaints of angry patients and family members. They are frontline people who spend hours dealing with people's most intimate physical needs and raw fears and emotions every day.

Nurses' inclination toward the practical aspects of the world and many hours of experience working in patient care settings contribute to a knack for improvising various things to meet the needs of the situation. For example nurses have a long history of interest and experience in solving problems related to practical aspects of hygiene and human waste collection.

Nurses have researched skin care issues and participated in inventing new products such as the Bag-bath™ (Incline Technologies, Inc.) to improve the skin care hygiene of bedridden patients. Ostomy nurses figure out unique solutions to help people go about their daily lives with as little interference as possible from their urine or stool ostomy diversion systems. Research interests and everyday practice interests are a good match with the kinds of skills that will be needed in space travel and long-term colonization of planets. In addition, nurses are accustomed to working long hours under stressful conditions.

A HEALTH CARE TEAM APPROACH - WITH A TEAM THAT MAKES THE MOST SENSE

Will any class of worker on a Mars expedition have all the answers? No. Nurses like doctors, engineers, farmers and builders, will all need to be connected to sources of information and receive specialized training for anticipated issues within the field, specific to space flight and space living conditions.

I think a most sensible use of health personnel in space involves a team approach that makes the most of the strengths of doctors and nurses, and one that is based on the premise that no one from any discipline will have all the answers. The contributions of nurses however, would be much broader than those exemplified by Nurse Chapel assisting Dr. McCoy in the Star Trek television series. Using the military model, flight surgeons and flight nurses work together collaboratively in many situations, both in close proximity and at a distance, where consultation takes place. Nurse-practitioners, who represent nurses with additional education and training in primary health care services, could handle many routine health needs of Mars colonists. Alternatively, nurses could serve in special roles such as a leader or member of a space "fitness team." This suggestion was offered by Dr. Phillip Harris, a management and space psychologist, at the Space Nursing Society convention of 1997.[16]

Zubrin contends that doctors should *not* be involved at least in the early Mars expeditions when engineers are more crucial because of their ability to work with and repair essential equipment.[17] While I can appreciate this argument for early missions of only four people, once settlements become established, systematically attending to the health needs of the people living for lengthy periods on the planet makes sense. Eventually, inhabitants will need assistance with aspects of life such as maintaining long term musculo-skeletal health, treating accidental injuries and giving birth. Harris suggests a significant role for nurses in the care of childbearing families and their children in space. Due to technological restrictions of the early missions, doing without a health care provider may be necessary.[16] But, once larger groups are present in space, attending to human health aspects en route and on the planet will be a prudent investment in the precious human cargo.

One of my major contentions is that a new class of health care workers or technicians is not needed as a substitute for nursing in space. There is no need to "reinvent the wheel" when the discipline of nursing has such a solid base on which to build the extra new components that are specific to health care in space.

CROSS-TRAINING SCIENTISTS TO DO HEALTH CARE - NOT THE ONLY SOLUTION

Zubrin suggests cross-training science mission personnel in basic medical skills.[17] Although this solution may work to a degree, the assumption that scientists and engineers can be cross trained to do all other essential functions on a mission has its drawbacks. I would like to point out that one of the great strengths of nurses is their ability to be compassionate and understanding when dealing with human beings at vulnerable times in their lives. Highly educated people who chose a career as a scientist to study the mysteries of Martian geology may not have the same inclination nor the sensitivity needed to deal with other people's urinary tract infections, anxieties and fears. To maximize efficiency and serve the crew better, it might be a better idea to cross train some nurses to work in the greenhouses or maintaining the life support and waste recycler systems. On the face, routine "medical technician" work may seem easy to teach and learn. However, truly caring for people's health is not quite so simple an operation. Someone who is doing it as a secondary line of work may not have the degree of investment in the results nor the skill to figure out what to do when things "don't go by the book."

Zubrin points out that astronauts are not thrilled with space doctors poking and prodding them with medical instruments while they are trying to accomplish their work, and noted the observation of the great Polar explorer Roald Amundsen that doctors were bad for morale.[17] This argument, that doctors would be narrowly focused on obtaining research data and that their continuous pessimistic proclamations might not be helpful to morale has some merit. On the other hand, dealing with those issues does not require the total elimination of doctors, nurses or other dedicated medical personnel from space missions.

By raising the point, Zubrin challenges those in health care professions to propose a model for health care in space that meets the evolving needs of the participants. In the early missions of the NASA space program, it is understandable that astronauts were called upon to participate in medical research, yet continuous participation in research involves stress. There are ways to handle human subjects research issues in space travel, such as intermittent or selective participation in research. Eventually, as more and more people travel in space, there will be less need for basic research and more of a need for health care providers to handle day to day accidents and illnesses along with safety and preventive measures.

Returning to Zubrin's reference to the opinion of Roald Amundsen about doctors, one only needs to read the book "Scott and Amundsen" to appreciate the many things that Amundsen did right on his expedition to the South Pole in contrast with the inept leadership of Sir Robert Scott.[18] The fact that Scott had doctors around was, in all fairness, not the only problem on his Antarctic expedition. A lesson we can take from the Amundsen style is an appreciation for pragmatism. When planning for health needs in space travel and habitation, we need to look at what we want to accomplish and how can it be done efficiently and pleasantly. The people who are putting together the mission can identify the values and services they want from a health care provider or team, and then select applicants who agree to work within the mission philosophy. For example, one might seek a health service grounded in the values of human dignity and self-care whenever possible. These ideals would set the tone and boundaries for a model of health care services for the particular space mission.

WHY ARE NURSES MISSING?

Dr. Harris, in the paper he delivered at the 6th National Conference on Nursing and Space Life Sciences, questioned why nurses are not included in various space programs.[16] I can offer several possible explanations from the perspective of a common citizen who watches the news. The NASA philosophy reflects an emphasis on safety and survival of astronauts. In addition, as space programs evolved in the U.S. and Russia, biomedical research projects were added to missions. With these objectives in mind, doctors represented an obvious choice for providing health care for astronauts and for conducting biomedical research.

Politics is another factor that undoubtedly plays a role in the formulation of mission objectives and selection of personnel. The hard sciences and biomedicine are disciplines represented by relatively large amounts of power and financial resources. Their presence in space is self-perpetuating so long as there is a critical mass of interest among the members of the various disciplines. Those disciplines have established niches within NASA, and issues tend to be framed from the perspectives of people in those disciplines. It is more difficult for new groups to "break in." As space operations expand and develop, I believe the missions will be served better by including a wider range of people in the collective endeavor, including those people who do not hold doctoral levels of education nor wield a great deal of economic or political clout.

CONCLUSION

Assuming a shifting focus toward establishing a longer term presence in space, a wider range of personnel is more desirable for many reasons. I have explained the how the qualifications, values and skills of nurses would contribute to quality of life and health of people on long-term space expeditions. I have one final reason to offer for the value of nurses to space exploration, and that is their effect on the perception of the public about space travel. With the picture of nurses on Mars, the image of everyday people going into space begins to come into focus. The nurse is more representative of the average citizen than the "rocket scientist" typically pictured by the public on space missions. There is a certain amount of awe people feel about a rocket scientist or brain surgeon, but almost everyone has a nurse as a family member or friend. With nurses in space, more people might start to think of space travel as a real project that they can support rather than a science fiction dream far off in the future.

Nurse have a heritage, a broad and useful knowledge base and qualities of character which could enable them to contribute quite positively to a Mars expedition. I encourage those who are planning the personnel needs of space expeditions to consider the many benefits of including nurses on the team.

REFERENCES

1. Malinski, V. (1986). Further ideas from Martha Rogers. In V. M. Malinski (Ed.), *Explorations on Martha Rogers' science of unitary human beings* (pp. 9-14). Norwalk, CT: Appleton-Century-Crofts.
2. Kalisch, P. & Kalisch, B. (1986). *The advance of American nursing*, 2nd Ed. Boston: Little Brown & Co.
3. Pfettscher, S. A., deGraff, K. R., Tomey, A. M., Mossman, C. L. & Slebodnik, M. (1998). Florence Nightingale. In A. M. Tomey & M. R. Alligood (Eds.) *Nursing theorists and their work* (pp. 69-85). St. Louis: Mosby.
4. American Nurses Association (1985). *Code for nurses with interpretive statements.* (Publication No. G-156) Washington, D.C.: American Nurses Publishing.
5. Giordano, B. P. (1997). "To be effective health care providers and patient advocates, we must keep the public's trust." *AORN Journal*, 65(3), 14.
6. Alligood, M. R. & Choi, E. C. (1998). Evolution of nursing theory development. In A. M. Tomey & M. R. Alligood (Eds.) *Nursing theorists and their work* (pp. 55-66). St. Louis: Mosby.
7. Rogers, M. E. (1970). *An introduction to the theoretical basis of nursing*. Philadelphia: F. A. Davis.
8. Rogers, M. E. (1986). Science of unitary human beings. In V. M. Malinski (Ed.), *Explorations on Martha Rogers' science of unitary human beings* (pp. 3-8). Norwalk, CT: Appleton-Century-Crofts.
9. Leininger, M. M. (1991). *Culture, care, diversity and universality: A theory of nursing*. New York: National League for Nursing Press.
10. Patton, T. J., Barnhart, D. A., Bennett, P. A., Porter, B. D., & Sloan, R. S. (1998). Jean Watson: Philosophy and Science of Caring. In A. M. Tomey & M. R. Alligood (Eds.) *Nursing theorists and their work* (pp. 142-156). St. Louis: Mosby.
11. Orem, D. (1995). *Nursing concepts of practice*, 5th ed. St. Louis: Mosby.

12. Andrews, H. A. & Roy, C. (1986). *Essentials of the Roy adaptation model*. Norwalk, CT: Appleton-Century-Crofts.
13. Plush, L. (1997, January). "Space research links to space nursing and earthbound healthcare." *Proceedings of the Life Support and Biosphere Science Conference*, Coronado Springs, CA.
14. Proctor, J. (1998, January). "The final frontier 'inner space': A nurse's role in distant oceanic para-medical practice, an example of environment health nursing in a hazardous remote location." Abstract of paper presented at the Life Support and Biosphere Science Conference, Coronado Springs, CA.
15. Czerwinski, B. (1996). "Adult feminine hygiene practices." *Applied Nursing Research*, 9(3), 123-129.
16. Harris, P. R. (1997, September). "The space nurse's role in Moon/Mars personnel deployment systems." Paper presented at the conference, "Pushing the Envelope II," sponsored by the University of Texas Medical Branch at Galveston, The Center for Aerospace Medicine and Physiology, Houston, TX.
17. Zubrin, R. & Wagner, R. (1996). *The Case for Mars: The plan to settle the red planet and why we must*. New York: Simon & Schuster.
18. Huntford, R. (1979). *Scott and Amundsen*. New York: Putnam.

MAN AND EXTENDED SPACE FLIGHT: MENTAL AND PHYSIOLOGICAL FACTORS

Bradley S. Tice[*]

The use of visual stimuli to promote a 'fight' or 'flight' reaction to engage psychological and physiological processes in human beings to facilitate both a pre and post exercise stimuli. This manipulation of the visual environment will result in desired behavioral traits that will increase the overall efficiency, motivation and gross-motor activities in humans for desirable health and fitness levels for long duration space flights and habitations.

INTRODUCTION

The case for putting a man on the planet Mars is the next big step for mankind. Once the moon held this much vaulted status but was transferred to Mars with the first moon landing in 1969. Although the question of 'human factors' is, at times, under-represented (Zubrin, 1996), the vast majority of research is inconclusive to the effects of a one to two year mission to Mars on the human body (C.O.A.S.S., 1960; S.P. & M., 1989; Lorr, Garshnek, and Cadoux, 1989; and Stine, 1997).

An area that can be tested is the area of psycho-dynamics of human behavior on earth on humans performing tasks similar to those found in long duration space habitation and flight. One of the key factors for such a mission is the physical health of the astronauts.

PRIMAL OVERDRIVE

Doctors have determined that progressive exercise of an anaerobic nature, i.e. weight training, is vital to the maintenance of the bodies' support and movement structures, i.e. muscular system and bones, that must stressed to counteract the effects of micro-gravity environments. Because the demands of space place a huge task on the human body that the efficiency of large systems of the human body, the normal or standard criteria for physical training and health are of little value when applied to such a long term mission in space.

What is a more ideal model for future astronauts is a more specialized body and a more active profile of the individual in regards to high level physical training. An ideal group of test subjects are the high level athletes from both the amateur and professional classes. But this is just the start of the new paradigm in that a new model of training must be designed to incorporate both psychological and physiological traits that compliment each other for a unified whole to the concept and application of high level resistance training in space. I have designed and used for many years what has been termed 'primal overdrive' that is basically a systematic system of specific visual cueing by way of 'motivational' films and videos that produce an increased response in the viewer that has physiological traits that aid in the motivation, efficiency and power expenditure of the person watching the 'stimuli'. A general increase in muscular

---

[*] Dr. Tice is Director and Institute Professor of Chemistry, Advanced Human Design, Medical Research & Development, P.O. Box 2214, Cupertino, California 95015-2214.

tone and increased respiration, and blood flow to extremities is the result of watching these stimulating images before an exercise session. This has a direct application to conditions in long duration space flight where motivation, attention and quality and intensity of the resistance training is of optimal importance (Tice, 1992a; Tice, 1992b; Tice, 1993a, and Tice, 1993b).

LADDER EFFECT

A similar system is used to promote a post-exercise state that will facilitate attention away from the bodies painful state of post-exercise recovery as well as the increase in the processes of that metabolic recovery. Because it involves the same method of visual presentation, although with a differing type of image content, the reverse of 'primal overdrive' is, in effect, the opposite result and hence the term ladder effect to denote the sequential and stepped processes that is the reverse of the pre-exercise stimuli promoter (Tice, 1998). While this paper is not intended as a full blown explanation of all the factors used in such a training system, some general points of interest can be made for this system for the training of future astronauts to Mars.

KEY FEATURES

Psychological States

1. Motivation to a high level for long durations.
2. Innate design (especially in male subjects).
3. Promotes a general sense of 'wellness' and body awareness.
4. Normal Earth activity, watching television or movies.

Physical States

1. Promotes a high level stimuli to physical ready state of the human body.
2. Effects both pre and post exercise environments.
3. Speeds up recovery in a post-exercise mode.
4. Functions as an ancillary facilitator to the primary anaerobic resistance exercise paradigm.

SUMMARY

The use of visual cues and related motivational stimuli that can be used to promote desired states in exercising astronauts that will have lasting attributes for the extended stay in space environs in the future years ahead. Both the primal overdrive and the ladder effect are a part of a much larger and dynamic health and fitness paradigm that matches normal or usual terrestrial activities with an increase in attention and motivational factors that will have a direct bearing on the overall quality of exercise activities in space.

REFERENCES

1. Broad, W. J., "Space doctors decide pumping iron is key to astronauts health," *The New York Times*, Section C1 & C3, Tuesday December 10, 1996.
2. Committee on Aeronautical and Space Sciences, United States Senate. "Space research in the life sciences: An inventory of related programs, resources, and facilities," July 15, 1960, Washington D.C.: United States Government Printing Office.

3. Lorr, D. B., Garshnek, V., Cadoux, C., *Working in Orbit and Beyond: The Challenges for Space Medicine*. Vol. 72, *Science and Technology Series*, American Astronautical Society, San Diego: Univelt, Inc.
4. (1989) *Space Physiology and Medicine*. Philadelphia: Lea & Febiger.
5. Stine, G. H. (1997) *Living in Space*. New York: M. Evans and company, Inc.
6. Tice, P. S. (1992a) "An abstract of behavioral science on anaerobic exercise stimulators." Unpublished Manuscript. February 25, 1992.
7. Tice, B. S. (1992b) "Emotional response to visual stimuli as a motivator for grip strength." Final project for HuP 250 San Jose State University. Fall semester 1992.
8. Tice, B. S. (1993a) "Visual stimuli and gross-motor response in human males." Masters of Science degree (interdisciplinary) in Sport Psychology (unfinished degree). San Jose State University.
9. Tice, B. S. (1993b) "Literary survey of imaging and motor skills." Term project for Psychology 896. Fall semester 1993. San Francisco State University.
10. Tice, B. S. (1998) "The ladder effect." Paper to be published in *The Journal of Advanced Human Design* in 1998.
11. Zubrin, R. (1996) *The Case for Mars: The Plan to Settle the Red Planet and Why We Must*. New York: The Free Press.

WHO SHOULD GO TO MARS

Paul VanSteensburg

Who should be in the first crews? The Mars Direct Plan describes the professions of the four crew members. This paper will propose, from a layman's point of view and in more detail, who should go on the first missions including such considerations as age, sex, types of personality as will as profession. Also, the paper will paint a picture of daily life experiences in space and on the surface of Mars and what are some of the personal hurdles that members of the crew must be prepared for? How do we verify that we have selected the right people for what will be a long, lonely, challenging and exhilarating journey?

By its nature, the first few missions to Mars will be different from any other journey. It will be long, lonely, challenging and exhilarating - a range of human emotion and experience. It is important that we take the time to select the right people to meet this challenge. Relying solely on scientific research and past crew experience in space may not be enough.

This is a very different journey. Unlike long duration flights on a space station, the Mars crew will be away from earth longer than the current human space duration record and at a distance well away from help or resupply. This journey will be unlike the Apollo missions which were well away from earth, but of short duration and very limited scope of exploration.

So who should we select for such a journey? How do we test to assure that we have the best selections? What are some of the personal challenges that they will face? The ideas I present are based on a mission architecture as proposed by Dr. Robert Zubrin in the Mars Direct Plan. With very minor exceptions, I support this common sense approach to explore and settle Mars.

One of the major issues of a journey to Mars is that of human interaction. When we decide on who should go on the first few missions to Mars, we need to temper research on human interaction with common sense; sometimes scientific research looses sight of this. For instance, I believe that the human interaction of the crews on the first missions is very much like that of a family. These folks will be together nearly 24 hours a day, 7 days a week for 130 weeks. And so I rely on my personal experience with my family of a wife and 4 children as well as my observations of other families. We also need to rely on other human interactions such as those in the workplace.

Some of the criteria that we must consider for crew selection are: the professions of the crew members, their age, their sex, their health and personality traits. Much of what I have to say about the criteria has to be innate in the people selected. It has to be part of their make-up. Training will only help one complete specific tasks.

Training could be analogous to a diet. People can usually discipline themselves to loose weight. However, a diet will only work long term if people make a conscious life-decision to change the way they think about and consume food. Most people fail at diets not because they can't loose the weight, but because they do not take that additional leap to change their attitude about food. Similarly, training will work for short duration and specific tasks, but unless the crew has much of the following criteria as part of their make-up, the training will not prepare them to successfully interact and be productive over a long period of time.

The professions of the crew members identified in the Mars Direct Plan are based on critical success factors. According to the Mars Direct Plan, we are going to Mars to do science. So, it is important to have a scientist onboard, probably a geologist. If you don't do science, the mission may be considered a failure. In order to do the science and do it well, you need properly working equipment. So, it is important to have a mechanic on board to keep the equipment working because without the equipment you can't do the science very well and therefore the mission may be considered a failure. In fact, these two professions are so critical that you need two scientists and two mechanics.

I propose that there is a third critical success factor needed to assure a successful mission and that is healthy live humans. Without humans the equipment doesn't run well, the science isn't done thoroughly and there may be disappointment, to say the least, if the people don't come back alive. Therefore, I propose that a doctor is needed on the crew. Not a flight surgeon-type as scornfully mentioned by Dr. Zubrin, but a general practitioner. Someone who can give a physical every six months as one may expect on Earth or who could set a broken bone, do minor surgery or diagnose and treat an illness.

So, I believe the crew should be made up of a scientist, a mechanic and a doctor. And these are so critical that the people selected have to be very experienced in their profession. In addition, they should have one of the other two professions as a special interest and they should be thoroughly cross-trained in that discipline so that there is a backup in each of these three disciplines.

The Mars Direct Plan calls for four crew members. The fourth person should be a generalist. Somebody who knows a lot about many things. This person would also be the commander. The generalist-commander is not anchored in any of the three disciplines and therefore can be unbiased in making decisions. In other words, the commander cannot be resented for making a particular decision because he favors one crew member's opinion over the others based on similar professional background. Instead, the commander relies on his experts for advice and then tempers it with personal knowledge and good sense. The generalist-commander probably has a broader perspective on life, the mission, its place in history and the role he and the crew are playing in this historic event.

But this commander is not an independent, take-charge, dictatorial kind of person, although this may be needed at times. The commander is not a Captain Kirk. For most of the journey the commander acts more like a business team leader. The role of the team leader is to facilitate the group's activities, guide and participate in the work of the team. The team leader assures that communication is maintained within the team and with those outside the team.

How old should the crew be? Assuming the plan to go to Mars will take about ten years from initiation to first mission, the ages I discuss are those at the time of recruitment. I have already stated that we need experienced people. The crew should not only be experienced in their professions, but they should also have a wealth of life experience. You will not have either type of experience sufficient to meet this challenge if you have just graduated from college or are even several years out of college. This type of innate ability requires many years of experience. Their professional decisions, quality of work and those issues interacting with their fellow crew members must be instinctive. Based on this, I believe the minimum age should be mid-thirties.

On the upper end of the age scale there are other considerations. With the passage of time, experience accumulates and this is good for crew members, but only to a point. We have all known people who by their mid-sixties are old and in poor health. But we have also known people who have been sharp, active and healthy well into their sixties and I believe this will be more probable as our health conscious society ages. But the reality is that as we get older the likelihood of having a life-threatening episode increases. Using mid-sixties as a probable age when we should be concerned with aging health issues, the maximum age at time of recruitment should be mid-fifties.

So we are looking for people between the approximate ages of 35 and 55. This will allow for ten years preparation for the mission, making their ages on Mars between 45 and 65.

The issue of the sex or gender of the crew or sex itself is one that should be considered. We all know the sex by itself is a very powerful human need. Jealousy sometimes goes along with sex and is itself a very powerful human reaction. There could be jealousy among the crew as well a jealously between crew members and loved ones left on earth. A mixed gender crew will certainly raise eyebrows of spouses as well as many others in our society, but more importantly relationships that go beyond the professional could jeopardize the performance of the crew especially if jealously is added to the mix or should I say mess.

An alternative would be to have a crew of one gender: all male or all female. However, given today's sexual climate there is the distinct possibility that this could also spark problems of sexuality and jealousy.

After giving this a great deal of consideration, I believe sex is such a complex set of circumstances and issues that one can never be sure of coming up with a good solution. Therefore, I believe that, no matter how you decide, this area of personal interaction has the potential of being disastrous. Aside from possibly selecting people with a "low sex drive", whatever that means, if the people selected meet all the other criteria, they will be of the right temperament and we will rely on their good judgment to deal with the matter of sexuality in a rational manner. Therefore, the gender make-up of the crew should be decided based solely on all the other criteria.

What are the health considerations for the crew? My thoughts on this are based on the fact that there is nobody in perfect health. They are healthy at the moment. Over the 130 weeks anything, from an accident to an unpredicted health problem, can occur. At age 43, I was the picture of "perfect health". One evening in October of that year, I began to get excruciating pains in my lower back. It was the most painful feeling of my life. My doctor diagnosed the problem over the phone and sent me to the hospital. I had a kidney stone. I had no warning of it and, in fact, I had never stayed in a hospital overnight until this episode. In November, I noticed swelling in my elbows. I went to the doctor again and was diagnosed with rheumatoid arthritis. In a matter of two months, I went from being the picture of "perfect health" to having two major medical conditions, one was corrected and one is under control with long-term medication, vitamins and exercise.

So, the best we can do is look for people who are in general good health, not necessarily perfect health. And we should be prepared that sometime during training or the journey a medical problem will present itself. But this should not necessarily jeopardize any person's participation in the mission unless it is serious. Similarly, we want people who are generally physically fit, not necessarily musclemen.

Most importantly, we want people who innately have a healthy life-style. They eat well, the do moderate exercise, they are clean and they take care of themselves.

What are some of the personality traits of the crew members? This area of consideration is perhaps the most important of all the criteria. It also touches on personal habits. Can you imagine two crew members stuck on the edge of a crevice in the pressurized rover while the other two crew members in the pick up truck attempt to wench them back without tipping them over the edge? As the cable tightens and begins to slip on the back of the rover and rover begins to sway perilously on the edge and tension builds, one of the rover crew members yells at the other, "will you stop picking your nose, your disgusting." The outburst and the reaction of the nose picker is such that the rover's sway increases and almost snaps the cable. While this may be somewhat humorous, sometimes it is the small annoying things that drive people crazy and can interfere with good judgment and the ability to work and play together. And the expression of that annoyance will burst forth at the worst possible moment.

The people selected should be:

- **determined** - willing to see a difficult task through to completion
- **dedicated** - committed to the successful completion of the entire mission
- **selfless** - putting the interests of others and the mission over their own personal interests; a big ego works well for short periods of time, but will be overbearing on a long journey
- **cooperative** - willing to pitch in on any task at any time
- **communicative** - this goes beyond professional communication; willing to socially communicate thoughts and feelings as well
- **adaptable** - willing to learn new or reject current thinking; open minded and open to change
- **orderly** - their environment and work methods must be neat for safety and efficiency, but not to the point of being annoying; one should be somewhere between an Oscar and a Felix
- **conservative** - materially not politically; they should innately practice the prudent use of resources in their daily lives.

The people should have:

- **a sense of humor** - an underestimated yet universal trait for any challenging task; willing to laugh at oneself as well as others; it is a tension reliever
- **few or no annoying habits** - no cracking of knuckles, excessive burping, etc.; you get the picture!
- **a good sense of direction** - sometimes very smart people don't know which way to turn when they get off an elevator even in familiar surroundings
- **good hygiene** - this is both a health and an interpersonal issue.

Having chosen the people, how do we test that we have selected the right people? Selection should be one of the initial steps of the Humans to Mars program. Shortly after selection, teams should be formed. The team members should work together as the program develops so that they are familiar with each other and the many aspects of the equipment and mission. Periodically, the teams should train in a closed environment such as a biosphere. Team members should be shifted from team to team or replaced as needed. This could happen because of personality differences or a drastic health or professional failing. In the end, we would have the years prior to launch to observe the teams' interaction and performance. As the time of launch nears, the best functioning team should be selected.

The journey itself will have many challenges.

The outbound journey will have an initial period of excitement. This may last for a week or so before boredom sets in. To counter the boredom, work shifts could be established; perhaps three per "day". There would always be a person on "the watch"; monitoring the ships systems and communication. Another could have housekeeping and service duty. One person may have a specific task or repair for the day as determined by the team or mission control. And another person may have the day off where a hobby or special interest is pursued. The night shift may only have a person on "watch" while others sleep. The duties and shifts may change among the crew from day-to-day or week-to-week. The point is that there will always be something for people to wake up to every day and while this may be routine, it forces people to get up and out of their staterooms. They will also feel as if they have not been "stuck" with one task.

Each day each crew member must take care of their personal hygiene and exercise. There should also be periodic emergency drills and one physical examination.

As the time approaches for landing the excitement will once again fill the cabin.

The second phase of the journey will be that of being on the new home planet. While routine may become part of daily life, the next year and one half will have few periods of boredom. Crew members should never venture outside alone and usually somebody should be left inside the hab. The first tasks will involve securing the hab, testing the equipment and checking the Earth Return Vehicle. One can only imagine the awe of the first sunsets, the beauty of the deep dark nights, the curiosity of experiencing the first dust devils, the thrill of each new "simple" discovery and the wonder of those few great discoveries. Each team member must be diligent at maintaining a careful daily log.

Exploration will begin as short journeys close to the hab. The team will become familiar with their surroundings, the handling of the equipment in low gravity and their ability to work in their Martian suits. As time passes, the Martians will venture well away from the hab. They will perfect methods of navigating and communicating over the horizon and perhaps establish depots of consumables to extend the range of their exploration. There will be mapping, cataloging of resources, experimenting with growing food and the search for evidence of life.

As the time nears for leaving, there will be both the excitement of returning to Earth and a sadness of leaving their new home. I can imagine that some will have a feeling of wanting to stay.

The potential for boredom on the journey back to the home planet is greater than the outbound journey. The team will also have to contend with the extended period of weightlessness and more confined quarters. What will mostly keep them going is the anticipation of the exciting completion of their journey and the longing to see loved ones and friends. It will be a long six months, but it may be the last peaceful moments they will experience as the greatest pioneers of their time.

BUSHNELL®
SPORTS OPTICS WORLDWIDE

OFFICIAL TELESCOPE SPONSOR
OF

THE  SOCIETY

BRING MARS INTO FOCUS

© Bushnell Corporation. ® denotes a registered trademark of Bushnell Corporation. • 9200 Cody, Overland Park, KS 66214 • (800) 423-3537 Fax (913) 752-3500 • http://www.bushnell.com

*National Geographic Books
is proud to sponsor the*

Founding Convention of the Mars Society and to introduce our spectacular new volume...

History was made during the summer of 1997 when Mars Pathfinder's arrival on the red planet was "greeted with the most attention since Apollo 11 touched down on the Moon" (*Time* magazine). National Geographic has captured that excitement in a book hailed by *Publishers Weekly* as "extraordinary... astonishing and unprecedented."

Combining dramatic digital images taken by Pathfinder and *Sojourner*, along with trailblazing images from Viking space missions of 21 years ago, this volume is a visual feast—including three-dimensional gatefold panoramic images and sweeping mosaics of the Mars surface.

Authoritative and compelling text by Paul Raeburn: former chief science correspondent for the Associated Press and currently senior science editor for *Business Week*.

Dr. Matthew Golombek: Mars Pathfinder project scientist and research scientist in the Earth and Space Sciences Division of the Jet Propulsion Laboratory at California Institute of Technology.

Available at this convention and in bookstores everywhere. Or call toll free **1-888-225-5647**, 24 hours a day.

- 224 oversize pages
- 135 fascinating photographs
- 3-D images and glasses included in the book